Kepler's New Star (1604)

Medieval and Early Modern Philosophy and Science

VOLUME 31

The titles published in this series are listed at *brill.com/memps*

Kepler's New Star (1604)

Context and Controversy

Edited by

Patrick J. Boner

BRILL

LEIDEN | BOSTON

Cover illustration: The "novus sydus," flanked by Jupiter and Saturn, as it appeared in the foot of the Serpent Bearer in October 1604. Elias Molerius, *De sydere nova* (Geneva: Jacob Stoer, 1606). Courtesy of the Zentralbibliothek Zürich.

Library of Congress Cataloging-in-Publication Data

Names: Boner, Patrick, editor.
Title: Kepler's new star (1604) : context and controversy / edited by Patrick J. Boner.
Description: Leiden ; Boston : Brill, [2021] | Series: Medieval and early modern philosophy and science, 2468–6808 ; volume 31 | Includes bibliographical references and index.
Identifiers: LCCN 2020042170 (print) | LCCN 2020042171 (ebook) | ISBN 9789004431638 (hardback) | ISBN 9789004437272 (ebook)
Subjects: LCSH: Kepler, Johannes, 1571–1630. De stella nova. | Supernovae—History. | Astronomy—History—17th century.
Classification: LCC QB843.S95 K47 2021 (print) | LCC QB843.S95 (ebook) | DDC 523.8/4465—dc23
LC record available at https://lccn.loc.gov/2020042170
LC ebook record available at https://lccn.loc.gov/2020042171

Typeface for the Latin, Greek, and Cyrillic scripts: "Brill". See and download: brill.com/brill-typeface.

ISSN 2468-6808
ISBN 978-90-04-43163-8 (hardback)
ISBN 978-90-04-43727-2 (e-book)

Contents

List of Figures VII
Abbreviations IX
Notes on Contributors X

Introduction 1

1 Straight Paths and Evanescent Bodies: The Physics and Dynamics of Celestial Novelties in Kepler's *De stella nova* 17
Dario Tessicini

2 Of Mites and Men (and Stars): Kepler on the Question of Star Sizes in *De stella nova* 41
Christopher M. Graney

3 The Measure of the Universe in *De stella nova* 63
Javier Luna

4 Celestial Novelty and the Science of the Stars: Kepler vs. Krabbe on Accuracy and Authority in Early Modern Germany 81
Patrick J. Boner

5 Stars, Crystals and Courts: Johannes Kepler and Anselmus Boëtius de Boodt 107
Jonathan Regier

6 Kepler's Astrological Play 129
Aviva Rothman

7 The Nova of 1600 in Cygnus and the Christianization of the Constellations 145
Miguel Á. Granada

8 Epicurean Astronomy? Atomistic and Corpuscular Stars in Kepler's Century 181
Pietro Daniel Omodeo

9 The Correspondence of Clavius, Dal Monte, Magini and Other Italian Astronomers on the Nova of 1604 204
Matteo Cosci

10 The Scientific Legacy of Kepler's 'Stella Nova' 259
William P. Blair

Bibliography 267
Index of Persons 293
Index of Places 297

Figures

1.1 Johannes Kepler, *De stella nova* (Prague: Paulus Sessius, 1606), p. 90 27

2.1 The more distant an object, as seen from the Earth (E), measures a certain apparent size (indicated by the shaded region), the greater its physical magnitude. A and B each have the same apparent size as seen from E, but B is over four times more distant than A, so B has over four times the diameter and over sixty-four times (4^3) the volume of A 42

2.2 Tycho's geoheliocentric hypothesis. With the Earth immobile at the center (V), Mercury, Venus, Mars, Jupiter, and Saturn circle the Sun (as in the Copernican hypothesis), while the Sun, Moon, and stars circle the Earth. Johann Georg Locher, *Disquisitiones mathematicae, de controversiis et novitatibus astronomicis* (Ingolstadt: Elisabeth Angermaier, 1614), p. 52. Source: ETH-Bibliothek Zürich 43

2.3 *Above*—the relative sizes of celestial bodies calculated by Brahe for (from left to right, upper row) the Sun, Mercury, Venus, Earth and Moon, Mars, Jupiter, Saturn, as well as for (lower row) a large star and a mid-range star in a geoheliocentric universe (where the stars lie just beyond Saturn, as in Fig. 2.2). The Sun, stars, and planets all fall into a fairly consistent range of sizes.
Below—the arrowed dots are the above figure, reproduced to scale and compared to the relative size of a mid-range star calculated by Brahe for the Copernican universe, where the stars are vastly more distant and far more enormous to explain their apparent sizes as seen from the Earth (as in Fig. 2.1). Brahe called the immense Copernican stars 'absurd' 45

2.4 Sketch of the Copernican system by Thomas Digges. Note his description of the heavens as the "palace of felicity," with lights "far excelling our Sun both in quantity and quality," the court of the angels and the dwelling of the elect. Leonard and Thomas Digges, *A Prognostication Everlasting* (London: Thomas Marsh, 1576). Source: History of Science Collections, University of Oklahoma Libraries 47

2.5 Kepler's illustration in Chapter 16 of *De stella nova* (p. 86), showing the globe of the Sun, the orbit of Earth, the orbit of Saturn, and the sphere of the fixed stars. Source: ETH-Bibliothek Zürich 53

2.6 A star as seen through a small-aperture telescope. This appearance of a sphere of measurable size is entirely spurious—an artifact of diffraction. However, early telescopic astronomers took such telescopic images to be the physical bodies of stars. John F. W. Herschel, *Treatises on Physical Astronomy, Light and Sound Contributed to the Encyclopædia Metropolitana* (1828), p. 491 & Plate 9. Source: ETH-Bibliothek Zürich 61

3.1 The parallax of the nova determined according to the distance of the Moon. Johannes Kepler, *De stella nova* (Prague: Paulus Sessius, 1606), p. 79 68

3.2 The parallax of the nova according to the heliocentric perspective. Johannes Kepler, *De stella nova* (Prague: Paulus Sessius, 1606), p. 82 69

3.3 The circular descent of the new star according to Johannes Kepler, *De stella nova* (Prague: Paulus Sessius, 1606), p. 91 75

4.1 Original appearance of the comet according to David Herlicius, *Astronomische und historische Erklerung des newen Sterns oder ungeschwäntzten Cometen* (Szczecin: Joachim Rhete, 1604), 1r 91

7.1 Nova of 1600 in Blaeu's globe of 1602. Peter van der Krogt, *Globi Neerlandici: The Production of Globes in the Low Countries* (Utrecht: HES, 1993), p. 198 146

7.2 Daniel Mögling, *Collegium Fraternitatis*, 1618. José Bouman and Cis van Heertum, eds., *Divine Wisdom, Divine Nature: The Message of the Rosicrucian Manifestoes in the Visual Language of the Seventeenth Century* (Amsterdam: In de Pelikaan, 2014), p. 98 151

7.3 Constellation of Cygnus. Johann Bayer, *Uranometria* (Augsburg: Christophorus Mangus, 1603). 155

7.4 Tycho's figure of the comet of 1580 with respect to Cygnus. *TBOO*, 13, 324 163

7.5 Christianization of the planets. Julius Schiller, *Coelum stellatum christianum* (Augsburg: Andreas Aperger, 1627). Source: Biblioteca de reserva, Universitat de Barcelona 175

7.6 Transformation of Cygnus into the Cross of Christ. Julius Schiller, *Coelum stellatum christianum* (Augsburg: Andreas Aperger, 1627). Source: Biblioteca de reserva, Universitat de Barcelona 177

9.1 Andrea Argoli (1570–1657) in his old age. Lorenzo Crasso, *Elogii degli huomini letterati*, vol. 2 (Venice: Combi & La Noù, 1666), p. 26 214

9.2 Relative position of the nova of 1572 in Cassiopeia according to Christoph Clavius, *In Sphaeram Ionannis De Sacrobosco commentarius* (Lyon: Jean de Gabiano, 1607 [5th ed.]), p. 218 223

9.3 Franciscus Ingolus, *De stella anni 1604 tractatus* (1604–1605), Biblioteca Apostolica Vaticana, MS Ott.lat.2484 pt. 1, c. 143v (selection): astrological diagrams. Image reproduced by concession of the Biblioteca Apostolica Vaticana (all rights reserved) 238

9.4 Tycho's posthumous star chart for Ophiuchus. Tycho Brahe, *Astronomiae instauratae progymnasmata* (Prague, 1602), cc. 267v–268r (composite image) 253

10.1 Three views of the remnant of Kepler's supernova, from NASA's Great Observatories: (left) an optical light view from Hubble Space Telescope; (middle) an infrared light view from the Spitzer Space Telescope; (right) the appearance in X-ray light from the Chandra X-ray Observatory. Hubble's camera shows details in the densest knots of circumstellar material, as shown in the enlargement. Spitzer shows hot dust that has been heated by the expanding supernova shock wave. The chemical composition of the X-ray emitting gas is what clinched the identification of Kepler's supernova as a Type Ia. Figure courtesy of the author 260

Abbreviations

OGG	GALILEI, Galileo, *Le opere di Galileo Galilei*. Ed. Antonio Favaro. 20 vols. Florence: G. Barbèra, 1890–1909.
JKGW	KEPLER, Johannes, *Gesammelte Werke*. Ed. Max Caspar, Walther von Dyck, *et al.* 22 vols. Munich: C. H. Beck, 1937–.
TBOO	BRAHE, Tycho, *Opera omnia*. Ed. J. L. E. Dreyer *et al.* 15 vols. Copenhagen: Swets & Zeitlinger, 1972.

Notes on Contributors

William P. Blair
(Johns Hopkins University) is an astrophysicist and research professor in the Henry A. Rowland Department of Physics and Astronomy at the Johns Hopkins University. He came to Johns Hopkins in 1984 to work on the Hopkins Ultraviolet Telescope, an instrument that was flown twice on the space shuttle. He was also previously the Chief of Operations for NASA's Far Ultraviolet Spectroscopic Explorer satellite, which was active from 1999–2007. He is currently the Project Scientist for User Support for NASA's James Webb Space Telescope project, headquartered at the Space Telescope Science Institute in Baltimore. His research on supernova remnants and their host galaxies has spanned several decades, including specifically work on the remnant of SN1604, also known as Kepler's supernova remnant.

Patrick J. Boner
(Catholic University of America) is a visiting scholar in the School of Philosophy at the Catholic University of America and an affiliate of the Department of History of Science at the University of Oklahoma. He has published extensively on the history of early modern science, including the book *Kepler's Cosmological Synthesis: Astrology, Mechanism and the Soul* (Leiden, 2013). He is currently preparing the first-ever English translation of Johannes Kepler's *On the New Star* (Prague, 1606), the primary source of the present volume.

Matteo Cosci
(University of Padua) is a Marie Curie research fellow and the author of *Verità e comparazione in Aristotele* (Venice, 2014) and co-editor (with Marco Sgarbi) of *The Aftermath of Syllogism: Aristotelian Logical Argument from Avicenna to Hegel* (London-New York, 2018). He is also the Assistant Editor of the online journal *Philosophical Readings*. He is especially interested in Aristotle and the history of Aristotelianism (with a particular focus on Paduan Aristotelianism and vernacular Aristotelianism in the Renaissance) up to Galileo. Cosci is currently carrying out a Marie Curie research project entitled "The Ophiucus Supernova: Post-Aristotelian Stargazing in the European Context (1604–1654)," for which he has completed several articles.

Miguel Á. Granada
(University of Barcelona) teaches History of Renaissance Philosophy. He has published widely on Renaissance philosophy and science, mainly on Giordano

Bruno and the cosmological revolution, in particular regarding the work of Bernardino Telesio, Tycho Brahe, Michael Maestlin, Christoph Rothmann, Helisaeus Roeslin, and Johannes Kepler. Among his more recent publications are: *Christoph Rothmann's Discourse on the Comet of 1585: An Edition with Accompanying Essays*, in collaboration with Adam Mosley and Nicholas Jardine (Leiden, 2014); *Unifying Heaven and Earth: Essays in the History of Early Modern Cosmology*, ed. by M. Á. Granada, P. J. Boner, and D. Tessicini (Barcelona, 2016); Juan Cedillo Díaz, *Ydea astronomica de la fabrica del mundo y movimiento de los cuerpos celestiales: Traducción de 'De revolutionibus I–III, de Nicolás Copérnico*, ed. by M. Á. Granada and F. Gómez Crespo (Barcelona, 2019); *Giordano Bruno: 'De immenso': Letture critiche*, ed. by M. Á. Granada, and D. Tessicini (Pisa-Rome, 2020).

Christopher M. Graney
(Jefferson Community & Technical College) is an adjunct scholar with the Specola Vaticana, the Vatican's astronomical observatory in Rome and Tucson. He is also a professor of physics and astronomy at Jefferson Community & Technical College in Louisville, Kentucky, and has been for thirty years now.

Javier Luna
(Metropolitan University of Mexico) writes about the history of early modern science and technology. He is especially interested in the history of astronomy and cosmology, as well as the intellectual history of the Americas and Europe during the sixteenth and seventeenth centuries.

Pietro Daniel Omodeo
(Ca' Foscari University of Venice) is a cultural historian of science, a professor of historical epistemology, and a permanent guest of the Max Planck Institute for the History of Science. He is the principal investigator of the ERC research endeavour "Institutions and Metaphysics of Cosmology in the Epistemic Networks of Seventeenth-Century Europe" (Horizon 2020, GA 725883). His research focuses on early modern science and philosophy, as well as political epistemology. He has authored *Copernicus in the Cultural Debates of the Renaissance: Reception, Legacy, Transformation* (2014), (with Jürgen Renn) *Science in Court Society: Giovanni Battista Benedetti's* Diversarum speculationum mathematicarum et physicarum liber *(Turin, 1585)* (2019), and *Political Epistemology: The Problem of Ideology in Science Studies* (2019). Among other edited volumes, he has recently published *Bernardino Telesio and the Natural Sciences in the Renaissance* (2019) and (with Volkhard Wels) *Natural Knowledge and Aristotelianism at Early Modern Protestant Universities* (2019).

Jonathan Regier
(Ca' Foscari University of Venice) is a Marie Sklodowska-Curie fellow in the Department of Philosophy and Cultural Heritage at Ca' Foscari University of Venice. His work explores the intersection of medicine, natural philosophy, and religion in sixteenth-century Europe. He is the author of several studies, including "Reading Cardano with the Roman Inquisition: Astrology, Celestial Physics and the Force of Heresy" (*Isis*, 2019).

Aviva Rothman
(Case Western Reserve University) is an Assistant Professor in the Department of History at Case Western Reserve University. She received her PhD in History from Princeton University in 2012. She has served as a Collegiate Assistant Professor and a member of the Society of Fellows in the Liberal Arts at the University of Chicago, and as a Bradley Fellow at Carthage College. Her first book, *The Pursuit of Harmony: Kepler on Cosmos, Confession, and Community* (Chicago, 2017), focused on German astronomer Johannes Kepler and the connections he saw between the harmony of the heavens and the harmony of church and state. She has two book manuscripts in progress: an anthology of texts on the Copernican Revolution for the Penguin Classics series, and a monograph on the relationship between science and myth in the seventeenth and eighteenth centuries.

Dario Tessicini
(Università di Genova) works on the history of astronomy and cosmology in the sixteenth and seventeenth centuries. His recent publications include the edited volumes *Unifying Heaven and Earth: Essays in the History of Early Modern Cosmology* (with P. J. Boner and M. Á. Granada, 2016) and *Giordano Bruno,* De immenso*: Letture critiche* (with M. Á. Granada, 2020), as well as journal articles on the observation of new stars and comets in Italy before Galileo and on the history of the term 'cosmology.'

Introduction

The supernova of 1604 captured the imagination of early modern society. Soon after the superior planets conjoined in Sagittarius, the eyes of Europe were stunned to see a bright light suddenly appear in their vicinity. In his sweeping account of the 'new star,' Johannes Kepler (1571–1630) could scarcely describe the sense of awe and anticipation aroused by the spectacle. The great conjunction of Jupiter and Saturn in December 1603 had marked the start of an eight-hundred-year cycle begun by the Fiery Trigon, producing a flood of predictions that "made the minds of everyone ripe for revolutionary matters."[1] When Mars joined Jupiter and Saturn nine months later, astrologers and astronomers closely studied their assembly. "Some sought to correct the motions of the stars," Kepler recalled, while others aimed to prove their predictions.[2] Yet few could fully express the fear and wonder they felt when the new luminary began outshining the neighboring planets. In fact, it sparkled so intensely and twinkled so turbulently that "many said they had never seen anything in the heavens move more swiftly."[3] The weight of the star on the world of science would prove even more powerful.

Soon after the appearance of the splendid novelty, a broad network of scholars began circulating their ideas and observations in print and private correspondence. In his first report on the phenomenon, David Fabricius (1564–1617) recounted the moment he discovered it "sparkling or twinkling like a blazing light buffeted by the wind."[4] Fabricius had spotted the star shortly after sundown on 3 October, while he was closely observing the conjunction of Mars and Jupiter. Albin Moller (1541–1617) described "a silver comet star," which resembled a round burning lamp "without a beard or tail, flickering furiously and far more greatly than the planets passing by it."[5] Echoing the opinion of other authors, David Herlicius (1557–1636) struggled to make sense of the marvel by

1 *JKGW*, 1, 157, ll. 14–15.

2 *Ibid.*, ll. 21–27.

3 *Ibid.*, ll. 21–24. In an epigram to open his account, Kepler compared the bright star to the birth of his son and his revolutionary work on Mars. While Kepler determined the orbit of Mars, he wrote, "behold! Nature gives birth to a red rival. While I yield the riches of the red star, behold! My wife gives birth to a golden son." See *ibid.*, 156, ll. 3–6.

4 David Fabricius, *Kurtzer und Gründtlicher Bericht / Von Erscheinung und Bedeutung des grossen Newen WunderSterns* (Hamburg, 1605), 4r.

5 Albin Moller, *Gründlicher und warer Bericht / Von den newen Cometstern / so in der Lufft unter dem Himmel gesehen / nach dem Niedergang der Sonnen / in October und Novemb. des 1604. Jahrs* (Eisleben, 1605), 2v.

 | DOI:10.1163/9789004437272_002

comparing it to another from the annals of history. Only the wonder of 1572 came close. "Although many comets have appeared since the beginning of the world, or at least for several hundred years, which have been diligently recorded by scholars," Herlicius wrote, "this one is special and far stranger, like the one that appeared in 1572."[6] Still others swore that the luminary of 1604 was specially intended for astronomers.[7]

Interpretations of the 'new star' spread across the learned world, spanning every sphere of society. While many court astrologers reported to their patrons and provinces, university professors often presented their accounts in learned texts and public lectures.[8] At the University of Valencia, Antonio Juan Ripollés (1554–1632) oversaw an account written by his student, Marco Antonio Palau, who endorsed the views of Jerónimo Muñoz (1515–1591).[9] Palau determined that if the fixed stars scintillate as strongly "as they are far from us," the brightly flashing star lay "well beyond the region of air and fire in the heavens."[10] Palau

6 David Herlicius, *Astronomische und historische Erklerung des Newen Sterns oder ungeschwänzten Cometen, so Anno 1604 im ende des Septembris, und folgendem Octobri, auch Novembri erschienen* (Szczecin, 1604), 7v.

7 In his poem about the "new flame in the sky," Johannes Weidner (1540–1612) suggested that the star had been signaled to astronomers. See Johannes Weidner, *Acclamatio ad stellam novam prodigiosam, anno labente 1604, mensib. Octobri & Novembri conspectam, post Solis occasum, in dodecatemorio Sagittarii* (Göltiz, 1605), 2r: "For a sign so rare and noble Jupiter first gives to the mathematicians." Recalling the words of Weidner, Kepler agreed that the star had appeared among the planets, "as if it were one among their number, so that no one more than the learned, especially those who understood astronomy and the disposition of the planets, could fail to notice it and establish its novelty." See *JKGW*, 1, 272, ll. 33–40.

8 On the public lectures of Galileo at the University of Padua, see Michele Camerota and Patrizia Ruffo, "Le lezioni di Galileo sulla nuova stella del 1604 nel resoconto di Antonio Alberti (17 dicembre 1604)," *Galilaeana*, 2015, 12: 193–201. For a fresh account of the polemic that spurred the pseudonymous publication of an astrological work attributed to Galileo, see Matteo Cosci, "Galileo *alias* Astolfo Arnerio Marchiano e la disputa padovana sulla Stella Nuova," *Atti e memorie dell'Accademia Galileiana di Scienze, Lettere ed Arti in Padova*, 2019, 131: 35–84.

9 According to Víctor Navarro Brotons, Palau proves the influence of Muñoz and "the persistence of [his] teachings and works on the nova of 1572 in Valencia." See *idem*, "Las novedades celestes en España entre 1572 y 1618," in *Novas y cometas entre 1572 y 1618: Revolución cosmológica y renovación política y religiosa*, ed. Miguel A. Granada (Barcelona: Universitat de Barcelona, 2012), pp. 15–41, on p. 38. On the career of Ripollés as a lecturer of astrology at the University of Valencia, see Tayra M. C. Lanuza Navarro, "From Intense Teaching to Neglect: The Decline of Astrology at the University of Valencia and the Role of the Spanish *Novatores*," *Early Science and Medicine*, 22 (2017): 410–437, pp. 420–422.

10 Marco Antonio Palau, *Discurso astronomico sobre el cometa: Estrella que de nuevo ha aparecido en el Zodiaco, en 18. grados del signo de Sagitario, cerca de la Ecliptica, con poca latitud septentrional, a los 8. de Octubre 1604* (Valencia, 1604), 4v.

further claimed that he could find no hint of parallax and concluded that the star derived from a denser part of the ether, "under the influence of Jupiter and Mars, at the time of opposition to the Sun and Moon."[11] Muñoz had come to a similar conclusion on the 'new star' of 1572.[12] No comet in the course of history had "anything in common with ours," Palau reported, "except the comet of 1572 that occurred in Cassiopeia, which was nothing other than a star in the firmament."[13] The precedent of stellar novelty had been firmly established.

Indeed, there was good reason to compare the great marvels of 1572 and 1604. For some, the comet or 'new star' of 1572 had commenced a series of cosmic signs that culminated with a similar spectacle in 1604.[14] According to Helisaeus Roeslin (1545–1616), the two luminaries appeared in the fiery signs of Aries and Sagittarius while they were longitudinally conjoined with Jupiter.[15] In what has been called his "naturalistic turn,"[16] Kepler focused on the physical nature and origin of the novelties, comparing their light and location in the heavens. He determined, among other things, "that both were in the highest ether, among the fixed stars in the Milky Way."[17] However, it is worth noting that the circumstances of the second star, appearing at a more salient point in space and time among the superior planets, gave Kepler the same pause as his peers. In a martial metaphor, Kepler described how the luminary of 1572 had emerged "like an unforeseen enemy," "beyond the limits of the zodiac in a lonely part of the heavens," away from the paths of the planets.[18] In the same poetic flourish, Kepler portrayed the more regal appearance of the 'new star' of 1604:

11 *Ibid.*, 5r.

12 Among many other excellent studies by the same author, see Víctor Navarro Brotons, *Jerónimo Muñoz: Matemáticas, cosmología y humanismo en la época del Renacimiento* (Valencia: Universitat de Valencia, 2019), pp. 128–139.

13 Palau, *Discurso astronomico* (ref. 10), 4r.

14 Robin B. Barnes, *Astrology and Reformation* (Oxford: Oxford University Press, 2016), p. 272.

15 Miguel A. Granada, "Kepler v. Roeslin on the Interpretation of Kepler's Nova: (1) 1604–1606," *Journal for the History of Astronomy*, 2005, 36: 299–319.

16 Robert S. Westman, *The Copernican Question: Prognostication, Skepticism, and Celestial Order* (Berkeley: University of California Press, 2011), p. 384. On the philosophical nature of his study of the 'new star,' see Gérard Simon, *Kepler astronome astrologue* (Paris: Gallimard, 1979), pp. 54–55.

17 *JKGW*, 1, 270, ll. 30–32.

18 *Ibid.*, 272, ll. 9–11.

> Our star, widely anticipated for a long time, appeared on the appointed day with great solemnity and triumphal pomp, in the same manner as a powerful monarch who, before calling on the capital of his kingdom, designates the places of his court members through markers sent far in advance [of his arrival].[19]

The appearance of the star inspired more than admiration. Panic in public circles resonated with apocalyptic predictions based on biblical, magical, and mystical sources. In his frightful account of forthcoming events, Wilhelm Neuheuser claimed that the star punctuated, rather than portended, a period of global transformation. "The new evening star appears to us during the time of great change [*Hauptveränderung*]," Neuheuser proclaimed, "which has already begun and will transform the entire world."[20] Summoning a familiar refrain in Lutheran Germany, Johann Sifard warned that the growing frequency of comets foretold the end of the fourth and final kingdom, the Holy Roman Empire.[21] Yet the prospect of Armageddon did not persuade every pastor, priest, and scholar. Many viewed the 'new star' more precisely as a call to penance and piety. In his brief account, Balthasar Büttner suggested that God sent "such an unspeakable fire" not only so that "man would marvel," but that Christians could witness "a sign of grace and great mercy," and atone for their sins.[22] In the face of fear, Büttner encouraged his readers to repent and strive to improve society.

---

19 *Ibid.*, 272, ll. 11–15: "[…] nostra vero, vulgo exspectata a longo tempore, cum multa solennitate et triumphali pompa, ad diem constitutum est ingressa; more praepotentis alicuius Monarchae, qui Metropolim regni invisurus, praemissis longe antea Metatoribus, loca comitatui designat." Despite any 'disadvantage,' it is fair to say that the new star of 1572 exercised an equal, if not greater, impact on early modern religion and science. See Lynn Thorndike, *A History of Magic and Experimental Science*, 6 (New York: Columbia University Press, 1941), p. 68. On the widely ranging dispute that erupted in Italy over the new star of 1572, see Dario Tessicini, "Il dibattito italiano sulla nuova stella del 1572 (con il testo del *Discorso intorno a la stella* di Giuseppe Valdagno, ms Ambrosiana R 95 sup.)," in *Novas y cometas entre 1572 y 1618: Revolución cosmológica y renovación política y religiosa*, ed. Miguel A. Granada (Barcelona: Universitat de Barcelona, 2012), pp. 43–93.

20 Wilhelm Eo Neuheuser, *Tractatus de nova stella: oder von dem newen Abent Stern Scheinende* (Strasbourg, 1604), 1r. On the world vision of Neuheuser, "on the verge of sudden and radical transformation," see Barnes, *Astrology and Reformation* (ref. 14), p. 272.

21 Johann Sifard, *Cometen Spiegel* (Leipzig, 1605), 3–4; cf. Ken Kurihara, *Celestial Wonders in Reformation Germany* (Abingdon: Routledge, 2016).

22 Balthasar Büttner, *Cometen Stundbüchlein* (Eisleben, 1605), p. 26.

Contrary to the eschatological expectations of his contemporaries,[23] Kepler conceived his age as an era of renewal that roundly improved upon the past. In his opinion, the great conjunctions of Jupiter and Saturn, seen by so many to measure and motivate the course of history, coincided remarkably with major improvements in the modern world. In his "joyous celebration" of recent erudition and innovation,[24] Kepler applauded the ground-breaking achievements that marked "the astonishing upheaval [*commutatio mirabilis*]" since the great conjunction of 1563.[25] In addition to progress in commerce, international conquest, and navigation, Kepler lauded the mass publication of modern texts as a leading source of change. "A greater number of writings in every subject is printed annually, especially since 1563," Kepler exclaimed, "than those written in the last thousand years around the world."[26] As a result of the record-setting numbers, Kepler referred to revolutionary advances in astronomy, law, medicine, and theology. "I truly believe that at last the world is alive, indeed seething," he wrote, "and that the stimuli of those remarkable conjunctions did not act in vain."[27]

Kepler did not express the same enthusiasm for the influence of the Fiery Trigon. He scorned any "foolish astrologers" who foretold the future according to "the natural force [*vis naturalis*] of the Fiery Trigon" that began in 1603.[28] The age of discovery had anticipated the Fiery Trigon by more than a century, and the world had witnessed so many accomplishments in science and technology that he could hardly imagine what, if anything, could surpass the swift pace of progress in the wake of such a widely revered event. What could the astrologers predict, Kepler pondered, "unless they should suppose that we shall discover another planet or learn how to fly and travel to the

23 See, for example, Miguel A. Granada, "Cálculos cronológicos, novedades cosmológicas y expectativas escatológicas en la Europa del siglo XVI," *Rinascimento*, 1998, 37: 357–435.

24 Nicholas Jardine, *The Birth of History and Philosophy of Science: Kepler's* A Defence of Tycho against Ursus *with Essays on Its Provenance and Significance* (Cambridge: Cambridge University Press, 1984), p. 272.

25 *JKGW*, 1, 330, ll. 2–3.

26 *Ibid.*, 331, l. 39–332, l. 1.

27 *Ibid.*, 332, ll. 4–5.; cf. Jardine, *The Birth of History and Philosophy of Science* (ref. 24), p. 278. On the optimistic outlook of Kepler and his positive view of "the possibilities of [his] time," see Ulinka Rublack, *The Astronomer & the Witch: Johannes Kepler's Fight for His Mother* (Oxford: Oxford University Press, 2015), pp. 9–11; cf. Aviva Rothman, *The Pursuit of Harmony: Kepler on Cosmos, Confession, and Community* (Chicago: Chicago University Press, 2017).

28 *JKGW*, 1, 332, ll. 6–9.

Moon or the surface of another sphere?"[29] By confining his scope of study to what Kepler called "nature alone [*sola natura*]," he broadened his role beyond mathematical astronomy to explore the physical causes and characteristics of the heavens.[30] Coupled with his attempt to cultivate his position at court in the same way as his illustrious predecessor, Tycho Brahe (1546–1601), Kepler defined his duties to fit the fusion of astronomy and natural philosophy.[31]

Before turning to the broader content of his book, *De stella nova* (1606), we may consider how Kepler perceived the place of the 'new star' in history. Among the many subjects that exercised his attention, chronology inspired Kepler to revise sacred and secular records in synchronicity with the stars.[32] Won over by the work of Laurentius Suslyga (1570–1640), Kepler seized the chance to show that if the star of Bethlehem had, in fact, been "the great conjunction of the three superior planets" two years before the birth of Christ, it could be compared with the 'new star' of 1604.[33] In a pioneering text appended to *De stella nova*, Kepler concluded that "at least four years, or perhaps as many

29 *Ibid.*, 332, l. 9–12.

30 *Ibid.*, 333, l. 14.

31 On his role in uniting mathematical astronomy and natural philosophy with the "most authoritative rationale," see Edward Grant, *A History of Natural Philosophy: From the Ancient World to the Nineteenth Century* (Cambridge: Cambridge University Press, 2007), p. 312; cf. Miguel A. Granada, "Novelties in the Heavens between 1572 and 1604 and Kepler's Unified View of Nature," *Journal for the History of Astronomy*, 2009, 40: 393–402. On the quest for unity at the court of Holy Roman Emperor Rudolph II "as a means of representing the divine scheme of creation," see Bruce T. Moran, "Patronage and Institutions: Courts, Universities, and Academies in Germany: An Overview, 1550–1750," in *Patronage and Institutions: Science, Technology, and Medicine at the European Court, 1500–1750*, ed. Bruce T. Moran (Bury St Edmunds: Boydell Press, 1991), pp. 167–183, on p. 175.

32 See Anthony Grafton, "Chronology, Controversy, and Community in the Republic of Letters: The Case of Kepler," in *Worlds Made by Words: Scholarship and Community in the Modern West* (Cambridge, MA: Harvard University Press, 2009), pp. 114–136.

33 *JKGW*, 1, 359, ll. 13–20; cf. Michael W. Burke-Gaffney, *Kepler and the Jesuits* (Milwaukee: Bruce, 1944), pp. 33–34. In his study of the 'new star' of 1572, Tycho assessed how other scholars associated it with the star of Bethlehem. If the star of Bethlehem "appeared to just a few men above the only place on earth where Christ was born," Tycho pondered, could the 'new star' of 1572, "visible to everyone around the world in the highest ether," signal the second coming of Christ? Tycho quickly concluded that the answer could be known only to God, "in whose hands lie time and the age of the world, concealed entirely from our own cognition [*a nostra cognitione omnino seclusa*]." See Tycho Brahe, *Astronomiae instauratae progymnasmata* (Prague, 1602), p. 330.

as five," were missing from the common era.[34] The turning point came with a lunar eclipse mentioned by the Jewish historian Josephus in connection with the final days of Herod, which Kepler dated three years differently.[35] (In addition to the death of Herod, Kepler was compelled by Matthew's account of the Murder of the Innocents to move the birth of Christ forward by one or two more years.)[36] Confident in his chronology, Kepler closed his account less certain about the significance of the star.

0.1 Context and Controversy

De stella nova witnesses a wide array of battles that Kepler fought on many fronts. The 'new star' of 1604 represented a rare opportunity to resolve a number of questions that stood in the way of the new cosmology. Beginning with a multi-chapter response to Giovanni Pico della Mirandola (1463–1494) on astrology, Kepler replied to a broad range of scholars on a series of relevant subjects.[37] The multivalent nature of his account, in discourse with so many sources, has prompted some to describe it as "a group of disputations" that remain "difficult, unruly, and, at times, indigestible for modern readers."[38] Others have deemed *De stella nova* "much less 'modern'" than the more 'scientific' works of Kepler, particularly his far more famous *Astronomia nova* (1609).[39] Yet Kepler wrote most of *De stella nova* shortly after he had completed the bulk of the *Astronomia nova* by the spring of 1605, and it is fair to consider the two

34 *JKGW*, 1, 388, ll. 3–4.

35 C. Philipp E. Nothaft, *Dating the Passion of Christ: The Life of Jesus and the Emergence of Scientific Chronology* (200–1600) (Leiden: Brill, 2012), p. 275.

36 Ludwig Günther, *Kepler und die Theologie: Ein Stück Religions- und Sittengeschichte aus dem XVI. und XVII. Jahrhundert* (Giessen: Alfred Töpelmann, 1905), pp. 99–100.

37 See Sheila J. Rabin, "Kepler's Attitude Toward Pico and the Anti-Astrology Polemic," *Renaissance Quarterly*, 1997, 50: 750–770.

38 Westman, *The Copernican Question* (ref. 16), pp. 403–404. On the "confusing amalgam" of ideas and influences that characterizes the work of Kepler, see Albert van der Schoot, "Kepler's Search for Form and Proportion," *Renaissance Studies*, 2001, 15: 59–78, on p. 66.

39 Owen Gingerich, "Kepler Then and Now," *Perspectives on Science*, 2002, 10: 228–240, on p. 237. John North has portrayed *De stella nova* as "a work with a certain popular appeal," made up of "musings" on physics and cosmology. See *idem*, *Cosmos: An Illustrated History of Astronomy and Cosmology* (Chicago: University of Chicago Press, 2008), p. 347. The most reliable modern translation of *De stella nova* may be found in German; see Johannes Kepler, *Über den Neuen Stern im Fuß des Schlangenträgers*, trans. Otto and Eva Schönberger and Eberhard Knobloch (Würzburg: Königshausen & Neumann, 2006).

complementary.[40] As many of the contributions to this volume will show, we should regard *De stella nova* as a continuation of his revolutionary campaign to found the heliocentric hypothesis physically and philosophically.

In concert with his campaign, Kepler confronted a crisis in astronomy sweeping other spheres of early modern scholarship. With the explosion of observational activity that soon followed the sudden appearance of the 'new star,' Kepler and his contemporaries faced what we may call "a challenge of integration."[41] How could the astronomical community establish the accuracy of observational instruments and techniques, coordinate observers, and correlate their results? As long as the definition and deployment of evidence remained under review, scholars struggled to settle on a communal methodology.[42] According to Kepler, some astronomers measured accuracy and authority insufficiently, relying on the wrong conventions to assess criticism and controversy. Before they could argue over their opinions, in other words, astronomers had to find common ground for evaluating their evidence. At the court of Holy Roman Emperor Rudolph II (1552–1612), Kepler enjoyed access to a virtual torrent of tales and texts about the 'new star.' From his place in Prague, Kepler endorsed the finest observations and forswore the work of those who performed or reasoned poorly.

As Imperial Mathematician, Kepler sought to preside over the practice of astronomy among his peers. At stake was not only accurate knowledge of the 'new star,' but the very pillars of precision and professional integrity. Kepler began his book with a "history of the new star," recounting the reports of observers from Giovanni Antonio Magini (1555–1617) in Bologna to "letters sent from Spain to Savoy announcing the appearance of a new star [*novi sideris exortus*]."[43] Along with expert accounts, Kepler lashed out at claims that he labeled "impossible" and explained how members of the public might have

---

40 On his association of the 'new star' as a "red rival" to his revolutionary work on Mars, see ref. 3. For another form of continuity between *De stella nova* and the *Astronomia nova*, see Francesco Barreca and Patrick J. Boner, "A Perfect Similitude: Science and Politics in Kepler's Dedicatory Letters to *De stella nova* and the *Astronomia nova*," in *Celestial Novelties on the Eve of the Scientific Revolution, 1540–1630*, ed. Dario Tessicini and Patrick J. Boner (Florence: Olschki, 2013), pp. 209–229.

41 See Lorraine Daston, "The Empire of Observation, 1600–1800," in *Histories of Scientific Observation*, ed. Lorraine Daston and Elizabeth Lunbeck (Chicago: University of Chicago Press, 2011), pp. 81–113, on p. 88.

42 On communal methodology as "a major determinant of perception of relevance," see Nicholas Jardine, *The Scenes of Inquiry: On the Reality of Questions in the Sciences* (Oxford: Clarendon Press, 1991), p. 78.

43 *JKGW*, 1, 157–165.

mistaken the appearance of the 'new star' near the horizon for Venus. In a sense, Kepler assumed the same role as Tycho, whose broader study of the 'new star' of 1572 Kepler had edited soon after the author abruptly died in 1601.[44] Tycho had scoured the available sources and submitted his study as part of a larger 'restoration' of astronomy.[45] Similar to his successor, Tycho tied the nature of the novelty to the divine order of the cosmos.[46] At the same time, Kepler and Tycho countered philosophical and religious convention by explaining celestial mutability through earthly analogies.[47]

We may compare the form and style of *De stella nova* to another work from the imperial court. Thaddaeus Hagecius (1525–1600), personal physician of the Holy Roman Emperor, authored an "exchange [*dialexis*]" about the 'new star' of 1572.[48] Hagecius presented his volume as a kind of capstone to the contributions of Cornelius Gemma (1535–1578) and Jerónimo Muñoz, whom he praised for their elegance and erudition. Hagecius suggested that, by selecting the texts of other scholars "who did not differ from our own opinion," he could arrive at a more faithful account of the phenomenon that filled the gaps for posterity.[49] Collating the observations of Muñoz in a brief "history of the new star appearing in the constellation of Cassiopeia," Hagecius turned to reciting and refuting the opinions of his rivals.[50] He began with Annibale Raimondo (1505–1591), an astrologer physician who argued that the star was not only "ancient," but

---

44 See Brahe, *Astronomiae instauratae progymnasmata* (ref. 33).

45 See Adam Mosely, *Bearing the Heavens: Tycho Brahe and the Astronomical Community of the Late Sixteenth Century* (Cambridge: Cambridge University Press, 2007), p. 159.

46 On the attempt to explore "divine activity in the world physically and astronomically," see Westman, *The Copernican Question* (ref. 16), p. 401.

47 To explain the disappearance of the 'new star' of 1572, Tycho compared it to "the metallic bodies born in the bowels of the Earth." Just as the 'new star' had dissolved in the ether, Tycho reasoned, so metals were consumed by fire, "some faster, some slower," depending on their measure of "prime matter [*prima materia*]." See Brahe, *Progymnasmata* (ref. 33), p. 410; cf. Alain-Philippe Segonds, "Astronomie terrestre/Astronomie celeste chez Tycho Brahe," in *Nouveau ciel, nouvelle terre: La révolution copernicienne dans l'Allemagne de la Réforme*, ed. Miguel A. Granada and Édouard Mehl (Paris: Belles Lettres, 2009), pp. 109–142.

48 Thaddaeus Hagecius, *Dialexis de novae et prius incognitae stellae inusitatae magnitudinis et splendidissimi luminis apparitione* (Frankfurt, 1574).

49 *Ibid.*, p. 7. On the influential study of Gemma, treasured by Hagecius as "truly a gem [*vere gemmeus*]," see Dario Tessicini, "Cornelius Gemma and the New Star of 1572," in *Change and Continuity in Early Modern Cosmology*, ed. Patrick J. Boner (Dordrecht: Springer, 2011), pp. 51–65.

50 *Ibid.*, pp. 12–20. Hagecius provided Tycho with copies of letters from Muñoz concerning the 'new star' of 1572; see Navarro Brotons, *Jerónimo Muñoz* (ref. 12), p. 199.

"concealed since the very act of creation."[51] Hagecius feared that Raimondo had "fooled many others by his opinion," and scorned him with the same contempt that Kepler would later criticize some of his own contemporaries.[52]

Although the 'new star' of 1604 caught the attention of more observers than the marvel of 1572,[53] Kepler could not completely appreciate the view. Plagued by poor vision, he observed the star in the company of other courtiers in the imperial garden in Prague. (When Venus began appearing so brightly in the fall of 1604 that "it cast a shadow," Kepler marveled that even he could see it during the day.)[54] In addition to the expert eyes of Jost Bürgi (1552–1632), Kepler referred to the observations of Fabricius, whom he considered the most accurate observer after the death of Tycho. "Greatly hindered" by his eyesight, Kepler frequently relied on Fabricius to confirm or correct his own observations.[55] To prove that the 'new star' was beyond the Moon, for example, Kepler consulted the work of Fabricius to determine that it could not span as many as six degrees between their respective places of residence in Bohemia and East Frisia. "Since our observations do not differ by six degrees," Kepler reasoned, "the star could not have been so lowly [*humilis*]."[56] It was far above the Moon, in fact, but Kepler measured each step of his account against the accuracy of other astronomers.[57]

Above all, Kepler was guided in his study by Tycho. Kepler employed the highly precise instruments of his predecessor to observe the phenomenon, occasionally in the company of Frans Gansneb Tengnagel (1576–1622), son-in-law of Tycho.[58] When Kepler determined the location of the 'new star' according to the positions of the planets and the stars, he adopted the same method that

51 *Ibid.*, p. 21. On the role of Raimondo in Italy, see Tessicini, "Il dibattito italiano sulla nuova stella del 1572 (ref. 19)."

52 Hagecius, *Dialexis* (ref. 48), p. 25: "It is no surprise that Raimondo is a poor logician [*malus logicus*], but we may also find that he is an inept astronomer [*ineptus astronomus*]."

53 On the slow response of some to the 'new star' of 1572, see Victor E. Thoren, *The Lord of Uraniborg: A Biography of Tycho Brahe* (Cambridge: Cambridge University Press, 1990), pp. 61–63.

54 *JKGW*, 1, 159, ll. 35–41.

55 For his full appraisal of Fabricius, see *ibid.*, 1, 210–211.

56 *Ibid.*, 1, 237, ll.7–13.

57 On the shortcomings of "sheer empiricism" and the early modern endeavor to reinforce it with analytical reasoning, see A. Rupert Hall, *The Scientific Revolution 1500–1800* (Boston: Beacon Press, 1962), p. 175.

58 On his bitter dispute with Tengnagel over the instruments and observations of Tycho, see James R. Voelkel, *The Composition of Kepler's* Astronomia nova (Princeton: Princeton University Press, 2001), pp. 142–169.

Tycho had pioneered in the *Progymnasmata* (1602). In short, Kepler mapped the 'new star' by plotting the distance from two or more points at a given time. After measuring the latitude on 11/21 October 1604 according to the location of Mars and α Ophiuci, the brightest star in the Serpent Bearer, Kepler collated his observations of Mars, Jupiter, Saturn, and two stars in Sagittarius on 7/17 October 1604 to establish the longitude. Dissatisfied with the result, Kepler calculated the longitude of the 'new star' over a series of several dates before settling on 17°43′35″ Sagittarius.[59] The first thing he did next was compare his figures with those of Fabricius. Yet it was precisely with Fabricius and Tycho, his principal sources of technical expertise, that Kepler clashed over the explanation of the 'new star' in the celestial ether.[60]

Finally, there can be no denying the deeply religious importance of the 'new star' for Kepler.[61] When he calculated the appearance of the novelty over the New World, Kepler recalled the words of Christ at the beginning of his ministry, "A people dwelling in darkness see the light [*populus sedens in tenebris videt lucem*]."[62] By comparing the 'new star' to the light of salvation that fulfilled the prophecy of Isaiah, Kepler evoked the prospect of conversion among "many peoples" across Brazil and Peru.[63] Despite the divine nature of the 'new star,' however, Kepler responded critically to many of his religious contemporaries. He referred to Fabricius as a perfect example of a faithful theologian who falsely extrapolated from his own field onto astronomy and physics.[64] At the heart of his disapproval, Kepler held to a natural explanation of the 'new star' that relied on the regular course of decay and renewal extending from the earth to the ether.[65] The result echoed the invisible principles of seminal

59 *JKGW*, 1, 211–216.

60 Kepler devoted Chapter 20 of *De stella nova* to disproving the theory of Fabricius that the star had been formed on the fourth day of creation and suddenly illuminated in the fall of 1604. See *ibid.*, 248–251; cf. Miguel A. Granada, "Johannes Kepler and David Fabricius: Their Discussion on the Nova of 1604," in *Change and Continuity* (ref. 49), pp. 66–92.

61 See Lynn Thorndike, *A History of Magic and Experimental Science*, 7 (New York: Columbia University Press, 1958), pp. 23–27.

62 *JKGW*, 1, 217, ll. 16–22; cf. Matthew 4:16 and Isaiah 9:2.

63 Ibid., 217, l. 19.

64 *JKGW*, 1, 249, ll. 30–32; cf. Jürgen Hübner, *Die Theologie Johannes Keplers zwischen Orthodoxie und Naturwissenschaft* (Tübingen: Mohr, 1975), pp. 244–247. See also Granada (ref. 60).

65 On the "spontaneous generation" of the 'new star' according to Kepler, see Massimo Bucciantini, *Galileo e Keplero: Filosofia, cosmologia e teologia nell'Età della Controriforma* (Turin: Einaudi, 2007), p. 139. On the equally important assumption that the science of optics could secure knowledge of astronomical objects "as valid as the observation of

reason that Augustine had employed to explain the spontaneous generation of living beings.[66] While the final meaning of the 'new star' remained a mystery, Kepler ultimately considered it knowable and natural.

0.2 Overview: Kepler's New Star

Many of the papers in this volume were originally presented at the conference "Kepler's Star: Context and Controversy," hosted by the Department of Philosophy at the University of Barcelona in the spring of 2017. The conference was organized as part of the research project, "Cosmología, teología y antropología en la primera fase de la Revolución Científica (1543–1633)," funded by the Spanish Ministry of Economy and Competitiveness (Project FFI2015-64498-P (AEI/FEDER UE) for the triennium 2016–2018). Together, the contributors examine a series of central questions raised or reinforced by the 'new star' among astronomers and other scholars over the seventeenth century. As prodigious as it was provocative, the stunning luminary inspired many to entertain novel ideas about the constitution of the cosmos and the course of history. At the core of our collection lies what is widely considered the definitive account of the 'new star,' *De stella nova in pede Serpentarii*, published in Prague in 1606.[67] The author, Johannes Kepler, intended to resolve a number of concerns about the great novelty. At the same time, his work spoke to more enduring questions about the design and structure of the world, which Kepler answered firmly in favor of the sun-centered system.[68] Each of the following essays takes a fresh look at the philosophical consequences of the 'new star' in the eyes of Kepler and his contemporaries.

The volume begins with an analysis of what Kepler feared to be a fatal mistake in the interpretation of the famous novelty. In the first essay, Dario Tessicini explores the role of Aristotelian elements in the eclectic cosmology of Kepler, focusing on his claim that the fixed stars were "neither infinitely dense

---

things at hand," see Raz Chen-Morris, *Measuring Shadows: Kepler's Optics of Invisibility* (University Park: Penn State University Press, 2016).

66 For an overview of seminal principles from Stoic physics to the sixteenth century, see Hiro Hirai, "*Logoi Spermatikoi* and the Concept of Seeds in the Mineralogy and Cosmogony of Paracelsus," *Revue d'histoire des sciences*, 2008, 61: 245–264.

67 Kepler reported that "no further trace of the star remained" by March 1606, but that it might have expired as early as October 1605. See *JKGW*, 1, 164, ll. 14–21.

68 On the size of the universe and the status of the Sun, see Judith V. Field, *Kepler's Geometrical Cosmology* (London: Athlone, 1988), pp. 19–21.

nor infinitely distant."[69] Tessicini shows how Kepler, contrary to a number of authors who argued that the 'new star' of 1572 became visible by moving directly toward the Earth from "the infinite altitude [*altitudo infinita*]" of the ether, called on Aristotle to combat the endless extension of the universe.[70] Christopher Graney turns to another widely held concern that ran counter to the world view of Kepler. Troubled by the "incredible mass" of the fixed stars according to Copernicus,[71] Tycho and other astronomers objected to what they viewed as a vile form of heresy in the heliocentric hypothesis. Graney provides a full translation of Chapter 16 of *De stella nova* covering the critical response of Kepler, which contains an argument borrowed from his mentor, Michael Maestlin (1550–1631), regarding the "incredible swiftness" of the stars.[72] As Graney notes, the size of the stars remained a source of concern long after Tycho and Kepler.

In our third essay, we encounter another question whose consequences challenged the principles of the prevailing cosmology. How far away was the 'new star,' Kepler asked, and what did it say about the "mind-numbing" distance of the sphere of fixed stars that stood as the final frontier of his finite world?[73] Javier Luna evaluates the answer and ensuing assault that Kepler launched against Giordano Bruno (1548–1600) and William Gilbert (1544–1603) over the scale and size of the world system. Together with the studies of Tessicini and Graney, Luna sheds light on the early evolution of Copernican theory and how Kepler attempted to tailor it according to his own ontology. The quest to fulfill the vision of Copernicus, fostered by Kepler through a "physically grounded representation of the cosmos," reflected the aesthetic values of beauty, harmony, and order.[74] Tessicini, Graney, and Luna build on a growing body of literature that traces the sources of such principles in early modern society and

69 *JKGW*, 1, 252, ll. 40–41.

70 *Ibid.*, l. 4.

71 *Ibid.*, 234, ll. 11–12.

72 *Ibid.*, ll. 8–9. Maestlin also wrote an incomplete manuscript on the 'new star' that was never published. For the first edition of this manuscript with an English translation, see Miguel A. Granada, "Michael Maestlin and His Unpublished Treatise on the Nova of 1604," *Journal for the History of Astronomy*, 2014, 45: 91–122.

73 *JKGW*, 1, 232, ll. 13–15.

74 Nicholas Jardine, "The Places of Astronomy in Early Modern Culture," *Journal for the History of Astronomy*, 1998, 29: 49–62, on p. 58. On the nature of early modern knowledge as "simply another form of display," see Paula Findlen, *Possessing Nature: Museums, Collecting, and Scientific Culture in Early Modern Italy* (Berkeley: University of California Press, 1994), p. 201.

their currency in the learned community.[75] In the end, Kepler chose not to calculate the colossal magnitude of the 'new star' for fear of "public ridicule."[76]

We turn to other friends and foes of Kepler in the following chapters. In the fourth essay, Patrick Boner discusses a telling dispute that Kepler waged with another court astronomer, Johannes Krabbe (1553–1616). Claiming that the 'new star' was a comet, Krabbe excluded "any form of deception" from his observations, including atmospheric refraction.[77] At a time of "great consensus among mathematicians and philosophers regarding the doctrine of parallax," Kepler sought to correct the remaining errors of observation and overturn the Renaissance view of vision as a secure foundation of knowledge.[78] In the following contribution, Jonathan Regier reveals the many similarities and sources of confluence between Kepler and his court contemporary, Anselmus Boëtius de Boodt (1550–1632). Comparing their cosmologies, Regier shows how De Boodt and Kepler shared a common sense of physical and spiritual coherence linking the heavens and earth. Regier concludes that De Boodt and Kepler were members of an intellectually fertile court in Prague flanked by physicians who supported a "medicalized view of nature" that resonated with the work of Marsilio Ficino (1433–1499).

The next two chapters deepen our knowledge of *De stella nova* by engaging with defining themes of the work. In her insightful study, Aviva Rothman reconsiders the role of culture as a target of reform that prompted Kepler to rid astrology of many principles and preserve a pristine core or 'kernel.' Rothman challenges the received view of his famous reform as "problematic in many ways" and prone to presentist views of early modern allegory, allusion, and

---

75 On the role of aesthetic responses as "reliable guides to the deepest truths about the fabric of the world," see Nicholas Jardine, "Kepler, God, and the Virtues of Copernican Hypotheses," in *Nouveau ciel, nouvelle terre: La révolution copernicienne dans l'Allemagne de la Réforme*, ed. Miguel A. Granada and Édouard Mehl (Paris: Belles Lettres, 2009), pp. 269–281.

76 *JKGW*, 1, 238, ll. 6–8. On his later attempt to transform the 'indefinite magnitude' of the world positively as "a principle of astronomy," see Édouard Mehl, "Kepler's Second Copernican Campaign: The Search for an Annual Stellar Parallax after the Roman Decree (1616)," in *Galileo Banned: The Entangled Matter of the Anti-Copernican Decree of 1616*, ed. Natacha Fabbri and Federica Favino (Florence: Olschki, 2018), pp. 191–209.

77 *JKGW*, 1, 198, ll. 6–7.

78 See Chen-Morris, *Measuring Shadows* (ref. 65), pp. 160–162. In response to one opponent of parallax, Kepler expressed his disbelief that, at a time of "great consensus among mathematicians and philosophers," the critic in question was not even a commoner, "but a philosopher" of noble origin at the epicenter of learned activity in Padua; see *JKGW*, 1, 229, ll. 8–13.

play. In her final analysis, Rothman argues that astrology ultimately remained cultural for Kepler, as it relied on a symbolic system of religiously inspired references. In the following essay, we turn to another 'new star' that Kepler discussed in detail at the end of the first part of *De stella nova*. Alerted by Melchior Jöstel (1559–1611) of a third-magnitude star in the neck of the Swan, Kepler set out to prove that it was new and not overlooked by astronomers until now. Following a "fancy" of Kepler, Miguel Granada traces the reconfiguration of the constellations according to three contemporaries who sought to Christianize the heavens. Granada suggests that, though Kepler only flirted with converting the Swan to the figure of Christ, he probably influenced others in Germany.

The final three studies extend our account in space and time down to the present day. Pietro Omodeo focuses on the 'philosophers of chance' who roused the ire of Kepler and influenced the later theories of René Descartes (1596–1650) and Pierre Gassendi (1592–1655). Tracing the thought of lesser known figures such as Nicholas Hill (1570–1610), Omodeo concludes with a closer look at the Cartesian astrology of Johannes Placentinus (d. 1683).[79] In the following contribution, Matteo Cosci surveys the many scholars in Italy who exchanged letters about the 'new star' of 1604. While the scope of his study centers on Cristoph Clavius (1538–1612), Cosci examines the records of several other sources, including one of the earliest observers of the 'new star' whom he identifies for the first time as Andrea Argoli (1570–1657). Cosci concludes with the work of Magini, another focal point in "the sprawling network of scientific exchange" whose observations and planetary tables proved indispensable for Kepler.[80] Together, Cosci and Omodeo give us a better sense of the burst of activity that swept learned society soon after the 'new star' appeared, as well as the questions that lingered long after it vanished from view.

William Blair closes the volume with an epilogue on the scientific legacy of the 'new star.' Based on his cutting-edge research on the remnant, Blair relates some of the "closely held secrets" about the last supernova visible to the naked eye. He describes how our knowledge of the supernova of 1604 has evolved with a more accurate explanation of the massive explosion mechanism behind it. However, the uncertainty surrounding the cause of such a spectacular

79 For a relevant study by the same author, see Pietro D. Omodeo, "Central European Polemics over Descartes: Johannes Placentinus and His Academic Opponents at Frankfurt on Oder (1653–1656)," *History of Universities*, 2016, 29: 29–64.

80 Kepler noted that the star "fooled" Magini when he first saw it on 2/12 October. See *JKGW*, 1, 159, ll. 7–9.

blast of energy means that we still do not know exactly how or why a white dwarf should explode at any given moment in time. As we look forward to further information about the conditions contributing to such a remarkable phenomenon, the supernova of 1604 continues to illuminate our understanding of stellar evolution. Though Kepler fiercely defended the finality and finitude of the cosmos, the idea of stars as subject to change was not entirely foreign to his philosophy. Messengers of mutability, new stars referred to the rise and fall of kingdoms for Kepler.[81] It is fitting that the 'star' named after him today should play such a pivotal role in revealing new details about the nature of the universe.

81 *Ibid.*, 341, ll. 13–16.

CHAPTER 1

Straight Paths and Evanescent Bodies: The Physics and Dynamics of Celestial Novelties in Kepler's *De stella nova*

Dario Tessicini

1.1 Introduction

In *De stella nova*'s Chapter 21, Kepler rejects the hypothesis that the new star of 1604 owed its appearance to a straight-line approach to Earth.* The argument is found in the first part of the chapter, just after a brief introduction, and fits within an articulate refutation whose aim is to distinguish the interpretation of the new star from the arguments in support of the infinite extension of the universe, an hypothesis that Kepler saw as a perverse "abuse" of Copernican heliocentrism and astronomical authority.[1] Closely familiar with Aristotle's refutation of rectilinear motion in the celestial region, in turn linked to infinitism, Kepler draws some of his arguments from the *Physics*. However, since in Aristotelian theory natural motions are determined by the physical properties of bodies, we may wonder whether Kepler's use of Aristotelian arguments against rectilinear motion beyond the sublunary sphere bears any consequences for his own celestial physics.

This issue has already sparked debate among historians and philosophers of science. In his extensive analysis of Kepler's arguments against the infinite universe in *From the Closed World to the Infinite Universe*, Alexandre Koyré read *De stella nova*'s Chapter 21 as a crucial episode in the clash between the "new" infinitist metaphysics, its potential to form the basis for empirical science, and Kepler's "new astronomy," which rejected infinitism for metaphysical reasons, while relying on observational evidence and geometrical demonstrations.[2] Koyré argues that Kepler's resumption of Aristotle is "archaic," a remnant of

* The present study appears as part of the Research Project FFI 2015–64498-P, "Cosmología, teología y antropología en la primera fase de la Revolución Científica (1543–1633)," financed by the Spanish Ministry of Science and Innovation.

1 Johannes Kepler, *De stella nova*, JKGW, 1, 253.20.

2 Alexandre Koyré, *From the Closed World to the Infinite Universe* (Baltimore: Johns Hopkins University Press, 1957), pp. 58–59.

 | DOI:10.1163/9789004437272_003

the old cosmology that Kepler exploits to achieve his aim of defending the finitude of the cosmos and the singularity of the solar system:

> Kepler's objection against infinity is, of course, not new: it is essentially that of Aristotle. Yet it is by no means negligible, and modern science seems rather to have discarded than to have solved the problem. [...] All that is not new, nor specific to Kepler: it is the traditional teaching of Aristotelian scholasticism. Thus we have to admit that Johannes Kepler, the great and truly revolutionary thinker, was, nevertheless, bound by tradition. In his conception of being, of motion, though not of science, Kepler, in the last analysis, remains an Aristotelian.[3]

Grounded on the conviction that spatial infinitism is one of the trademarks of modern science, Koyré finds it difficult to reconcile the coexistence of revolutionary and traditional traits in Kepler's cosmology.[4] In an attempt to establish the authority of the 'new astronomy' and science over the 'new metaphysics,' Kepler resorts to 'traditional' Aristotelian metaphysics. In contrast with Kepler's "science," which Koyré intends presumably as the astronomical and geometrical arguments of *De stella nova*, his theories on being and motion appear as residues of Aristotelian philosophy that remain at work within Kepler's thought as an 'archaic' agent.

The unresolved tension in Koyré's reading of *De stella nova* directly or indirectly influenced two generations of interpreters. On the one hand, the idea of the infinite extension of the universe, supported by Giordano Bruno, one of Kepler's polemical targets, and apparently confirmed by Galileo's early telescopic discoveries, has been studied in order to stress the clash with Kepler's theologically motivated version of heliocentrism.[5] The paradigmatic changes in the observational evidence that occurred after the publication of *De stella nova* undermined Kepler's arguments on the distribution of the stars in the universe, to the extent that the long and complex demonstrations of Chapter 21 became practically ineffective following Galileo's discovery that the stars, when seen through a telescope, did not increase in diameter according to the same

3 *Ibid.*, pp. 87.

4 For a recent critique of Koyré's point of view and a revision of the question of spatial representation, cf. David Marshall Miller, *Representing Space in the Scientific Revolution* (Cambridge: Cambridge University Press, 2014).

5 On Kepler's long-lasting preoccupation, starting from *De stella nova*, with the consequences of Bruno's cosmological conclusions on the size of the universe and its internal organization, see Antonella Del Prete, *Universo infinito e pluralità dei mondi: Teorie cosmologiche in età moderna* (Naples: Città del Sole, 1998), pp. 96–114.

proportion as the planets. This phenomenon vitiated Kepler's argument that an observer located away from the solar system would see nearby stars three to five times larger than when the same stars were seen from our Earth. By the time Kepler wrote the first book of the *Epitome astronomiae copernicanae* (1620), one of his principal arguments against the infinite universe had shrunk to a very short paragraph.[6] Conversely, though, successive telescopic discoveries seemed to tilt the balance of the argument in Kepler's favor, or at least did not provide additional ammunition for the supporters of the infinite universe and the plurality of worlds.

On the other hand, *De stella nova*'s astrology and celestial physics have been considered a serious subject of scholarly investigation ever since Gérard Simon's seminal work, *Kepler astronome astrologue* (1979).[7] While Koyré included philosophical and theological motives in his study of Kepler, astrological reform and animism remained beyond the range of his interest. Simon's analysis, under the influence of Foucault's work on the formation of discursive spaces, integrates Kepler's study of astrological aspects and his doctrine of the soul with the main astronomical and cosmological motives of his work.[8] The legacy of Koyré remains an active agent in Simon's work, although reshaped in such a way that the idea of Kepler as a "Janus bifrons" comes to signify how the modern and archaic faces cannot be separated, even when the archaic comprises Kepler's reform of physical astrology.[9]

Further broadening the boundaries of the Copernican narrative, Robert Westman's *The Copernican Question* (2011) presents Kepler's reform of astrology on center stage. Astrological motives, stemming from Pico's assault at the end of the fifteenth century, are not only integrated, but represent the leading drivers of cosmological reform. No wonder, then, that *De stella nova* is considered a milestone, "a new moment in the evolution of the Copernican question,"[10] marking the point where theology, theoretical astronomy and astral causality meet. In fact, Kepler's own brand of heliocentrism has one of its original

---

6 For an analysis that stresses the role of observations over metaphysical concerns, cf. Judith V. Field, *Kepler's Geometrical Cosmology* (London: Athlone Press, 1988), pp. 20–21.

7 Gérard Simon, *Kepler astronome astrologue* (Paris: Gallimard, 1979).

8 Simon is concerned with the role of analogy, a concept he derives from Foucault's *Les mots et les choses*, in Kepler's conception of a unified physics; see Simon (ref. 7), p. 42.

9 Simon (ref. 7), p. 449: "Si selon le mot de Koyré, Kepler est un véritable *Janus bifrons*, il l'est d'abord en ce que la face moderne et la face archaique de son génie sont rigoureusement indissociables." On the same issue, see Nick Jardine, "Koyré's Kepler/Kepler's Koyré", *History of Science*, 2000, 38: 363–376, and Patrick J. Boner, *Kepler's Cosmological Synthesis: Astrology, Mechanism and the Soul* (Leiden-Boston: Brill, 2013), pp. 15f.

10 Robert S. Westman, *The Copernican Question: Prognostication, Skepticism and Celestial Order* (Berkeley: University of California Press, 2011), p. 394.

moments in *De stella nova*, a work that, echoing Patrick Boner, "represents a central part of Kepler's larger cosmological program,"[11] entailing both the defense and reform of astrology, as well as the elaboration of a unified physics that bridges the gap between the terrestrial and the celestial. Along these lines, the following chapter intends to contribute to the study of Kepler's cosmology in *De stella nova* by probing the issue of how the alleged motions of the recent celestial phenomenon—its 'dynamics'—stimulated Kepler's reflection on relevant aspects of celestial physics and its philosophical sources.

1.2 Straight Motions and the New Star

The title of *De stella nova*'s Chapter 21, "Whether by motion the star was admitted into the depths of the aether, and whether the sphere of the fixed stars extends infinitely,"[12] addresses the question of how the appearance of the new star led some interpreters to underscore an infinite expansion of the universe. The argument claims that the star became visible by variations in its distance from Earth, that is, a motion that brought it closer to our planet and then back toward its original location. Kepler's interpretation of this hypothesis is that the distance eventually covered by the celestial body in order to become visible to observers on Earth presupposes an infinitely extended body.[13]

Kepler's arguments against cosmological infinity fall under three principal headings. The first argument (pp. 251–253) discusses the question from a natural-philosophical perspective, rehearsing Aristotle's rejection of the existence of an infinite body in the *Physics*. The details of this argument will be examined below. The second section of Chapter 21 (pp. 253–256) deals with the astronomical arguments, drawn from the celestial appearances contrary to infinite extension, and in favor of the singularity of our own solar system as well as its central position in the universe:

> First, I say that it is certainly possible to demonstrate from astronomy that the sphere of the fixed stars is finite and there is a clear lower boundary, and that what they claim is not true, [namely] that there is no difference

11 Boner (ref. 9), p. 75.

12 *JKGW*, 1, 251.26–29: "An motu recepta sic haec stella in profundum aetheris, et an fixarum sphaera in infinitum procurrat? Caput XXI."

13 *Ibid.*, 251–252.

> in the appearance of this inferior world, with its Sun, from any of the fixed stars, that is, from region to region or place to place.[14]

Kepler's assumption is that the stars' apparent diameters are proportional to their real ones, and that stars that appear to be of similar size can be assumed to be equidistant from the Earth. This initial hypothesis is put to the test in the following part of the argument, where Kepler considers the eventual consequences of Bruno's tenet that the stars are distributed uniformly in an infinite universe.[15]

The third and shortest section of the chapter (pp. 256–257) takes into consideration metaphysical arguments only ("Quod reliquum est, Metaphysica superstruat"). Here, Kepler exposes the contradictions arising from the assumption of an infinite universe with respect to the observations. For one thing, there would be a logical contradiction if the Milky Way were assumed to have a center, as appearances confirm, and be without one, as the concept of infinity requires. Similar arguments are applied by Kepler to refute the existence of stars infinitely distant from the Earth, since 1) such stars would possess bodies of infinite size, and 2) it would imply the existence of an infinite space without stars. Finally, Kepler concludes that "an infinite body cannot even be comprehended by thought."[16]

From the above conclusion, it appears that the primary aim of the chapter is to attack the arguments in support of the infinite universe that were furthered by some of Kepler's friends and correspondents, in particular the Baron and Imperial Councilor Johannes Matthaeus Wacker von Wackenfels (1550–1619), and the nomadic English nobleman Edmund Bruce.[17] Both were admirers

---

14 *Ibid.*, 253.28–31: "Ajo primum, ex Astronomia doceri posse certissime, fixarum regionem deorsum evidenti limite terminatam: neque verum esse, quod ponunt, nihil differre aspectu mundum hunc inferiorem, suo cum Sole, a qualibet stellarum fixarum, hoc est regionem a regione, sive loco."

15 A detailed analysis of Kepler's argument in this section of *De stella nova*, Chapter 21, may be found in William H. Donahue, "Astronomy in the Service of Cosmology?: Kepler's Arguments for a Finite Universe," in *Astronomy as a Model for the Sciences in Early Modern Times*, ed. Menso Folkerts and Andreas Kühne (Augsburg: Dr. Erwin Rauner Verlag, 2006), pp. 297–306.

16 *Ibid.*, 257.11–12. On this final set of arguments, see Koyré (ref. 2), pp. 71–72; and Jean Seidengart, *Dieu, l'univers et la sphère infinie: Penser l'infinité cosmique à l'aube de la science classique* (Paris: Albin Michel, 2006), pp. 356–357.

17 On the historical and intellectual context of this exchange of ideas, see Massimo Bucciantini, *Galileo e Keplero: filosofia, cosmologia e teologia nell'Età della Controriforma* (Torino: Einaudi, 2000), pp. 96–101; Westman (ref. 10), p. 400; Miguel Ángel Granada, "Kepler and Bruno on the Infinity of the Universe and of Solar Systems," *Journal for the History of Astronomy*, 2008, 39: 469–493.

of Bruno's philosophy and supported some of his most radical cosmological ideas, namely, (1) that the Earth is one of the planets and the universe an infinite body, (2) that the space the universe occupies is undifferentiated and that therefore each of its parts answers to the same physical laws, (3) that the elements of our earthly region are the same throughout the universe and, in turn, (4) that the universe is populated by infinite worlds, each of which are similar to our own heliocentric system.[18]

These issues called into question fundamental cosmological principles, interweaving both the Aristotelian philosophy and Copernicus' theoretical assumptions. Chapter 21's opening lines appropriate Aristotle's assessment of the Pythagorean cosmology in *On the Heavens*, II, 13, a chapter that Kepler knew very well, and that he considered one of the earliest historical records of the heliocentric theory. The relevance of the Aristotelian text can hardly be underestimated. In the years immediately preceding the composition of *De stella nova*, Kepler twice examined this chapter of *On the Heavens*. First, the Pythagoreans' heliocentric theory is discussed in a letter to Herwart von Hohenburg (30 May 1599); then, a section of Aristotle's text (translated into Latin by Kepler himself) makes its way into the *Apologia Tychonis contra Ursum*. After the publication of *De stella nova*, Kepler translated Aristotle's chapter into German and wrote lengthy notes on it.[19] However, contrary to the above occurences, in which Kepler reads Aristotle's passage in the context of the debate on planetary order, *De stella nova*'s reprise of *On the Heavens*, II, 13, focuses on a methodological critique against a "sect of philosophizers" who, rather than deriving the physical causes from observable evidence and experience, conceive a cosmology ("mundi constitutio") of their own, in an 'out-of-the-blue' manner revealing their zeal more than reason.[20] Once captivated by

18 See Bruce's letter to Kepler dated 6 November 1603, in *JKGW*, 14, no. 450, and Bucciantini (ref. 17), pp. 92–116, where the letter is discussed on pp. 96–97.

19 See the letter to Herwart von Hohenburg in *JKGW*, 13, no. 123; the *Apologia* in *JKGW*, 20, 1, 32–33; Kepler's German translation of *On the Heavens* may be found in *JKGW*, 20, 1, 150–167: 150. See also *ibid.* (ref. 1), 161, for Kepler's interpretation of the relevant passage. On the composition of this text, see *ibid.*, 495, and Nicholas Jardine, *The Birth of History and Philosophy of Science: Kepler's* A Defence of Tycho against Ursus *with Essays on Its Provenance and Significance* (Cambridge: Cambridge University Press, 1984), p. 161.

20 *JKGW*, 1, 251.30–252: "Sed est alia secta, eorum qui (ut Aristoteles de Pythagoreorum sententia, a Copernico repetita, immerito quidem pronunciat) *non initium ratiocinationis ex sensibus deducunt, neque causa rerum ad experimenta accomodant*: sed qui ex abrupto, et quasi quodam enthusiasmo, concipiunt et depingunt intra sui cerebri parietes, aliquam de mundi constitutione opinionem: eamque ubi mordicus fuerint amplexi; iam porro ea quae sub experientiam cadunt, quaeque quotidie eveniunt, crinibus pertracta ad axiomata sua accommodant (my emphasis)." Cf. this passage with Kepler's translation

their own ideas, they are said to stubbornly bend the interpretation of phenomena to their own chosen principles.

Kepler's borrowing from Aristotle is an adaptation of the original text concerning the error of the Pythagorean philosophers, while the considerations on the method followed by the sect—"ex abrupto" and "quasi quodam enthusiasmo"—are his own additions. It is worth noting that the latter expression occurs in another context of the heliocentric reception, namely in a letter by Melanchthon to Camerarius of 25 July 1542, in which the reformer can barely forgive, and only on account of his age, that young Rheticus' mind was carried away "by a certain enthusiasm for the [heliocentric] philosophy."[21] If the reference is more than incidental, it reinforces the link between the theme of *De stella nova*'s Chapter 21 and the interpretation of heliocentrism. The allusion could even be regarded as an oblique reference to Wacker, in which Kepler, taking his cue from Melanchthon, disagrees benevolently with his friend by justifying, to a certain extent and perhaps not without a degree of condescension, his critical attitude toward the infinite extension of the universe.

Overall, the purpose of the initial section of Chapter 21 is to introduce the argument that will be rejected in the course of the chapter. The position mentioned in the chapter's title is explained more thoroughly in the following section:

> They believe that this new star, and if there were any others of the same kind, gradually descended from the deepest recesses of Nature, which, they affirm, extends to an infinite altitude, until according to the rules of optics it appeared to such an extent that it became visible to the human

---

of Aristotle in *JKGW*, 20, 1, 33: "Non quaerant rationes et causas, ad ea quae apparent, reddenda et efficienda, sed ad praeconceptas suas quasdam opiniones et ratiocinationes ipsa phaenomena violenter accomodent, et conciliare conentur." For an analysis of Kepler's translations of Aristotle, and his use of Perion's translation, see Nicholas Jardine and Alain Segonds, "Kepler as Reader and Translator of Aristotle," in *Philosophy in the Sixteenth and Seventeenth Centuries: Conversations with Aristotle*, ed. Sachiko Kusukawa and Constance Blackwell (Aldershot: Ashgate, 1999), pp. 205–32, on p. 216.

21 *Melanchthons Briefwechsel: Kritische und kommentierte Gesamtausgabe*, ed. Christine Mundhenk (Stuttgart-Bad Cannstatt, Frommann-Holzboog, 2010), 11, letter to Joachim Camerarius, 3013 (*Corpus Reformatorum*, 4, 848): "Indulsi aetati nostri Rhetici, ut ingenium quasi quodam enthusiasmo incitatum ad hanc philosophiam partem, in qua versatur proveheretur." On this letter, see Hans Blumenberg, *The Genesis of the Copernican World* (Cambridge, MA-London: MIT Press, 1987), pp. 222–223.

eye; then, retracing its steps backwards, returned to the infinite depth, growing smaller each day the higher it proceeded.[22]

The above text entails an expression that appears to be taken from Virgil's *Aeneid* (9, 391), "vestigia retro observata legit." Kepler's text reads, "vestigiis retro lectis," suggesting that the motion of the new star followed the same path in both its descent and its ascent. Virgil was considered by some as a vehicle of Pythagorean lore, and this reference echoes the Pythagorean motive of the chapter's opening lines.

Kepler does not provide any further details about the motion of the new star at this stage. However, there are two further points to be raised concerning this section of the chapter: first, the argument of *De stella nova* is organized according to the principles of Aristotelian physics regarding the relations between substances and motions; and secondly, the reference to Virgil echoed other lines of the same poet, whose interpretation linked a "backward motion" of the stars to the religious explanation of the *nova* as a divine prodigy.[23] Furthermore, Kepler rejected an argument put forward by several other scholars, including at least one of his friends and assiduous correspondent, the astronomer David Fabricius, whose explanation of the origin of new stars involved the exceptional intervention of divine power.[24] Fabricius was also Kepler's main interlocutor during the composition of his *Astronomia nova* (1609). He received Kepler's first letter announcing that the path of Mars was not circular but oval, and over the following years Fabricius discussed Kepler's new hypotheses concerning the planetary motions.[25]

22 *JKGW*, 1, 252.3–8: "Iis placet, novum hoc sidus, et si qua alia fuerunt hujusmodi, ex penitissimo Naturae sinu, quem in infinitam altitudinem exporrigi affirmant, paulatim descendisse; donec ratione optica in tanta quantitate apparens, oculos hominum in se converteret; indeque vestigijs retro lectis, rursum in infinitam altitudinem subductam, tanto quotidie minorem, quanto pervaserat altius." On this passage, cf. Koyré (ref. 2), p. 59, and Westman (ref. 10), p. 400.

23 Cf. Virgil, *Aeneid*, IV, 489: "sistere aquam fluviis, et vertere sidera retro." The line is quoted by Tycho Brahe in his early work on the new star of 1572: see Tycho Brahe, *De nova stella*, *TBOO*, 1, 19: "Reliquum igitur est, ut statuamus Dei totius machinae mundanae opificis, admirandum hoc esse ostentum, praeter omnem naturae ordinem, a seipso in initio constitutum: nunc demum advesparascenti mundo exhibitum. Divina enim maiestas liberrime agit, nec ullis obstricta est Naturae vinculis, sed cum vult, sistit aquam fluviis et vertit sidera retro." The same line also appears in Cornelius Gemma, *De prodigiosa specie naturaque cometae* (Antwerp: Christopher Plantin, 1578), p. 38.

24 On this episode of Kepler's *De stella nova*, see Granada (ref. 17), pp. 475–480.

25 On Fabricius' role in Kepler's work on planetary theory ahead of the publication of his *Astronomia nova*, see James R. Voelkel, *The Composition of Kepler's* Astronomia nova (Princeton: Princeton University Press, 2001), Chapter 8, pp. 170–210.

On the first point above: according to Aristotle, simple bodies have simple motions. These can be either rectilinear, a natural form of motion for the terrestrial elements, or circular, belonging to bodies solely composed of superlunary quintessence. Mixed or composite motions are proper to beings in the terrestrial region, such as animals and meteorological phenomena.[26] In turn, the dichotomy between circular and straight motion is the pivotal point of this first part of Chapter 21, in which Kepler argues that the hypothesis of the star's descent from, and ascent to, the depth of the heavens can come only from one of these two forms of motion. Each of the two options is presented by Kepler as supported by one of two sides.[27] However, there is no evidence, as far as I can find, that the first hypothesis positing the circular motion of the new star as a cause of its variation in distance is anything more than a theoretical option deriving from the principles of Aristotelian dynamics. In addition, Kepler's text is reminiscent of Aristotle's arguments against an infinitely extended body, in particular *Physics*, Book 8, Chapter 9, where the infinite is discussed in relation to circular and straight motions.[28] In Aristotle's words, "any one point as much as any other is alike beginning, middle and end," so that the body in circular motion is always at the beginning and end of its motion. Circular motion is the simplest and most perfect, as it is continuous and infinite, although it does not require an infinite space to perform its motion. Therefore, the sphere is both at rest and in motion at any given time.[29] These properties of circular motion and the sphere were appropriated by Copernicus in *De revolutionibus*, 1, 4:

> I shall now recall to mind that the motion of the heavenly bodies is circular, since the motion appropriate to a sphere is rotation in a circle. By this

26 Cf. Aristotle, *On the Heavens*, trans. W. K. C. Guthrie (Cambridge, MA: Harvard University Press, 1939), I, 2, pp. 10–17. An effective summary of Aristotle's doctrine of nature and motion is found in Simplicius, *On Aristotle's* On the Heavens *1.2–3*, trans. Ian Mueller (London-New York: Bloomsbury, 2011), p. 44: "1. There are two simple motions, in a circle and in a straight line; 2. a simple motion attaches to a simple body; 3. the motion of a simple body is simple; 4. one natural motion attaches to one thing; 5. for a single thing there is a single contrary; 6. heaven moves in a circle, as perception indicates." See also Edward Grant, *The Foundations of Modern Science in the Middle Ages: Their Religious, Institutional and Intellectual Contexts* (Cambridge: Cambridge University Press, 1996), p. 64.

27 See the parallel texts presenting the two positions in *JKGW*, 1, 252, ll. 8–9 and ll. 28–29: "Qui hoc affirmant, pars naturae caelorum consulentes, lege circulari hoc factum esse pronunciant"; "Fuerunt itaque qui ante hos triginta annos ista sobrie magis exponerent."

28 The main discussion and rejection of an infinite body takes place in *Physics*, V, 5–8. See Ursula Coope, "Aristotle on the Infinite," in *Oxford Handbook of Aristotle*, ed. Christopher Shields (Oxford: Oxford University Press, 2012), DOI: 10.1093/oxfordhb/9780195187489.013.0011.

29 Aristotle, *Physics*, in *The Complete Works of Aristotle: The Revised Oxford Translation*, ed. Jonathan Barnes (Princeton: Princeton University Press, 1984), VIII, 9, 265b1–2, p. 157.

> very act the sphere expresses its form as the simplest body, wherein neither beginning nor end can be found, nor can the one be distinguished from the other, while the sphere itself traverses the same points to return upon itself.[30]

Thus, Kepler's adoption of Aristotle's arguments against linear motion can be understood both as functional to refuting the infinitist consequences of some interpretations of the new star, as well as part of the interpretative context of Copernican heliocentrism, at least as far as the principles of motion are concerned.

The explanation of the appearance and disappearance of the new star through circular motion—a "lex circularis," as Kepler calls it—requires the ingenious solution of the celestial body being transported on a circle whose center is located in the celestial region somewhere far away from the Earth. A representation of the star's descent and ascent is provided by Kepler with reference to an earlier figure in Chapter 17 of *De stella nova* (Fig. 1.1).

In its original context, the purpose of the illustration is to demonstrate that the new star can only be located among the fixed stars, even when one assumes a geocentric planetary order. A planet (or any other celestial body) moving along the circle PQXI (with A as the Earth and BCDE the Sun's orbit) will be seen to follow a retrograde motion along FKZY. Since the new star does not appear to move at all, either it does not possess a motion comparable to that of the Sun and the planets, or its position is so far away from the Earth that its motion is not observable. This means that the circle on which the new star moves needs to appear smaller than 2′ (which is equal to the star's diameter), a size that would require the fixed star to be far more distant from the Earth than assumed by Tycho.[31]

In Chapter 21, the meaning of the figure is only slightly reframed to allocate for the hypothetical motion of the fixed star. Drawing from Aristotle's *Physics*, the principle of circular motion of the celestial body is that its descent is inextricably linked to its ascent. The moment between the two motions is what Kepler names the "nexus," which ultimately proceeds from the definition of circular motion as seen from above: the inversion of the motion's direction is relative to the observer, similar to how each point on the circumference can

30 Nicholas Copernicus, *On the Revolutions*, trans. and commentary by Edward Rosen (Baltimore: Johns Hopkins University Press, 1992), 1, 4, p. 10. A recent analysis of this passage in relation to Aristotle's representation of space may be found in Miller (ref. 4), pp. 56–57.

31 *JKGW*, 1, 239–240.

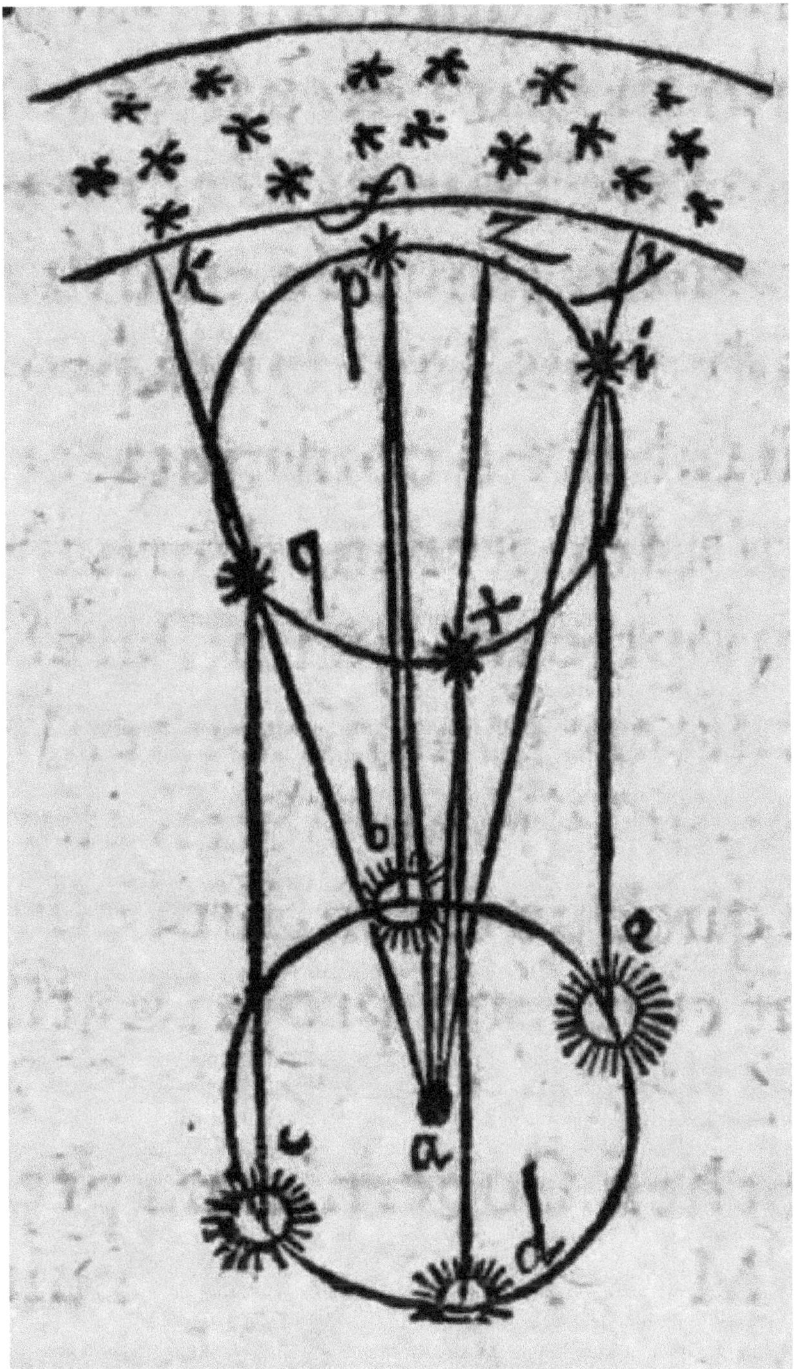

FIGURE 1.1 Johannes Kepler, *De stella nova* (Prague: Paulus Sessius, 1606), p. 90.

at the same time be the beginning, the centre or the end of the motion.[32] In the wake of the demonstration from Chapter 17, however, Kepler argues that since the star does not appear to have any motion in latitude, the diameter of the circle over which it supposedly moved should be so small that it would be undetectable to the human eye, or at least smaller than the visible diameter of the star.[33] However, the latter case leads to the paradoxical consequence that the new star would have traced a circle whose diameter was contained within its body. As a final objection, Kepler observes that the visibility of the new star was greatest at its first appearance and did not improve gradually, as would have been expected if the star had gradually drawn closer to the observer.[34]

The second hypothesis for the new star's variation in distance suggests that the celestial body became visible through a direct, straight-line motion towards the Earth. This explanation was first conceived at the time of the appearance of the new star of 1572, when it was supported by a heterogeneous group of interpreters that included the English "magus" John Dee, the mathematical practitioner Thomas Digges, the professor of medicine from Louvain, Cornelius Gemma, and the professor of mathematics at the University of Frankfurt on the Oder, Elias Camerarius.[35] The opinions of these authors were amply explored and critically assessed by Tycho Brahe in his posthumous *Astronomiae instauratae progymnasmata* (1602). Kepler oversaw the publication of Brahe's work and probably had its arguments in mind when he discussed the possibility of the new star's rectilinear motion. Brahe argued that the cause of the new star's disappearance could be attributed to distance only if an immense celestial space was admitted.[36] This possibility was inconceivable to Brahe, as

---

32 *Ibid.*, 252.9–11: "ut ita causa sit, cur ex descensu, nexus, et quodammodo prognatus fuerit contrarius ascensus. Id enim quomodo fiat, in rotis est obvium."

33 *Ibid.*

34 *Ibid.*, 25–26: "Adde quod non, ut paulatim visa est minui, sic paulatim et incrementa sumpsit, sed subito effulsit maxima."

35 A variant of this opinion was held by Cornelius Gemma: see Dario Tessicini, "Cornelius Gemma and the New Star of 1572," in *Change and Continuity in Early Modern Cosmology*, ed. Patrick J. Boner, (Dordrecht: Springer, 2011), pp. 51–66. On Camerarius' work on the new star, the *Observatio and descriptio novi sideris*, see Michael Weichenhan, *"Ergo perit coelum …": Die Supernova des Jahres 1572 und die Überwindung des aristotelischen Kosmologie* (Stuttgart: Steiner, 2004), pp. 578–580.

36 Tycho Brahe, *Astronomiae instauratae progymnasmata*, in *idem*, TBOO, 4, 204, referring to a report from Christoph Rothmann that John Dee had argued that "stellam hanc neotericam, intra quidem Mundi Aetherei vastissimam capacitatem, contentam extitisse: attamen non in eodem loco, quo ad remotionem a Terris, perpetuo permansisse, sed paulatim ab humiliori situ in sublimiorem, per lineam rectam ascendendo, se recepisse."

it would have entailed the existence of a vast void between Saturn and the sphere of the fixed stars, shattering the harmonic structure of the universe.[37]

Taking as his point of reference the new star of 1572 as interpreted through the *Progymnasmata*, Kepler briefly considers how 'his' new star could have travelled by a straight-line motion back to the ethereal depths of the heavens. Two objections are raised in response. The first is similar to what was formulated against the hypothesis of circular motion. It focuses on the moments of the star's appearance and disappearance by asking why the phenomenon became suddenly visible but disappeared gradually.[38] The absurd consequence of this assumption, under the premises of Kepler's argument, is that the motion of descent of the new star towards the Earth was faster than its motion in the opposite direction, with no feasible explanation for it. The second objection questions the changes in the direction of the star's motion: namely, what is the reason for the star's inversion of direction ("quae causa conversi cursus")?[39] In the case of the star's circular orbit, the change of direction can be explained by the continuous nature of its motion. This was one of the reasons why Aristotle deemed circularity superior to straight paths, and thus appropriate for celestial bodies: "Now rotatory motion can be eternal; but no other motion, whether locomotion or motion of any other kind, can be so, since in all of them rest must occur and with the occurrence of rest the motion has perished."[40] By contrast, linear motion in a finite universe cannot continue indefinitely, but must eventually come to an end. Aristotle argues at this point that "the line traversed in rectilinear motion cannot be infinite; for there is no such thing as an infinite straight line."[41] Since there is no infinite space, infinite motion on a straight line is impossible. Kepler's argument follows in the wake of Aristotle on this point, claiming that the new star had first descended, and then, perhaps after stopping, reversed its motion (in the opposite direction). Kepler's explanation and final comment of this section states that "no natural body behaves in this way" ("Naturalium enim rerum haec proprietas non est"). This particular and seemingly isolated reference to the properties of natural bodies marks, in my

37 On the double defense deployed by Kepler against those who, on the one hand, "believed the figures were absurd" and, on the other, "found evidence for an infinite universe," see the contribution by Javier Luna to this volume.

38 *JKGW*, 1, 252.29–31: "Ii vero et causam dicant, cur subito per immane illud spacium demissa (quia subito effulsit), inde non subito, sed paulatim revolaverit, et quae causa conversi cursus." Earlier on, Kepler questioned (25–26): "Adde quod non, ut paulatim visa est minui, sic paulatim et incrementa sumpsit, sed subito effulsit maxima."

39 *Ibid.*

40 Aristotle, *Physics*, (ref. 29), VIII, 9, 265a25–27, p. 156.

41 *Ibid.*, VIII, 9, 265a16–17, p. 156.

view, the juncture between the dynamics and ontology of Kepler's discourse on the new star.

To understand Kepler's line of thought, we need to return our attention to Aristotle's *Physics*. In *Physics*, Book 8, Chapter 8 (263a1), Aristotle posits that a change of direction for a body in linear motion equates to coming to a halt and then initiating another motion in the new direction: "Therefore that which turns back in traversing a rectilinear course must in so doing come to a stand. Consequently, there cannot be a continuous rectilinear motion that is eternal."[42] This appears to be the same dynamics by which Kepler interprets the star's rectilinear motion according to the hypothesis explored above.

Moreover, *Physics*, Book 8, Chapter 9 (265a19–21) states that "rectilinear motion on a finite [straight] line is composite if it turns back, i.e., two motions, while if it does not turn back it is incomplete and perishable." Therefore, a body moving on a straight line and reversing (or changing) its direction does not perform one continuous motion, but multiple motions separated by stops. Conversely, if the body does not turn back, its motion will necessarily come to a halt. The premise of the above assertions is found in the physics of natural bodies. For Aristotle, natural bodies have only one natural motion. The purpose is for each body to travel to its own natural place. Since "the movement of each body to its own place is motion towards its own form," once a body is at rest, it has realized its form and therefore remains at rest unless moved by external forces.[43] Kepler's objection that "no natural body behaves in this way" can be understood in the wake of Aristotelian arguments on linear motion and their motivations. Similar to Aristotle's theoretical case, Kepler interprets the linear motion of the new star as composed of two motions in opposite directions that are separated by the body's coming to a halt and starting again in the other direction. Once the theory of linear motion has been interpreted according to this conceptual framework, Kepler denies the existence of any physical principle that can explain the phenomenon's appearance, although at this point in Chapter 21 there is no commitment as to which natural "proprietas" Kepler is referring to. Clearly, Kepler's aim is to refute one unpalatable hypothesis, while the implications of extending terrestrial physics into the celestial realm will be discussed further on in *De stella nova*. At this point, the focus of the argument shifts from the natural to the extra-natural realm. Could there be a metaphysical explanation for the origin of the new star? Kepler briefly considers the hypothesis of God's intervention in the creation of the new star and

42 *Ibid.*, VIII, 8, 263a1–2, p. 152.

43 See Peter K. Machamer, "Aristotle on Natural Place and Natural Motion," *Isis*, 1978, 69: 377–387, on p. 380.

in its downward motion ("ab eodem demissam ex alto"), but only to dismiss it immediately, noting the difficulties of understanding how the double motion could be the effect of the same potency, and lamenting the "inextricable labyrinths" in which the supporters of the infinite universe are plunged.[44]

Kepler's explicit polemical targets in this section are Giordano Bruno and William Gilbert. In a much-commented passage from Chapter 21, Kepler voices his apprehension of the consequences of their infinitist philosophies.[45] Gilbert rooted his cosmology in the theological principle that God's infinite power can only produce an infinite creature. In his universe, there is no place for the sphere of the fixed stars, whose diurnal motion is taken up by a rotating Earth. Once unbounded, the stars are conceived by Gilbert as dispersed in the unlimited space extending beyond Saturn, in a way that resembles the world vision described in Thomas Digges' *Perfit Description of the Caelestial Orbes* (1576). Gilbert's full adherence to the heliocentric theory is a disputed matter, however, since in his system the Earth does not perform an annual motion.[46]

Bruno's cosmological conception further pushes the conceptual boundaries of Copernican heliocentrism by admitting the infinite extension of the physical universe, positing its homogeneous nature, and populating its space with innumerable worlds modeled on our solar system. The prospect of a universe with no absolute center or periphery is what unsettles Kepler's soul:

> This very cogitation carries with it I know not what secret, hidden horror; indeed, one finds oneself wandering in this immensity, whose limits, whose center, and therefore any determinate places are denied.[47]

As Miguel Ángel Granada points out, there are precise textual references in this passage that demonstrate Kepler's reading of Bruno's *De immenso*. In Book 4,

44 On the question of the relevance of theological explanations for celestial phenomena for Kepler and other Protestant authors, see at least Charlotte Methuen, "This Comet or New Star": Theology and the Interpretation of the New Star of 1572," *Perspectives on Science*, 1997, 5: 499–515. For other instances and sources, cf. *Christoph Rothmann's Discourse on the Comet of 1585: An Edition and Translation with Accompanying Essays*, ed. Miguel Á. Granada, Adam Mosley and Nicholas Jardine (Leiden-Boston: Brill, 2014), p. 173.

45 Analyses of this passage can be found in Koyré (ref. 2), pp. 61–62; Field (ref. 6), pp. 18–19; Seidengart (ref. 16), pp. 357–358; and Granada, (ref. 17), pp. 476–480.

46 On Gilbert's cosmology and *philosophia magnetica*, see Gad Freudenthal, "Theory of Matter and Cosmology in William Gilbert's *De Magnete*," *Isis*, 1983, 74: 22–37.

47 *JKGW*, 1, 253.15–17: "Quae sola cogitatio, nescio quid horroris occulti prae se fert; dum errare sese quis deprehendit in hoc immenso; cuius termini, cuius medium, ideoque et certa loca, negantur." Translated in Koyré (ref. 2), p. 61. See also Granada (ref. 17), pp. 477–478, for this passage's reference to Bruno's *De immenso*.

Chapter 3, Bruno invites his reader to imagine an "ascent to the heavens and the true contemplation of the world."[48] Once the Earth is observed from the orbit of the Moon, features of the terrestrial landscape gradually fade away and the Earth itself begins to resemble the Moon as it appears from our earthly point of view. Forests, rivers and lakes, cities and islands, even straight and curved lines begin to blur and become indistinct from a distance.[49]

Bruno's evocative wording runs parallel to Kepler's own terminology, although the positions of the viewer and the subject appear to be inverted by the two authors. In Bruno, the "ascensus" confuses the shapes of distant terrestrial objects, so that the boundaries between the curved and the straight become indistinguishable to the viewer. This observation is subsequently reinforced in the same chapter, when Bruno later argues that linear and circular motions cannot be detected from great distances. On the contrary, observers believe them simply to be stationary.[50] In turn, Kepler's presentation of the arguments in favor of distance variation as the cause of the new star's appearance and disappearance seems to be motivated by similar questions concerning the nature of observing distant objects, although he arrives at a radically different answer than Bruno.[51]

Furthermore, the very hypothesis that the new star became visible by its own motion could also be attributed to Bruno. According to Bruno, the possibility of observing comets (considered to be seldom-seen planets) and new stars depends, among other circumstances, on their distance from the observer. In *De immenso*, Book 3, Chapter 7, Bruno takes aim at Aristotelian celestial perfection and immutability, as well as the simple motions of the heavens. He argues against the Aristotelian philosophers' demotion of the sense of sight,

48 Giordano Bruno, *De immenso, et innumerabilibus*, in *Jordani Bruni Nolani Opera Latine conscripta* (Naples: Morano, 1884; facsimile ed., Stuttgart: Frommann, 1962) I, 2, p. 15.

49 *Ibid.*, p. 16: "Dic: ubi sylvarum species? ubi flumina, montes,/ Stagna, lacus, urbes, brumae discrimen, et aestus?/ Ut tantum species candentis mansit, et atri?/ Ut maculat clarum Oceanum nigra insulam passim?/ *Quae recti nusquam aut curvi geometrica amussis,/ Atqui ex iis quid confusum discriminat ambo* (my emphasis)."

50 *Ibid.*, p. 21: "Hoc, quotidiana iugeque experientia, sensus docet, qui quae propinquius gyrantia et in directum deambulantia moveri novit, eminus eodem ordine mota atque mensura, fixa manere iudicabit."

51 On this point, see the accurate analysis by Field (ref. 6), p. 18, where she remarks that "Kepler's philosophical argument against the idea of an infinite universe is directed against the limitations it would place on the value of observation. If the Universe is immense or infinite, in the sense that we must believe it to contain objects we cannot observe, such as stars that are sometimes too far away for us to see them (such stars had been invoked to explain the observation of new stars in 1572 and 1604), then we cannot know how to construct theories to explain what we observe."

and therefore of observations that do not reinforce their paradigms. A "nova lux" in the heavens is taken as an example of their oversight. Its appearance, attributed to the star's motion of "accessus" and "recessus" from the Earth, challenges the Aristotelian theory that only perfect circular motions exist in the celestial region.[52]

Ultimately, in Chapter 21 the issue of linear motion, presented as the principal case for the hypothesis that distance variation caused the appearance and disappearance of the new star, is closely implicated in the interpretation of the cosmological principles underpinning heliocentrism. As a follow-up to the first lines of the chapter, in which the Aristotelian assault on the Pythagoreans was addressed to the modern supporters of the infinite universe, the entire passage on Bruno and Gilbert is enclosed by references to the cosmological debate on celestial motion and the size of the universe. Just before addressing the two natural philosophers, Kepler recounts the principal lines of argument that derive the infinite size of the sphere of fixed stars from Copernicus' abolition of their diurnal motion:

> For we shall, if possible, take this immensity away from them, so that the ascent comes crashing down. That immensity is revived by the ancient scholia of pagan philosophers, in particular by this argument: that Aristotle had demonstrated from motion [i.e., the diurnal motion of the fixed stars] that the cosmos is finite, while Copernicus, having removed that motion, may allow the sphere of the fixed stars to be infinite.[53]

At the end-point of the argument, following his discussion of Bruno and Gilbert, Kepler returns to the terminology from the beginning of the chapter, wagging his finger at the "sect" of philosophers who "abused Copernican authority and, in fact, all of astronomy."[54] While defending the extension of the universe as a consequence of Copernican heliocentrism, Kepler resists its infinite extension and the threat posed by the "ascensus" of the new star by deploying an arsenal

52 Bruno (ref. 48), I, 1, p. 371.

53 *JKGW*, 1, 252.40–253.3: "Adeoque eripiamus illis, si potest fieri, hanc immensitatem: tunc enim ascensus ille per se cadet. Reducitur illa ex antiquis gentilium Philosophorum scholis, hoc praecipue argumento; quod Aristoteles ex motu finitum mundum demonstraverit: Copernicus vero fixarum Sphaeram adempto motu, infinitam esse patiatur." Koyré (ref. 2), p. 60, translates, mistakenly in my view, "ascensus" with "assertion" ("then, indeed, the assertion will fall of itself"). I believe Kepler intends to refer to the possibility of the infinite ascent of the new star "into the depth of the heavens," a thesis that is clearly impossible once the immense or infinite size of the universe is refuted.

54 *JKGW*, 1, 253.20–21; cf. Granada (ref. 17), p. 479, for a translation and interpretation of this passage.

of philosophical, astronomical and theological arguments in defense of the principles of his own cosmology. Essentially, the weapons of his world vision are "based on archetypal metaphysics, a bounded universe well-proportioned according to the dimensions of the five regular solids."[55]

1.3 The Matter of Celestial Novelties

The location, origin and material nature of the new star are discussed in chapters preceding and following Chapter 21. After demonstrating the celestial position of the new phenomenon (Chs. 13–17), Kepler investigates its light and scintillation (Ch. 18), all of which provide further evidence for the hypothesis that the new luminary is celestial. Kepler follows Scaliger's *Exotericarum exercitationum* (1557) in considering light as innate to the body of the stars. The way in which light is emitted by the stars is exemplified by the analogy of the motion of the heart and the arteries.[56] Chapter 19 is the first of four chapters (together with 20, 22–24) devoted to the material nature of the new luminary. Two hypotheses are discussed in the first instance: whether the new star is a conflagration of matter ("flamma"), and therefore an evanescent body, or a perennial one ("corpus, sua constans materia").[57] If the latter is true, the scintillation either depends on a "palpitation" of the body or its rotation. Kepler prefers the hypothesis that the new star is a conflagration whose fuel, once completely consumed, causes the disappearance of its light.[58] The chapter features a protracted refutation of Francesco Patrizi's thesis on the immaterial nature of the new star.[59] Instead, Kepler's explanation implies the abandonment of celestial immutability and promotes the presence of elemental matter (fire and water) in the heavens.

A crucial step in the extension of earthly physics to the celestial realm is the role of Copernicus' heliocentric theory in reframing the status of the relation(s)

55 Cf. Westman (ref. 10), pp. 393–401, and the discussion of Kepler's use of *modus tollens* against the arguments of Bruno and Gilbert on pp. 399–400.

56 *JKGW*, 1, p. 241.24–29: "Placet Scaligero, et ipse vehementer approbo, *naturalem esse in sideris corpore vim luminis sese spargentis non sine motu* [...] Exemplum puto satis aptum in motu cordis, arteriarumque; ut hoc quoque cordi cum Sole, vitae cum lumine, praeter ea, quae cap. 1 et 6 Optices dixi, commune sit (my emphasis)."

57 *Ibid.*, 245–246.

58 *Ibid.*, 246.5–6: "Equidem faveo sententiae primae, flammam fuisse; quia ut flamma, consumpta est, quasi deficiente alimento."

59 An extended analysis of Chapter 19 and its intellectual context can be found in Patrick J. Boner, "A Tenuous Tandem: Patrizi and Kepler on the Origins of Stars," *Journal for the History of Astronomy*, 40, 2009: 381–391.

between the Earth and celestial bodies. By considering the Earth as one of the planets ("unus planetarum"), Copernicus elevated it to the same ontological status as the planets, thus conferring what Kepler calls a "jus civitatis in coelo," or a right of 'heavenly citizenship' among the celestial bodies. That is to say, the distinction between the terrestrial and celestial realms is all but gone. Any remaining differences are dictated by distance, not by substance.[60]

Based on Copernicus' analogy between the Earth and the planets, the relationship between the celestial and terrestrial is re-drawn in terms of their connection or affinity. A marginal heading placed next to this passage in the text reads: "Cognatio coelestium cum sublunaribus."[61] The term *cognatio* is revelatory of Kepler's selective appropriation of Copernican motives, although once again they are not entirely distinct from Aristotle, as we will soon see below. In the first book of *De revolutionibus*, *cognatio* first appears in Chapter 8, where the term refers to the 'connection' between the rotating Earth and the elements that follow its motion.[62] In order to counter the age-old argument that clouds and projectiles would be left behind by a moving Earth, Copernicus considers that the water, "part of the air" and the bodies that are in it, move together with the earth, their link being described as the *cognatio* between them, and more particularly between the earth and the bodies in the lower region of air, that is, extending to the top of the mountains, as explained in *De revolutionibus*, Book 1, Chapter 2. Following Aristotle's *Meteorology* (Book 1, Chapter 7), Copernicus claims that the upper region of air is transported by the circular motion of the heavens, as evidenced by the rising and setting of comets, or "bearded stars." Despite their sudden generation in the sphere of air, they are unaffected by terrestrial motion. Crucially, the disconnection between comets and the Earth is attributed by Copernicus solely to their distance, similar to how Kepler conceives the difference between the terrestrial and the celestial, as just seen above.[63]

---

60 *JKGW*, 1, 246.19–24: "[…] non tam essentiae dissimilitudine, quam intervallis, distincta esse illa coelestia ab his sublunaribus […] Ut Copernicum etiam mittam, qui Telluri ipsi jus civitatis in coelo dedit." Cf. Patrick J. Boner, "Life in the Liquid Fields: Kepler, Tycho and Gilbert on the Nature of the Heavens and Earth," *History of Science*, 2008, 46: 275–297, on pp. 277–278.

61 *JKGW*, 1, 246.19–20.

62 Copernicus (ref. 30), p. 18: "We would only say that not merely the earth and the watery element joined with it have this motion, but also no small part of the air and whatever is linked in the same way to the earth [*et quaecumque eodem modo terrae cognationem habent*]."

63 *Ibid.*: cf. on this passage and its relation to Aristotle and Arabic sources, particulary the work of al-Tusi, F. Jamil Ragep, "Tusi and Copernicus: The Earth's Motion in Context," *Science in Context*, 2001, 14: 145–163, on pp. 146–148.

In addition, in *De revolutionibus* Book 1, Chapter 10, the term *cognatio* refers to the affinity between the Earth and the Moon as part of the arguments for the Earth as "one of the planets" and for the symmetry and *nexus* of the cosmological order of the planets. As in the previous passage, there is a reference to Aristotle ("Aristoteles de animalibus ait"), according to whom "maximam Luna cum terra cognationem habet." Although the reference is likely taken from Averroes' *De substantia orbis*, which misunderstands an Aristotelian statement from *De generatione animalium*,[64] the idea that the physical similarity between the Earth and the Moon was endorsed by Aristotle himself strengthens the authority of Copernicus' argument on the *ratio* of his cosmological order. Thus, the Earth's displacement from the center of the cosmos leads to its integration into the *systema mundi*, since both motion and substance render the Earth one of the planets.

Kepler appropriates Copernicus' terminology, but expands the notion of "affinity" to include the recent celestial novelty, as well as comets. Similar to our earthly region, heavenly matter allows for substantial change and the existence of temporary bodies. Chapter 20 is dedicated entirely to defending the natural generation of the new star, contrary to David Fabricius' theological explanation of the celestial body as a perennial being.[65] In turn, Kepler traces the origin of the new star back to alterations in the celestial region. These are conceived with reference to both meteorology and physiology in the wake of an organicist approach to natural phenomena. Thus, Kepler claims that the new star was created by a "congregation" of celestial matter from the sphere of the fixed stars, a natural process of spontaneous generation similar to how smaller forms of life come from larger organisms.[66] Once the matter from which it was generated is exhausted, the heavenly formation dissolves.[67] Kepler is aware that the concept runs contrary to Aristotle, but he invokes a principle of

64 Cf. Copernicus (ref. 30), p. 360: Rosen notes how Copernicus' misunderstanding of Aristotle's *Generation of Animals* (4, 10: 777b.24–27) derives from Averroes.

65 On this debate, cf. Miguel Á. Granada, "Johannes Kepler and David Fabricius: Their Discussion on the Nova of 1604," in *Change and Continuity in Early Modern Cosmology*, ed. Patrick J. Boner (Dordrecht: Springer, 2011) pp. 67–92.

66 *JKGW*, 1, 257.31–35: "Ut vero de materia prius agam, et pertractam hactenus disputationem finiam: supra dixi, materiam ex terra non fuisse. Ex coelo igitur fuit desumpta materia; siquidem aliquo pacto stellae generatio (etsi monstrosa, ut anguillae genitura in alvo humano) potest naturae viribus transcribi. Nam citra materiam natura nihil procreat."

67 *Ibid.*, 267.3–7: "In prasens autem quinque prioribus probationibus sufficiat hoc reddidisse verisimile, materiam coeli esse alterabilem: proinde nihil absurdi consequi, si quis dicat, *Novae stellae materiam ex ipsa fixarum sphaera coactam; stellaque extincta in eandem vicissim esse refusam* (my emphasis)."

plausible agreement: if Aristotle had observed changes in the heavens like those seen recently, he would have revised his doctrine.

Kepler regards the generation of celestial bodies as the result of a "natural faculty," which is envisaged as a vital heat that, similar to terrestrial organisms, produces excretions that in turn represent part of the life cycle.[68] The biological and meteorological model of explanation adopted by Kepler for the new star does not entail an astronomical study of its motion, since its matter simply congregates in one place and grows in a way similar to the organization of the organic matter of plants and "animalcula" during their generative process. This type of generation is common to other celestial phenomena, such as, crucially, comets, which are in fact mentioned in *De stella nova* and occasionally serve as a point of comparison—the difference between new stars and comets being that the former are generated in the region of the fixed stars, while the latter originate far below among the planets.[69]

In one particular instance, Kepler argues in Chapter 22 for the celestial nature of the new star by considering analogies with comets:

> And if, in fact, comets are in the heavens, and if, as can be demonstrated by examples, they proceed from the farthest part of the ether down to Earth, and occasionally pass it by through rectilinear motion, and if new stars arise among the fixed stars, what prevents [us] from saying that they are made of the ethereal substance?[70]

This passage forms part of a discussion on the nature of celestial novelties whose focal point involves Tycho Brahe's theory on the generation of the new star of 1572, according to which the celestial body was formed by dense matter in the Milky Way.[71] While on the one hand Kepler praises this opinion ("pul-

---

68 Cf. Boner (ref. 9), pp. 91–98, on Kepler's discussion of the generation of the new star and the relevance of Aristotelian biology. See also Granada (ref. 65), note 76, referring to Kepler's correspondence with Fabricius, and the use of Stoic sources in the adoption of the spiritual principle operating in the universe.

69 A detailed account of the generation of comets and novae in *De stella nova* can be found in Boner (ref. 60).

70 *JKGW*, 1, 257.35–258.3: "Et quidem si cometae in coelo sunt; si, quod exemplis demonstrari potest, ex profundissimo aethere, tellurem verso feruntur, et aliquando proxime hanc motu rectu praetereunt; si novae inter ipsas fixas oriuntur stellas: quid prohibet affirmare, illas ex aetheria constare essentia?"

71 *Ibid.*, 258.4–8: "Hic iam supra mentio facta est pulcherrimae sententiae, quam Braheus dixit: stellas huiusmodi procreari ex via lactea. Et visus est confirmare rem sensu oculorum; ostensa lacuna, eo loco, quo sidus anni 1572 effulsit; quasi portio illa materiae in corpus sideris coacta fuerit; quae postmodum fortasse conflagraverit, aut dissipata fuerit."

cherrima sententia"), since it establishes the principle that generation occurs in the celestial region, he also makes some critical considerations that depart from Tycho's theses. Referring to Fabricius' observations of the new star of 1596, Kepler notes that the phenomenon did not appear in the vicinity of the Milky Way, and therefore that it may not have originated from that particular part of the heavens. Thus, Kepler believes contrary to Tycho that new stars can be generated in any part of the heavens. In addition, while new stars may be infrequent phenomena, they are formed naturally from a process of accumulation of celestial matter.[72] In this context, the rectilinear motion of comets, although mentioned briefly, recalls Kepler's discussion in Chapter 21; at the very least, the wording used in this instance is reminiscent of the title of that previous chapter. Comets originate "ex profundissimo aethere," they approach the Earth, pass it by, and continue their trajectory away from it.

Taking comets as a comparative case study helps to shed light on Kepler's underlying assumptions on the straight-line paths of celestial bodies. Since 1602, he believed that comets moved along straight lines, their motions linked to both the Copernican theory and the nature of celestial matter.[73] In a letter to Herwart von Hohenburg from October 1602, Kepler argued that "whoever follows the Copernican hypothesis may defend this concept: comets are merely ethereal projectiles which clearly move almost uniformly in straight lines."[74] While accepting Tycho's explanation for comets and their celestial nature, Kepler adds that they originate "e profundissimo caelo," and move towards the Earth ("versus terra ruentes"), as well as past it ("et proxime illam transeuntes in partem alteram"), that is, back towards the deep recesses of the heavens.

72 *Ibid.*, 259.10–16.

73 Cf. Édouard Mehl, "Théorie physique et optique des comètes de Kepler à Descartes," in *Novas y cometas entre 1572 y 1618: Revolución cosmológica y renovación política y religiosa*, ed. Miguel Á. Granada (Barcelona: Universitat de Barcelona, 2012), pp. 255–274.

74 See the entire passage in Kepler to Herwart von Hohenburg, 7 October 1602, *JKGW*, 14, no. 228, ll. 53–67: "Cometarum [...] sententia tamen haec est: Si quis Copernici sequatur hypotheses, illum hoc defendere posse. Cometas nihil esse aliud quam trajectiones aetherias: moveri scilicet in linea recta pene aequabiliter. Ac tantum abesse ut confiniantur sublunares quin potius iam non tantum per parallaxes diametri terrae, sed etiam ex ipsis observationibus crassiore modo acceptis in aetherem reponantur, et quaqua versum impetu capto ruant. Hinc tute ipse iudica, qualis sit materia caelestis, si ubique cometas transmittit et aliquando ibi est cometa, ubi Terra, Luna, vel Martis etc. stella paulo ante fuit. Ita hic nihil a Tychone differo, nisi quod cometis circulos adimo, per quos ipsis Tycho pene divinitatem conciliavit. Multum itaque caelorum perfectioni decederet, si constaret etiam trajectiones ibi esse, easque non ex terra ortas, sed quasdam e profundissimo caelo versus terram ruentes, et proxime illam transeuntes in partem alteram: adeo nihil commune habent cum terra." Cf. James A. Ruffner, "The Curved and the Straight: Cometary Theory from Kepler to Hevelius," *Journal for the History of Astronomy*, 1971, 2: 178–194, on p. 181.

In *De cometis libelli tres* (1619), Kepler conveys a more comprehensive vision of his cometary research, covering at once astronomical, physical and astrological aspects. Straight-line motions receive considerable attention in both the first and second parts of his text, beginning with one of the principles ("assumpta") that introduce the discussion of the comets of 1607 and 1618: "No different than projectiles, comets constantly move through space by a continuous straight line."[75] Later on in the text, Kepler is prompted by the works of Isaac Habrecht (1589–1633) and Philipp Müller (1585–1659) to further articulate his position in response to their support for the circular motion of comets:

> Considering the difference between everlasting and transitory bodies, there should follow a similar difference in their motion; circular motions, and thus revolutions, shall belong to everlasting bodies, whether the Sun moves or the Earth in its place, while rectilinear motions shall belong to evanescent bodies. For both types of bodies have their own constitution, which attaches to their state arising from their respective form; eternity for the circle, and that of mortality for the straight line, which can by no means be infinite. We therefore join the ancients in our judgement that rectilinear motions are suitable for comets.[76]

The model of relations between bodies and motions envisaged by Kepler appears to align with Aristotelian physics, at least for what concerns the dichotomy between eternal and temporary bodies. The "forma" of the body is linked to the kind of motion it performs, similar to how Aristotelian doctrine conceived the relations between bodies and their natural motion. Kepler appears to appropriate the same principle, but in place of the ontological distinction between ethereal and elementary bodies, he predicates a distinction based on the duration of the bodies. As in *De stella nova*, his preoccupation is that the rectilinear motion of an eternal body requires an infinite universe, while on the contrary, the motion of a body destined for dissolution will, in time, come to an end.

---

75 *JKGW*, 8, 142.24–25: "Cometam non secus ac traiectionem aliquam ferri per spacia mundi in linea recta in directum continue." Cf. *ibid.*, 205–209, on the rectilinear motion of the comet of 1618.

76 *JKGW*, 7, 217.3–9: "Et cum differentia perrennium temporariorumque corporum sequatur in eorundem etiam motus; circulares utique revolutiones erunt perennium corporum, seu Sol moveatur, seu eius loco Tellus; rectilinei vero vanescentium: habent enim utrique conditionis suae causam in forma quisque sua; aeternitatis in circulo, mortalitatis in linea recta, quae infinita utique nequit esse. Iungimus igitur cum veteribus suffragia nostra, cometis rectlineos motus aptos esse." Cf. Ruffner (ref. 74), p. 181.

1.4 Conclusion

Kepler's discussion and ultimate rejection in *De stella nova* of the hypothesis that the new star became visible through a rectilinear motion from the heavens calls into question not only his commitment to an archetypal cosmology and its consequences, including the centrality and singularity of our solar system, but also the nature of the relations between celestial bodies and their motions, his interpretation of the Copernican theory, and, last but not least, his cometary theory. Consequently, the arguments used to counter the challenge posed by the supporters of an infinite extension of the universe cannot be seen solely as a clash between 'old' and 'new' philosophical systems or astronomical models. As Rhonda Martens suggests, Aristotle's natural philosophy was often "taken as the alpha, though by no means the omega, of approaches to natural philosophy."[77] Kepler's celestial straight paths might be taken as a case in point, since the appropriation and re-purposing of different Aristotelian materials is used both to defuse the challenge posed by cosmological infinitism to Kepler's archetypal cosmology and to support the rectilinear motions of 'transitory' celestial bodies. The difficulty of reconciling Kepler's rejection of straight-line motion for the new star with his cometary theory, whose conception runs parallel to the reconsideration of the age-old paradigm of celestial circular uniformity, might be understood, if not dissolved, by considering the elaboration of Kepler's cometary theory as part of his celestial physics. In the wake of the consequences of the heliocentric theory, especially the possibility of extending terrestrial physics to the celestial realm once the Earth is 'promoted' to the planetary ranks, Kepler adapts for his own purposes the Aristotelian principle that the type of motion natural to a body depends on its physical nature. Thus, it appears that his aversion to rectilinear motion in the heavens is limited to eternal celestial bodies, excluding temporary bodies such as comets. In both cases, Kepler seeks to defend his own interpretation of the Copernican heliocentric cosmos, expediently applying Aristotle's arguments at every opportunity. In the second case, the consequences are far less Aristotelian.

77 Rhonda Martens, *Kepler's Philosophy and the New Astronomy* (Princeton: Princeton University Press, 2000), p. 99. In her analysis, Martens refers to Charles Schmitt's study of Renaissance Aristotelianism.

CHAPTER 2

Of Mites and Men (and Stars): Kepler on the Question of Star Sizes in *De stella nova*

Christopher M. Graney

2.1 Introduction

In his *De stella nova* (1606), Johannes Kepler attempted to answer Tycho Brahe's objection that the heliocentric hypothesis, as outlined by Nicolaus Copernicus, required all the fixed stars to dwarf the Sun. Brahe considered such a prospect a profound problem that denied any sense of proportion. In this paper, I briefly explore Brahe's argument as well as Kepler's response, which compares the magnitudes of men, mites, and snakes, all with respect to the dimensionless nature of their divine creator. In addition, I present a complete translation of Chapter 16 of *De stella nova*, where Kepler articulates his argument.

2.2 Brahe and Star Sizes

Among his many objectives in *De stella nova*, Kepler set out to resolve an argument raised by Brahe against the heliocentric theory of Copernicus. The argument was as follows: in the heliocentric theory, the fixed stars had to be very distant in order to explain why Earth's annual motion around the Sun produced no corresponding visible annual changes in their position—no 'annual parallax.' In addition, stars were not seen to grow brighter when the Earth moved toward them as it revolved around the Sun, nor were they seen to grow dimmer when it moved away from them. The explanation for this absence of change was that the annual path of the Earth was like a point in comparison to the distance of the stars. Earth's orbit was negligible in size, and thus Earth's motion was negligible in effect, with regard to the stars. But Brahe noted that stars have a measurable apparent size as seen from the Earth. He had measured these sizes, determining that the more prominent or 'first magnitude' stars measured a little less than a tenth of the apparent diameter of the Moon—a little less than three minutes of arc, since the Moon has an apparent diameter of approximately thirty minutes, or one-half of one degree. Brahe's measurements of the apparent diameters of stars were generally consistent with those of Ptolemy

 | DOI:10.1163/9789004437272_004

FIGURE 2.1 The more distant an object, seen from the Earth (E) to measure a certain apparent size (indicated by the shaded region), the greater its physical magnitude. A and B each have the same apparent size as seen from E, but B is over four times more distant than A, so B has over four times the diameter and over sixty-four times (4^3) the volume of A.

and others.[1] An object of a given apparent size has a physical size that is some fixed fraction of its distance (Fig. 2.1). At the vast distances required for the stars in the heliocentric hypothesis, those apparent sizes measured by Brahe translated into enormous physical dimensions. Were Copernicus correct, every one of the stars would dwarf the Sun. The Sun, then, would be a uniquely small body in a universe of giants.[2]

Johann Georg Locher and his mentor Christoph Scheiner (1573–1650) would neatly summarize Brahe's 'giant star' objection in their book, *Disquisitiones mathematicae* (1614). They wrote that, according to Copernicus, the orbit of the Earth was like a point within the universe of stars; but the stars, possessing measurable sizes, were larger than points; therefore, in the Copernican hypothesis every star must be larger than the orbit of the Earth and, of course, vastly larger than the Sun:

> [E]ven the smallest star visible to the eye is much larger than the whole circle of Earth's orbit. This is because such a star has a measurable size, as does the circumference of the sky. The ratio of the size of the star to the size of the firmament of fixed stars is therefore perceptible. But according to the Copernican opinion, the ratio of the size of the circle of Earth's orbit to the size of the firmament is imperceptible. For in the

1 Albert Van Helden, *Measuring the Universe: Cosmic Dimensions from Aristarchus to Halley* (Chicago: University of Chicago Press, 1985), pp. 27, 30, 32, 50–52. As Van Helden notes, Brahe spoke to a sense of cosmic scale "that was deeply etched into the collective consciousness of educated Europeans. Therefore his objection was a very serious one; no answer was in sight [p. 52]."

2 See Christopher M. Graney, *Setting Aside All Authority: Giovanni Battista Riccioli and the Science Against Copernicus in the Age of Galileo* (Notre Dame: University of Notre Dame Press, 2015), pp. 32–38.

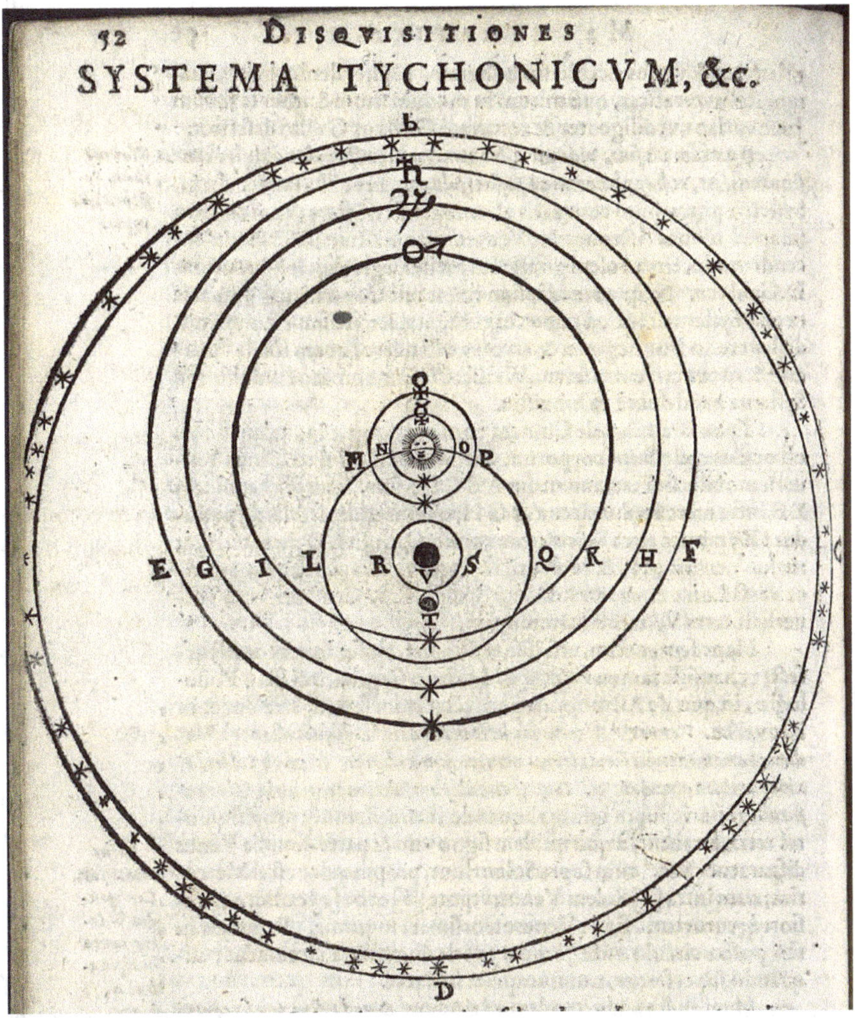

FIGURE 2.2 Brahe's geoheliocentric hypothesis. With the Earth immobile at the center (V), Mercury, Venus, Mars, Jupiter, and Saturn circle the Sun (as in the Copernican hypothesis), while the Sun, Moon, and stars circle the Earth. Johann Georg Locher, *Disquisitiones mathematicae, de controversiis et novitatibus astronomicis* (Ingolstadt, 1614), p. 52.
SOURCE: ETH-BIBLIOTHEK ZÜRICH

Copernican opinion the size of the Earth's orbital circle holds the same proportion to the firmament as the size of Earth itself holds to the firmament in the common geocentric opinion. Yet experience shows the Earth to be of imperceptible size compared to the firmament. Thus in the Copernican opinion it is the circle of Earth's orbit that is of imperceptible

> size compared to the firmament—and therefore smaller than the smallest perceptible star.[3]

The giant stars of the Copernican hypothesis differed significantly from the more commensurate star sizes found in Brahe's own hypothesis, a hybrid geocentric (or *geo-heliocentric*) hypothesis in which the Sun, Moon, and stars circled an immobile Earth, but the planets circled the Sun (Fig. 2.2).

Brahe's hypothesis was observationally and mathematically equivalent to the Copernican hypothesis, insofar as the Sun, Moon, and planets were concerned. However, since the Earth was immobile, there was no expectation of parallax, and thus no need for the stars to be so distant in order to explain the absence of observable parallax. Brahe had the stars located a bit beyond Saturn.[4] And, since the stars were roughly similar to Saturn both in terms of distance and in terms of their appearance in the night sky, they had to be similar to Saturn in physical size as well. According to Brahe, sizes were commensurate in the universe, with the Moon being the smallest body, the Sun being the largest, and the Earth, stars, and planets all falling in between. Albert Van Helden suggests that this scheme lent a sense of "planetary hierarchy" found in the Copernican system as well.[5] However, in the Copernican hypothesis every last star dwarfed the Sun, Moon, and planets (Fig. 2.3).[6]

In Brahe's view, giant stars were an absurdity,[7] and following Brahe a number of people raised the star size issue against the Copernican hypothesis. These authors included Simon Marius (1573–1625), who in 1614 first noted that stars seen with the telescope still had circular shapes and measurable sizes; Locher and Scheiner, also in 1614, as mentioned previously; Francesco Ingoli (1578–1649), who raised the star size objection in an essay to Galileo of 1616, which Galileo believed to have been influential in the rejection of the Copernican hypothesis by the Congregation of the Index in Rome later that year; Peter Crüger (1580–1639), who in his *Cupediae astrosophicae* (1631) remarked that, granting

3 Christopher M. Graney, Mathematical Disquisitions*: The Booklet of Theses Immortalized by Galileo* (Notre Dame: University of Notre Dame Press, 2017), p. 30.

4 Van Helden (ref. 1), p. 50.

5 *Ibid.*, p. 51. On the size of the Earth's sphere as the "common measure" for the other planetary spheres according to Copernicus, see Fernand Hallyn, *The Poetic Structure of the World*, trans. Donald M. Leslie (New York: Zone Books, 1990), p. 97.

6 Graney (ref. 2), pp. 32–38. On the scale of proportion according to "a fixed unit" in the Vitruvian sense, see Giora Hon and Bernard R. Goldstein, "*Symmetry* in Copernicus and Galileo," *Journal for the History of Astronomy*, 2004, 35: 273–292. Hon and Goldstein argue that Copernicus was moved by "considerations of symmetry in the Vitruvian sense" to conceive a world system whose parts cohered "to form a beautiful whole."

7 Graney (ref. 2), p. 37, 80.

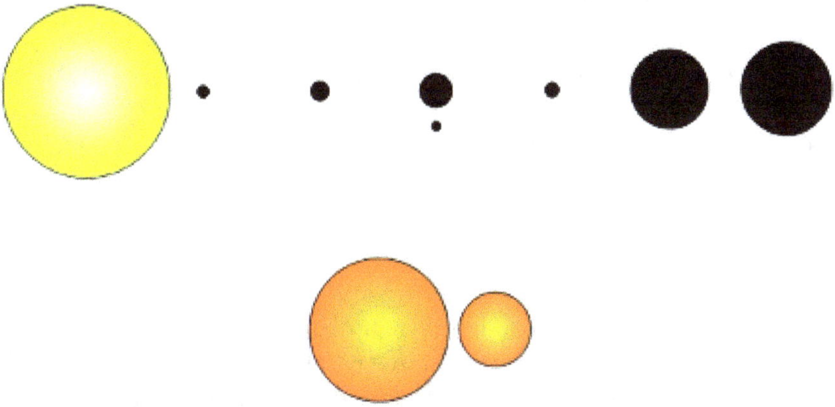

FIGURE 2.3 *Above*—the relative sizes of celestial bodies calculated by Brahe for (from left to right, upper row) the Sun, Mercury, Venus, Earth and Moon, Mars, Jupiter, Saturn, as well as for (lower row) a large star and a mid-range star in a geoheliocentric universe (where the stars lie just beyond Saturn, as in Fig. 2.2). The Sun, stars, and planets all fall into a fairly consistent range of sizes. *Below*—the arrowed dots are the above figure, reproduced to scale and compared to the relative size of a mid-range star calculated by Brahe for the Copernican universe, where the stars are vastly more distant and far more enormous to explain their apparent sizes as seen from the Earth (as in Fig. 2.1). Brahe called the immense Copernican stars 'absurd.'

the "impeccable" logic of Brahe,[8] "I therefore do not understand how the Pythagorean or Copernican world system can survive"; and Giovanni Battista Riccioli (1598–1671), who analyzed 126 arguments for and against Copernicus in his *Almagestum novum* (1651), and considered star sizes to be one of the few decisive arguments.[9]

8 Van Helden (ref. 1), p. 51.
9 Graney (ref. 2), pp. 50–52, 64–71, 129–139.

2.3 The Star Size Question in *De stella nova*

After exploring a number of other concerns, Kepler devoted Chapter 16 of *De stella nova* to the sizes of the fixed stars in the new cosmology. (A translation of Chapter 16 appears below.) This follows his discussion of their vast distance from the Earth in Chapter 15. Albert Van Helden has written that Brahe's measurements of the apparent sizes of the stars were unassailable, and his logic incontrovertible, so Copernicans simply had to accept Brahe's objection and agree that the stars were giant.[10] This Kepler does. He does not question that the stars are all vastly larger than the Sun. In fact, in Chapter 16 he grants that a star with an apparent diameter of three minutes (one-tenth the apparent diameter of the Moon) has the same physical size as the orbit of Saturn; he grants that Sirius, the most brilliant of stars, is larger than that, and the nova of 1604 larger still. It follows from Kepler's numbers that any star whose physical size was the same as the orbit of the Earth, one-tenth the diameter of Saturn's orbit, would possess an apparent diameter of three-tenths of a minute, or eighteen seconds, the apparent diameter that Brahe had determined for sixth-magnitude stars, those barely visible to the naked eye.[11] And as the physical diameter of the Sun is less than one-hundredth the diameter of Earth's orbit according to Kepler, clearly every last star in the sky utterly dwarfs the Sun. For Kepler, the Sun and its planets are surrounded by giants, and only by giants.

Kepler's answer to Brahe, then, is that the Sun and planets being surrounded by giants makes sense, or at least more sense than the geocentric (or geoheliocentric) alternative. Kepler argues that Brahe fixes upon size, but that what is commensurate in the Copernican hypothesis are speeds. Speeds in any geocentric hypothesis, Kepler says, are more incommensurate than sizes in the heliocentric hypothesis. His words echo the objection of his mentor, Michael Maestlin. In his preface to the *Narratio prima* that he appended to Kepler's *Mysterium cosmographicum* (1596), Maestlin condemned the "indescribable speed" of the heavens required by geocentrism.[12] Moreover, Kepler explains that a vast range of sizes exists in the physical world. He illustrates some of these: the longest snake vs. the smallest insect; human beings vs. the Earth; and the Earth vs. the universe. And finally, there is the limitless power and

10 Van Helden (ref. 1), p. 51.

11 Graney (ref. 2), p. 34. Brahe reported sixth-magnitude stars as having a diameter of twenty seconds.

12 Katherine A. Tredwell, "Michael Maestlin and the Fate of the *Narratio prima*," *Journal for the History of Astronomy*, 2004, 35: 305–325, on p. 309.

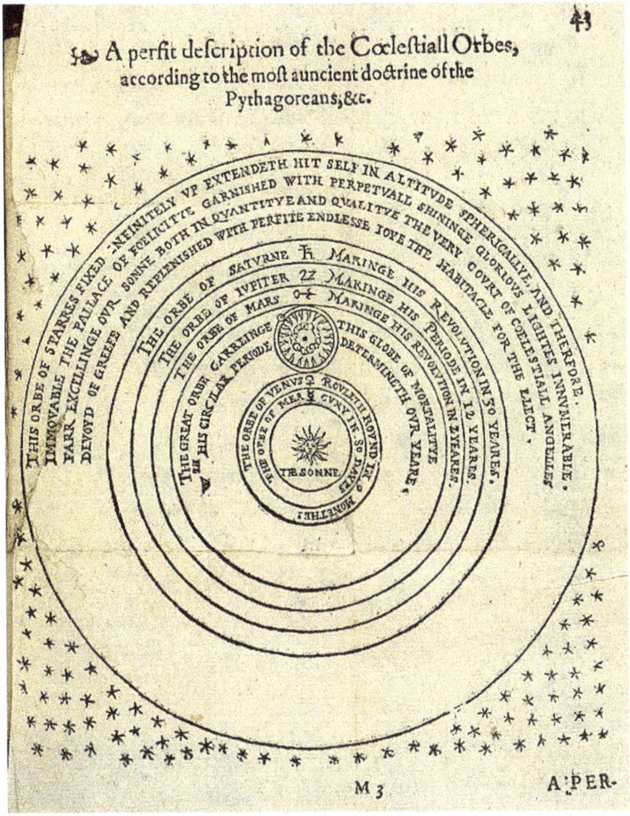

FIGURE 2.4 Sketch of the Copernican system by Thomas Digges. Note his description of the heavens as the "palace of felicity," with lights "far excelling our Sun both in quantity and quality," the court of the angels and the dwelling of the elect. Leonard and Thomas Digges, *A Prognostication Everlasting* (1576).
SOURCE: HISTORY OF SCIENCE COLLECTIONS, UNIVERSITY OF OKLAHOMA LIBRARIES.

creativity of God, for whom nothing is too great, and yet who also confers value upon the small.

By invoking divine power in response to the question of star size, Kepler stands among a number of Copernicans. These include: Thomas Digges (1546–1595), whose famous sketch of an infinite Copernican universe features a caption stating that the starry universe includes stars "far excelling our Sun both in quantity and quality"—the palace of felicity, the court of the angels, and the dwelling-place of the elect (Fig. 2.4); Christoph Rothmann (1560–1600), who also invoked the idea of a celestial palace in response to Brahe's star size argument, noting that "the greater the King, so much more greater and larger the palace befitting his Majesty," and asking what sort of palace is fitting for God; and Philip Lansberg (1561–1632), who envisioned the giant stars as God's own warriors, the palace guard of heaven itself—an idea he reinforced with numerous references to scripture.[13]

2.4 Chapter 16 of *De stella nova*

Let us now turn to Kepler's brief but significant Chapter 16, for nothing better conveys Kepler's views than his own words. Indeed, Kepler's writing here is not only clear and succinct, but at times almost poetic. In this annotated English translation, I have broken Kepler's original account into additional paragraphs. The original paragraph breaks are indicated by →; the added paragraph breaks are indicated by indentations.

Kepler begins with a general introduction. Again, this follows the fifteenth chapter's discussion on the distance of the 'new star' of 1604 according to the Copernican hypothesis.

> CHAPTER 16[14]
>
> CONCERNING THE IMMENSITY OF THE SPHERE OF THE FIXED STARS IN THE HYPOTHESES OF COPERNICUS: AND CONCERNING THE SIZE OF THE NEW STAR. [→]
>
> At this point I am sure that, among those who are reading, there are some who are ready to ridicule the insanity of Copernicus. Other readers, who embrace his opinion on account of different astronomical and physical arguments, are nevertheless troubled by this immensity, and are beginning to doubt whether those arguments (which are produced

13 Graney (ref. 2), pp. 76–85. On the "palace" portrayed by Rothmann in response to Brahe, see Van Helden (ref. 1), p. 52.

14 Johannes Kepler, *De stella nova in pede Serpentarii* (Prague: Paul Sessius, 1606), 83–89. Copies of Kepler's original book are freely available via Google Books and E-rara.ch (the platform for digitized rare books from Swiss libraries). For secondary sources that discuss Chapter 16, see Robert S. Westman, *The Copernican Question: Prognostication, Skepticism, and Celestial Order* (Berkeley: University of California Press, 2011), and Patrick J. Boner, "Kepler's Copernican Campaign and the New Star of 1604," in *Change and Continuity in Early Modern Cosmology*, ed. Patrick J. Boner (Dordrecht: Springer, 2011), 93–114. Neither of these discuss at any length Kepler's attention to the issue of the sizes of the fixed stars themselves. Rather, both focus on the question of the size of the universe as a whole. Boner (104) does include a quotation from Kepler that remarks upon the size of the 'new star' under the Copernican hypothesis as exceeding that of the entire planetary system.

out of other knowledge for the purpose of establishing the opinion of Copernicus) may be asserted altogether truly, or whether the occasion may be right for those arguments to be dissolved by some other reasoning, and for it to be shown how the Copernicans were led to error.[→]

Brahe added force to this objection by finding a lack of elegance [*concinnitas*] in the most perfect of works,[15] if the vastness of the sphere of one of the fixed stars be so insane; the meagerness of all the wandering stars so contemptible. How great the flaw in the human body, he says, if the finger, if the nose, might surpass by many times the bulk of the whole remainder of the body.

Therefore, in order perchance to dilute this objection, I shall say three things. First, I shall show how many more incredible things emerge from the opinion of the ancients and Ptolemy concerning the motion of the fixed stars. Next, I shall establish through different examples chosen from the world the existence of an inestimable range of proportion. And lastly, I shall point out a mutual compensation by which some equality arises in the most unequal of things.[→]

Having provided an introduction to Brahe's objection and an outline of what he intends to discuss in Chapter 16, Kepler now launches into a discussion of the speeds required for bodies in the heliocentric hypothesis. He also includes a short discussion of how the Sun powers the movement of the planets, propelling the nearer ones around more quickly than the farther ones.

Accordingly, the motion of the bodies of the universe in the hypotheses of Copernicus is such that in one hour the globe of Saturn traverses about 300 German miles; Jupiter, 400; Mars, 600; the center of Earth, 740;[16] Venus, 800; Mercury, 1,000—a beautiful proportion, whereby that which is closer to the quiescent Sun (the dispenser of all motion) is

15 On the role of aesthetic virtues such as *concinnitas* as signs of certain knowledge for Kepler, see Nick Jardine, "Kepler, God, and the Virtues of Copernican Hypotheses," in *Nouveau ciel, nouvelle terre: La revolution copernicienne dans l'Allemagne de la Réforme* (*1530–1630*), ed. Miguel Ángel Granada and Édouard Mehl (Paris: Les Belles Lettres, 2009), 269–281.

16 Kepler gives the Earth-Sun distance as 1,200 terrestrial semidiameters, which in turn he gives as 860 German miles (he states that a German mile is 5,000 paces of 5 feet per pace). Thus, the circumference of Earth's orbit is $2 \times \pi \times (1200 \times 860) = 6{,}484{,}247$ miles. A year being 365.25 days and a day being 24 hours, this yields a speed of $6{,}484{,}247 / (365.25 \times 24) = 740$. Speeds for other planets are calculated in the same way. For example, Saturn, having an orbital radius roughly ten times that of Earth, and an orbital period roughly thirty times, has an orbital speed of roughly a third that of Earth.

always swifter. And because the Earth not only is conveyed annually, but also is revolved daily, therefore in one hour the regions of the Earth that are near the equator traverse 240 German miles around the center,[17] the rest of the regions less (as they are nearer to the poles). Therefore, the 240 miles of the surface added to the 740 of the center make about 1,000. This is how much ethereal space some city situated at the equator traverses in one hour in the middle of the night. And 240 subtracted from 740 leaves 500 miles remaining—the journey through the ethereal vapor of one midday hour at the equator.[18]

Whence it should be considered, because the nighttime speed of the lands at the equator is twice that of the daytime, whether this somewhat aids in cooling the nights, just like a kind of ventilation, and whether indeed on account of this, the morning time receives the greater favor. For the parts of the Earth, on which the Sun now rises, at this moment are carried directly into the ethereal vapor. Meanwhile, the parts where the Sun sets are, at this time, as it were, pulled away from the ethereal vapor.

The Mόon remains for us to consider. In one hour, it traverses 476 miles around the center of the Earth, because it is distant from Earth by 60 terrestrial semidiameters.[19] Add 476 to the journey of the center of Earth, 740: the journey of the full Moon comes to 1,216 miles through the ethereal vapor. Subtract, in turn, 476 from the journey of the center, 740: the journey of the new Moon in one hour leaves behind 264 miles.

From Copernicus, you have commensurate speeds for all the stars, fixed and wandering [i.e. planets]. Indeed, the fixed stars are like a place, and are quiescent. The Sun is like a mover, and maintains the middle station of the immobile world. The body of the Sun, rotated in place, circularly stirs by this rotation, throughout the expanse of the universe, an immaterial species [*species immateriata*] (unless I merely concocted such a fitting circumstance by physical reasoning, in the commentaries concerning the motion of Mars),[20] a mover which the wandering stars all follow, more swiftly the closer they are to it.[→]

17 If the semidiameter of Earth is 860 German miles, the circumference of Earth at the equator is then $2 \times \pi \times 860 = 5{,}404$ miles. 5,404 / 24 hours = 225 rather than 240.

18 Earth's orbital and rotational speeds add together on its night side and subtract on its day side. Galileo would use this concept in his discussion of the ocean tides serving as evidence for the motion of Earth.

19 The Moon takes 27.3 days to circle the Earth: $2 \times \pi \times 60 \times 860$ miles / $(27.3 \times 24$ hours$)$ = 495 rather than 476.

20 Kepler is referring to the *New Astronomy* (1609), which he had largely completed by the fall of 1605. For a brief study of the concept of "*species immateriata*" in the work of

> Lest this passage through a thousand miles in one hour still seem truly incredible to you, however, I urge you to consider the proportion of the density of air to the density of the ether, which I have demonstrated in the *Optics* [1604]. There it is proven how the passage of a thousand miles through the ether in one hour may be more tranquil, by an incredible number of times, than the passage of one mile through our air in one hour.[→]

Next, Kepler turns to the geocentric hypothesis of Ptolemy and the speeds required by it. He shows that the speed of the most distant planet, Saturn, is over seven-thousand times greater in the hypothesis of Ptolemy than in that of Copernicus. He argues that multiplying the size of the Ptolemaic sphere of the fixed stars by that amount (which will multiply both the distance to the stars and the sizes of individual stars) results in even more bulk than is found in the Copernican universe.

> Go now to Ptolemy and the ancient opinion; you will find everything more incredible. In that [opinion], the semidiameter of the sphere of the fixed stars occupies 20,000 semidiameters of Earth. The circumference, therefore, will be 63,000[21]—truly a modest number [*multitudo modesta*], compared to the Copernican, but all of which is said to go round in one day. Therefore, 2,625 semidiameters (each of which contains 860 miles) are covered in one hour.[22] Behold here what to me is an immense distinction. In the view of Ptolemy, Saturn is closest to the fixed stars, such that it will almost touch them. Following Copernicus, in one hour it traverses 300 miles; following Ptolemy—2,257,500 miles.[23] Saturn must be believed to be 7,525 times swifter under Ptolemy than under Copernicus.[24] Whoever should attempt to comprehend this incredible velocity mentally is overcome just as much as, and indeed more severely than, someone who attempts to comprehend the Copernican immensity.
>
> Meanwhile, weigh carefully, O Philosopher, what proportion of accident to its subject is to be desired here; who can rightly desire much more than the proportion of part to part of the universe following the view

Kepler and his modern translators, see Sheila J. Rabin, "Was Kepler's *Species Immateriata* Substantial?," *Journal for the History of Astronomy*, 2005, 36: 49–56.

21 $20{,}000 \times \pi = 63{,}000$. As the circumference is determined by $\pi \times$ diameter, not semidiameter, this number is too small by half. As seen in the previous footnotes and as will be seen below, there are a variety of such typographical errors in Kepler's text of Chapter 16.

22 $63{,}000 / 24 = 2{,}625$.

23 $860 \times 2625 = 2{,}257{,}500$.

24 $2{,}257{,}500 / 300 = 7525$.

> of Copernicus? For accidents are not without a suitable subject. More credible is a great subject without motion than a great motion in a small subject. And so, if I have carried on regarding the Ptolemaic proportion of movements versus the Copernican proportion of bodies, I will have done so to the greater applause of the philosophers.
>
> Ptolemy multiplies the motion of Aristarchus or Copernicus 7,500 times. I shall multiply by that many times the Ptolemaic body of the fixed stars, 20,000 semidiameters, so they may rise to 150,500,000 times.[25] Dividing into this 1,200, which is the measure of the interval between the Sun and the Earth, yields 125,000 solar intervals—specifically 125,417.[26] But we do not require as much bulk in order to defend Copernicus as it appears the ancients required speed to defend Ptolemy. Let us commend the universe for its elegance of proportion [*concinnitas proportionis*]—not for its proportion of size, as Brahe was in the habit of desiring, but rather for its proportion of beauty and reasoning.

In the following passage, Kepler makes a quick change. He now turns from comparing dimensions and distances between Ptolemy and Copernicus to an argument regarding the Copernican system more specifically. Kepler sets forth the idea that the size of the planetary realm, demarcated by the orbit of Saturn, ought to bear the same proportion to the Sun as the size of the sphere of the fixed stars bears to it.

> The perfection of the universe is motion, which is as it were a certain life of it. Three things are required for motion: a mover, a movable, a place. The mover is the Sun. The movables are Mercury up through Saturn. The place is the outermost sphere of the fixed stars.
>
> But if I may express a physical matter mathematically, the movables are the proportional mean [*medium proportionale*] between mover and place. For they aspire to the quiescence of the place or of the encircling body. They receive their motion from the mover, which they resist, as if by a certain heaviness (what do you ridicule, O inexpert spinners of philosophical phrases, abounding with so many imaginary things, most destitute of the true?).[27] From this are derived their different periods of

25 20,000 × 7,525 = 150,500,000.

26 150,500,000 / 1200 = 125,417.

27 This comment is likely directed to those who would remind Kepler that, at least in the traditional Aristotelian view, celestial bodies were supposed to be made of an ethereal substance and would have no heaviness.

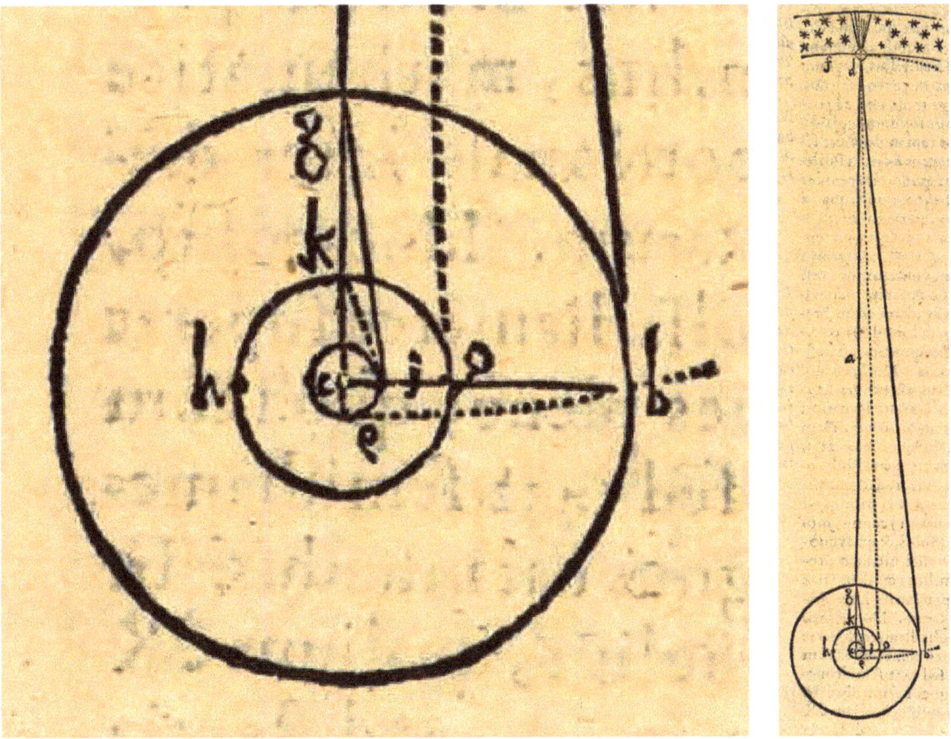

FIGURE 2.5 Kepler's illustration in Chapter 16 of *De stella nova* (86), showing the globe of the Sun, the orbit of Earth, the orbit of Saturn, and the sphere of the fixed stars.
SOURCE: ETH-BIBLIOTHEK ZÜRICH

> time, contrary to their nearness to the mover, which is unique and moves uniformly. Thus both the motion and the quiescence are mingled in a certain way in the movables. But if the movables are physically the proportional mean between mover and place, what is more probable, even mathematically, than for the diameter of the movables to be the proportional mean between the diameter of the mover Sun, and the diameter of the place or the fixed stars?

Presenting his principle of proportion between "the sphere of the fixed stars and the body of the Sun,"[28] Kepler proceeds to apply it to calculating the scale and size of stars.

28 This phrase occurs in a marginal note: Kepler (ref. 14), p. 85.

> With this posited, it is easy to investigate the extent [*quantitas*] of the fixed stars. For the diameter of the Sun surpasses the diameter of Earth by 5 ½ times. It may surpass it by fully 6, for as I say, I attribute not 1,200 but 1,432 semidiameters to the distance between the Sun and Earth. Yet I may be too generous with this measure, so that the universe may be made altogether large.
>
> In the following figure [Fig. 2.5], the semidiameter CE of the body of the Sun is 6, and the distance CK from the Sun to the Earth is 1,432. The loftier sphere of Saturn is GB. Now Saturn is 10 times the distance of the Sun—specifically, CG is 10 times CK. It is, in fact, somewhat less, but nevertheless this number will do. Therefore, Saturn is distant by 14,320 semidiameters of Earth, and this is the length of CG. And so now, as the semidiameter of the solar body CI[29] (6 semidiameters of Earth) to the semidiameter of the orb of Saturn CG or CB (14,320 semidiameters of Earth); so this CB to the semidiameter of the sphere of the fixed stars CD, formed by continuing CG to the fixed stars.[30] Upon this line semicircle EBD is drawn, carried across through marks E and B, the center of which is A. Reaching the fixed stars, CG will manifest to us the maximum number of semidiameters of Earth, 34,077,066 ⅔.[31] Yet this sum is still not ¼ that previous number [150,500,000], which we had taken up according to the example of multiplied speed in the ancient opinion.[→]
>
> And so why do the philosophers clamor to remove this speck of the immensity of the fixed stars from the eye of Copernicus, when at the same time they ignore in their own eye the enormous log, more than four times greater, of the insane speed of the fixed stars?[32]

Kepler next notes that Brahe's hypothesis results in a universe that is smaller than that of Ptolemy, improving things somewhat for the geocentric hypothesis—but, he says, not enough.

> Tycho Brahe has somewhat helped the old opinion here by reducing the height of Saturn by slightly more than a third, for he sets it at 7,000 semidiameters of the Earth,[33] while he lavishes the fixed stars with double this [distance]—the former, indeed, by necessary demonstration from his

---

29 Kepler's text reads CO, but this is certainly a typographical error.

30 CI / CB = CB / CD.

31 (14320 / 6) × 14320 = 34,177,067 (note the typographical error).

32 Cf. Matthew 7:3.

33 7,000 is a value lower than Brahe, in fact, gives; see Graney (ref. 2), p. 34.

> own supposed model of the universe [Fig. 2.2], the latter from plausible conjecture. But whether by ⅔ or by ⅓, either figure is surpassed by 3 or 4 times.[34] And what is the single star of Saturn to the monstrous multitude of the fixed stars?—which all according to the received hypothesis are moved by that swiftest motion; which all in the view of Copernicus rest immobile, only seven little bodies marching along.[35][→]

The next question is whether Kepler's various calculations suggest that an annual parallax might be detectable. His answer? No, of course. Based on his numbers, any annual parallax will be less than what the best of astronomers can measure.[36]

> Yet might it be that the universe, adorned by the proportion pointed out by me here, is narrow and so near that the fixed stars may produce some parallax?—that is, that the proportion of line CK or CO, by which the Sun and Earth stand apart, may become sensible compared to CD, the separation of the Sun and the fixed stars? Not at all. For that previous number [namely 34,000,000], is 20 times the distance that could produce a parallax of 2′ (a little less, in fact, just over 16 times); and so 34,000,000 shall claim a parallax of a little less than ⅟16 of 2′, that is, of approximately ⅛ of 1 [minute].[37]
>
> Or put more artfully and briefly, because the Sun occupies 30′, as seen from earth—that is, because angle CKI is 15′—and certainly Saturn is 10

34 Kepler's text says "4 or 5," but I take this to be an error, since the math is "3 or 4."

35 The five planets, Earth, and the Moon.

36 In a note to the second edition of Kepler's *Mysterium cosmographicum* (1621), Maestlin argued that annual parallax fell below the threshold of the best instruments of the day. See Harald Siebert, "The Early Search for Stellar Parallax: Galileo, Castelli, and Ramponi," *Journal for the History of Astronomy*, 2005, 36: 251–271, on p. 253.

37 Kepler's words here regarding the calculation of parallax are very difficult, and these two sentences are as much the product of my interpretation as of translation. Kepler seems to start with the distance that will produce a parallax of two arc minutes, or $^2/_{60}$ of a degree, "*distantia ... potuit facere duorum minutorum parallaxin.*" Tan ($^2/_{60}$) = 0.0005236, and the inverse of that, 1,719, is the distance to the stellar sphere (CD in Kepler's figure), measured in Earth-Sun distances (ESD), that will produce a two-minute parallax. Sixteen times this value is 27,502 ESD. If we suppose that Kepler used his earlier value of 1,200 semidiameters of the Earth as his ESD value for this calculation, then we get 1,200 × 27,502 = 33,000,000. As Kepler specifies that the multiplier is a little more than sixteen, specifying a parallax of approximately ⅛ of a minute as a little less than ⅟16 of two minutes ("*sedecimum huius paulo minus*"), we can take the multiplier as 16 ½, and we get 1,719 × 16.5 × 1,200 = 34,000,000. Happily, the sentences that follow provide a much clearer discussion regarding parallax.

times higher, then as seen from Saturn, the Sun occupies 1⁄10 of 30′—that is, 3′. Truly, this quantity is double the angle CGI. But it is posited that, as IC the Sun is to CG the semidiameter of the circle of Saturn, so CG or CB is to CD the semidiameter of the fixed stars. And so, connecting points B and D, triangles BCD and ICG, will be similar; hence, angles BDC[38] and IGC are equal. Therefore, even that circle of Saturn CB, as seen from the sphere of the fixed stars or point D, might produce an appearance of 1½′. And if CO, the circle of Earth [*telluris coelum*], should be 1⁄10 of the semidiameter[39] of Saturn CB, therefore angle ODC will be about 1⁄10 of angle BDC. Thus, the semidiameter of the annual orbit of the Earth CO is not more than 9″,[40] so it should cause a parallax of 1⁄7 of 1′.[41]

But there is no astronomer with the observational skill necessary to boast of achieving an observation of 1⁄7 or 1⁄8 of an arc minute. For instance, though the separation of two stars near point D may be less by 1⁄7 of a minute when the Earth is moving through O (when D is reckoned at a right angle to the Sun or to line OC), than when the Earth is moving through K (when D is opposite the Sun), the astronomer still will not distinguish these distances, observing them both at quadrature O and in opposition to the Sun K, since the bodies of the fixed stars themselves generally occupy and subtend one, two, three, or four minutes.[→]

Now comes Kepler's comparison of sizes of things on Earth—his argument that "enormous proportions are not rare among creatures."[42] Since tiny mites and giant snakes exist on Earth, he reasons, then a tiny Sun surrounded by giant stars should neither be distasteful nor disturbing. Here, Kepler's writing becomes almost poetic.

So therefore it is shown that no one who embraces anything as true from the hypotheses, and does not wish the whole of Astronomy to be utterly overturned, is prudent to present the immensity of the universe as an objection to Copernicus.[→]

Now it still remains to be seen by just what examples that immense proportion may become palatable to us. Here, I normally presented to Brahe as an objection the proportion between the mite that burrows into

---

38 The original reads "BCD." In the following sentence, the original reads "BDC."

39 The original reads "*diametri*," but CD and CB refer to semidiameters.

40 $1.5' / 10 = 0.15' = 0.15 \times 60'' = 9''$.

41 $60/7 = 8.57 \approx 9$.

42 This phrase occurs in a marginal note: Kepler (ref. 13), p. 87.

the skin of the hand and that 120-foot serpent described by Pliny. Hides of the serpent were preserved at Rome, and Brahe claimed that an even larger one had been seen in the North. The proportion of length extended to 100,000.[43]

But this example is a stunt. Take another. How small, I ask, is a man compared to the globe of Earth? Let us calculate. From the surface to the center there are 860 miles, each of which measures 5,000 paces. Therefore, there are 4,300,000 paces.[44] Five times that figure will give you feet, namely 21,500,000.[45] Grant to the height of a man seven whole feet, and divide the above number by that. Thus, if you arrange 3,100,000 men in succession along the same line, it will stretch from the surface of Earth to the center of it. So one diameter of the Earth shall equal more than 600 myriads of human height [i.e., 6,000,000 men].

Everyone grants this immense proportion of man to the globe of the Earth. But now they call incredible the proportion of Earth to the circle of Saturn [*coelum Saturni*], which is 1 to 12,000.[46] They call incredible the proportion of the realm of the movables [*coelorum mobilium*] to the immobile sphere of the fixed stars [*coelum fixarum immobile*], which is 1 to 3,000.[47] These both are much smaller than the above proportion. Certainly they do not consider proportion, but magnitude, since they may be small. Of course, the universe is not large to God, but we are small to the universe.

Yet consider again an analogy. Where magnitude waxes, there perfection wanes, and nobility follows diminution in bulk. The sphere of the fixed stars according to Copernicus is surely very large; but it is inert, without motion.

The movable world is next. Now this—all the more divine as it is smaller—has taken on motion that is as admirable as it is well ordered. Nevertheless, that place neither contains a nutritive faculty [*facultas vegetans*], nor does it reason or discourse. What it does (when moved) it did not learn, but it retains what was impressed upon it from the beginning.

---

43 120 feet / 100,000 = 0.0012 feet, which is roughly half of a millimeter—a reasonable estimate for a barely visible mite.

44 860 × 5000= 4,300,000.

45 4,300,000 × 5 = 21,500,000.

46 Although earlier Kepler stated the distance from the Earth to the Sun to be 1,432 terrestrial semidiameters rather than 1,200, here he uses 1,200: that multiplied by Saturn's tenfold greater distance from the Sun yields 12,000.

47 34,177,067 / 12,000 = 2,848 ≈ 3,000.

> What it is not, it will never be. What it is was not made by it—the same thing endures that was created.
>
> Then comes this our little ball, the little cottage of us all, which we call the Earth: the womb of the growing, herself informed by a certain internal faculty. The architect of marvelous works, she kindles daily so many little living things from herself—plants, fishes, insects—as she may easily scorn the rest of the bulk in view of this her nobility.
>
> Lastly, behold if you will the little bodies that we call the animals. What, in comparison to the universe, can be imagined that is smaller than these? But there now behold feeling [*sensus*], and voluntary motions—an infinite architecture of bodies.
>
> Behold if you will, among those [animals], these fine bits of dust, which are called men; to whom the Creator has granted such, that in a certain way they may beget themselves, clothe themselves, arm themselves, teach themselves infinite arts, and daily work toward the better; in whom is the image of God; who are, in a certain way, lords of the whole bulk [*domini totius molis*].
>
> And who among us would choose a body the breadth of the universe in exchange for no soul? Let us learn, therefore, what well pleases the Creator, who is the author of both coarse bulks and minute perfections. For He glories not in bulk, but ennobles those that He willed to be small.[→]
>
> In the end, through these intervals from the Earth to the Sun, from the Sun to Saturn, and from Saturn to the fixed stars, we may learn gradually to ascend toward recognizing the immensity of divine power.

Finally, we arrive at Kepler's closing remarks on "the great bulk of the new star."[48] Kepler notes that Sirius appears one-third larger to us on Earth than he said the orbit of Saturn (marking the frontier of the planetary realm) would appear to someone located at the sphere of the fixed stars. Thus, Sirius well exceeds the size of Saturn's orbit, as does the nova, which is at least the size of Sirius.

> I have included the above things on the vastness of the fixed stars raised against Copernicus all the more willingly because they pertain to the incredible magnitude that must be estimated for the new star. For if it occupied only four minutes (the size Sirius appears), then according to this hypothesis of Copernicus it was far larger than the entire realm of the movables [*tota machina mobilium*]. For earlier we attributed only three minutes to that realm if one were to look upon it from the fixed stars.

48 This phrase occurs in a marginal note: Kepler (ref. 14), p. 89.

> Indeed, I refrain from expressing such a magnitude in numbers, dreading that I may have already pushed things too far toward being condemned by the unlearned mob.

2.5 Kepler's Universe

Kepler does indeed refrain from expressing the size of Sirius and the nova by too many numbers. A few more numbers, however, provide a clearer picture of the sun-centered system that he has in mind. First, Kepler grants Sirius a diameter of four minutes, one-third greater than the three-minute diameter that he supposes Saturn's orbit would exhibit if seen from the sphere of the fixed stars. Since Saturn's orbit is ten times greater than the distance from the Earth to the Sun, the semidiameter of the globe of Sirius must measure over thirteen Earth-Sun distances. Thus, Kepler is assigning to Sirius the same size that Brahe had granted to the whole universe. It follows that the stars Kepler mentions as having apparent diameters of three minutes, two minutes, and one minute would all have physical sizes at least equal to or a significant proportion of Saturn's orbit;[49] all would be well larger than the orbit of the Earth. A star the size of Earth's orbit, just one-tenth of the diameter of Saturn's orbit, would be the smallest star visible to the naked eye. Obviously, the Sun would be orders of magnitude smaller than any visible star.

Consequently, the heliocentric universe of Johannes Kepler is one in which the Sun is a tiny, unique body at the center of a sphere of truly stupendous stars—stars whose bulk is best understood in comparison to geocentric universes, not to the Sun. The Sun is much closer in size to the Earth or even the Moon than it is to the smallest visible star in Kepler's universe. We can sympathize with Brahe for finding this absurd, but we can also marvel at Kepler's flexibility of thought in finding it imaginable. Kepler does not refute Brahe's star size argument; he makes it his own.[50] Through these sizes, he says, we can recognize "the immensity of divine power." Clearly, Kepler believed that all things were possible for God, even single stars as large as our planetary system.

49 The smallest star sizes result if the stars are assumed to lie on a sphere. If the starry universe is supposed to have depth, then the physical size of any star is larger, the more distant it is (Fig. 2.1).

50 Kepler used the idea of a heliocentric universe in which the Sun is a tiny, unique body at the center of a sphere of truly stupendous (and also very dim) stars to argue against Giordano Bruno's idea of a universe of other suns that were orbited by other Earths. See Christopher M. Graney, "The Starry Universe of Johannes Kepler," *Journal for the History of Astronomy*, 2019, 50: 155–173.

2.6 Conclusion: An Unpersuasive Response

We have seen how Kepler relied on divine power and celestial speeds in response to Brahe's star size objection to Copernicus. Arguments such as Kepler's would endure, but not necessarily persuade. In his 1651 analysis of arguments for and against the heliocentric hypothesis, Riccioli concluded that the vast majority of arguments in either direction were indecisive. He saw the star size objection as one of only two decisive arguments (both of which were in favor of the anti-Copernicans—the other being the absence of any detectable "Coriolis Effect," as it would be called today, caused by the rotation of the Earth).[51] By 1651, the telescope had long been brought to bear on the star size question, including by Riccioli himself.[52] And since, as Marius first noted, the small-aperture telescopes of the time showed stars as definite disks (not yet understood to be spurious artifacts of diffraction; see Fig. 2.6) with no detectable parallax, Brahe's objection still stood. Riccioli noted how that objection could be answered by appealing to what Kepler had called the "insane" speed of the stars, but he ultimately dismissed this answer. The rising and setting of the stars is caused either by the rotation of the Earth or by the rotation of the stars around the Earth, Riccioli reasoned, and whichever rotates will turn though one of its own circumferences per day. In either case, the rate of rotation will be proportionally identical; only in terms set by human perspective are the speeds different. And as for appealing to the power of God, the Jesuit Riccioli argued that, while such an answer could not be refuted, it did not satisfy the prudent. Besides, he said, if divine power could be called upon to explain the difficult aspects of a hypothesis, could not the vast speeds of the geocentric hypothesis also be explained by divine power?[53]

Thus, although Kepler set out in *De stella nova* to answer Brahe's star size objection to the heliocentric theory, that objection would still carry force almost five decades later. What would finally resolve Brahe's objection would not require any references to geocentric speeds, or comparisons of the sizes of snakes and mites, or appeals to divine power. Rather, it would involve the discovery that the apparent sizes of stars, whether measured visually or with a telescope, were the spurious products of optical systems, products which gave no indication of the true sizes of those stars. The first evidence suggesting the spurious nature of apparent stellar sizes—Jeremiah Horrocks' observations that stars winked out instantaneously when being occulted by the

51 Graney (ref. 2), pp. 115–128; Graney (ref. 3), p. xix.

52 Graney (ref. 2), pp. 53–61, 129–139.

53 *Ibid.*, 136–138.

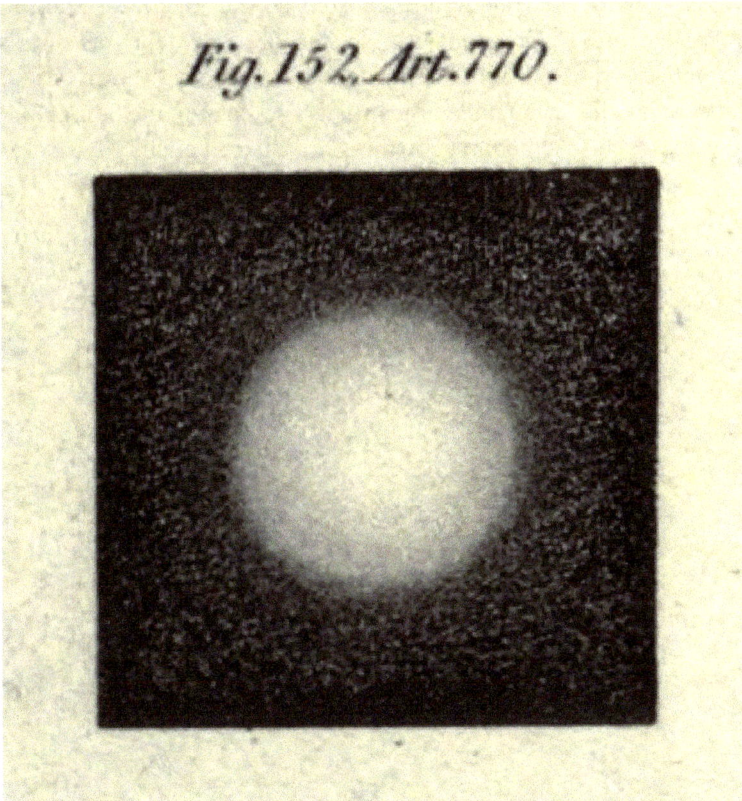

FIGURE 2.6 A star as seen through a small-aperture telescope. This appearance of a sphere of measurable size is entirely spurious—an artifact of diffraction. However, early telescopic astronomers took such telescopic images to be the physical bodies of stars. From John F. W. Herschel, *Treatises on Physical Astronomy, Light and Sound Contributed to the Encyclopædia Metropolitana* (1828), p. 491 & Plate 9.
SOURCE: ETH-BIBLIOTHEK ZÜRICH
Note: See Christopher M. Graney and Timothy P. Grayson, "On the Telescopic Disks of Stars: A Review and Analysis of Stellar Observations from the Early 17th through the Middle 19th Centuries," *Annals of Science*, 2011, 68: 351–373.

Moon, suggesting they were in fact far smaller than they appeared—was not published until a decade after Riccioli's *Almagestum novum*, six decades after *De stella nova*.[54] Such evidence would eventually show that stars were not all required to be giants in a sun-centered system. Indeed, more recent progress in

54 Graney (ref. 2), 150–151.

astronomy has shown that, while some giant stars do exist that dwarf the Sun, they are relatively rare. The large majority of stars are small, dim 'red dwarfs' that are far outclassed by the Sun. Having crafted such a creative response to Brahe to encourage the acceptance of giant stars surpassing the Sun, Kepler would surely have been disappointed by this starry state of affairs.

CHAPTER 3

The Measure of the Universe in *De stella nova*

Javier Luna

3.1 Introduction

In his book *De stella nova* (1606), Johannes Kepler contributed a new and much needed piece to the puzzle of heliocentric cosmology by offering one of the first public calculations of the possible distance of the stars for the new hypothesis according to the astronomical knowledge of his time as well as the accuracy of instruments available. This achievement was made possible by the combination of the most precise observations and the support of an ancient astronomical tradition. Tradition served as a double-edged sword, however. On the one hand, it offered astronomers geometric techniques and basic information for measuring astronomical bodies and distances while, on the other, passing on preconceptions and false data. To build the bridge between the geocentric and the nascent heliocentric worldviews, it was necessary to reconsider the data together with the forms of measure, filling the gaps with fresh assumptions and information. Kepler based his astronomical model on the assumption of the absolute centrality of the Sun and the finitude of the universe. These principles and others compelled him to oppose the cosmology of Giordano Bruno, who had insisted the universe was infinite and without a center in his study of space and other cosmological subjects.[1]

Kepler approaches the problem of measuring the universe according to a series of steps in *De stella nova*. Beginning with the explanation of the geometric foundation of parallax, he continues by showing that the nova cannot be as close as the Moon, the Sun or even Saturn. Kepler thus concludes that the nova was indeed a new star. Until this moment, Kepler's approach is valid both for the geocentric and heliocentric models, but from this point on he focuses exclusively on the sun-centered model. Considering that the annual motion of the Earth does not exhibit any stellar parallax, the only valid option for a Copernican would be that the stars are thousands of times farther away than what was previously thought. Kepler presents one calculation in *De stella nova* for the minimal distance to the stars and another for what he believes could

1 On Kepler's bitter opposition to Bruno and the "sect" of philosophers who "abused Copernican authority," see the contributions by Dario Tessicini and Pietro Omodeo to this volume.

 | DOI:10.1163/9789004437272_005

be the real distance. The immensity of these figures urges Kepler to confront, on the one hand, supporters of the geocentric view because they believed the figures were absurd, while, on the other, philosophers who thought they had found evidence for an infinite universe. This two-front assault forms the line of argument I will follow in my study. Of course, the subject of stellar distances in *De stella nova* has been discussed before. In his concise survey, Albert Van Helden presented the figure of Kepler's real distance and the mathematical means of obtaining it.[2] Before that, Judith Field dedicated a chapter and an appendix in her book to the problem.[3] However, Van Helden deals with the problem generally, never delving in detail to recount the specific steps of the process, while Field focuses primarily on another related problem, the size of the stars.

3.2 The Measurement of Space in the Astronomical Tradition

The history of the measurement of the universe has relied as much on philosophy as on astronomy, for the way it is measured depends on the intuition of the whole and the beliefs that can or cannot be supported in relation to the known phenomena. The account also depends on how the totality of time, space and matter is conceived. Equally important, the measurement relies on the instruments applied and their accuracy, as well as the limits of the empirical evidence and the creativity to employ it. The dramatic interplay between accuracy, evidence and ingenuity can be traced back to the pioneering works of Aristarchus and Hipparchus, which were done by applying accurate geometrical tools to false data; hence, the inaccuracy of their results. Nevertheless, these attempts yielded the first measures of the distance to the Moon and the Sun as well as their diameters. Their limitations, however, were not due to a lack of effort, but to the limits of naked-eye observation. The extremely large astronomical distances were an obstacle as much for the senses as for human understanding. For this reason, calculating the various dimensions of the Earth or the distance to the Moon has been a real challenge, and this is only a tiny portion of the distance to the Sun and the other planets, not to mention the stellar distances, which are on a completely different scale.

2 Albert Van Helden, *Measuring the Universe: Cosmic Dimensions from Aristarchus to Halley* (Chicago: University of Chicago Press, 1985), pp. 61–62.

3 Judith V. Field, *Kepler's Geometrical Cosmology* (Chicago: University of Chicago Press, 1988), pp. 18–29, 191–194.

To the naked eye, it is easy to assess that the Sun and Moon are the most luminous objects in the sky, and we can determine that both are of similar apparent size. At the same time, if we consider the large measure of lunar parallax and the occurrence of solar eclipses, it is clear that the Moon is closer to the Earth. The ancient astronomers also knew that the periods of the superior planets (Mars, Jupiter and Saturn) increase progressively, and that they all exhibit a retrogradation or an apparent temporary reverse motion. However, this is as far as solid evidence carried the day. The Sun, Mercury and Venus apparently have the same period; on the other hand, the Sun's parallax was extremely difficult to measure.[4] Therefore, there was no solid foundation for ordering the inferior planets or calculating their distance from the Sun.[5] To complete the picture, Claudius Ptolemy combined geometrical techniques with a wide range of philosophical principles,[6] in order to provide a complete astronomical system in a truly skillful manner. The result of this effort was a cosmic scheme that prevailed for nearly one and a half millennia:

> The scheme of the sizes and distances that was welded into the heads of European students was not, however, the product of the science as we know it. Rather, it was a mélange of philosophical notions, such as Aristotle's argument that the cosmos has no empty spaces; unverifiable naked-eye estimates, such as Ptolemy's statement that Mercury's apparent diameter is one-fifteenth of the Sun's diameter; and geometrical methods that produced spuriously precise results that had, in retrospect, more to do with numerology than with science. Its longevity resulted from the fact that it could not be falsified.[7]

4 The Sun's parallax for Ptolemy (Almagest, v, 17) was almost 3', the same figure for Copernicus (DR, IV, 21) and even for Tycho Brahe. On Brahe's acceptance of 3' for the Sun, see Victor E. Thoren, *The Lord of Uraniborg: A Biography of Tycho Brahe* (Cambridge: Cambridge University Press, 1990), p. 234.

5 On the role of planetary order as a central source of concern for Copernicus, see Robert S. Westman, *The Copernican Question: Astrology, Skepticism, and Celestial Order* (Berkeley: University of California Press, 2011).

6 Although the most common denomination for Ptolemy is 'Aristotelian,' he actually combines several philosophical systems in his quest to complete a broad astronomical panorama. "Depending upon the treatise, in fact, Ptolemy has been variously portrayed as a Platonist-Pythagorean, an Aristotelian, a Stoic, an Empiricist, and even a Positivist. The resulting impression is of a schizophrenic intellectual whose philosophical allegiances are sworn and resworn whenever the analytic circumstances demand." A. Mark Smith, "Ptolemy's Theory of Visual Perception: An English Translation of the *Optics* with Introduction and Commentary," *Transactions of the American Philosophical Society*, 1996, 86: iii–300, on p. 17.

7 Van Helden (ref. 2), 2.

Following the techniques of his predecessors Aristarchus and Hipparchus, Ptolemy calculated a distance ranging between 33 and 64 e.r. (earth radii) to the Moon; and between 1,160 and 1,260 e.r. to the Sun. Accepting Aristotle's *plenum* theory, Ptolemy then developed the principle of nesting spheres with which he calculated the epicycles and added the distances to the remaining planets, and especially the distance to the outer vault of the world, the so-called sphere of fixed stars, estimated at 19,865 e.r. In a simplified way, these distances were expressed as 60 e.r. between the Earth and the Moon, 1,200 e.r. to the Sun (or 20:1 between the two) and 20,000 e.r. to the sphere of the fixed stars.[8] Finally, the measure of the Earth's radius was determined to be equivalent to 250,000 Greek stadiums. These figures became the basis of astronomical studies as well as for education at higher levels; moreover, they became a milestone for the worldview of the general public, even for those who shared no knowledge of the exact figures.

For this reason, the Copernican system was not only a simple change of position between the Earth and the Sun; it shook every previous principle to the core. Now the Earth was no longer considered a motionless observatory, but a moving body, relocated to the third heaven together with the Moon, newly considered a satellite. This created a strange paradox for Aristotelian philosophy, since the Earth—allegedly a heavy body—was now considered to be joined in motion with the Moon, a body of an entirely different nature.[9] The position of the Sun—now motionless—was near the centre. As a consequence, the number of original planets was reduced to five, in addition to the Earth, and their circuits were no longer considered to be in mutual contact, since the annual motion of the Earth rendered the large epicycles no longer necessary to explain retrogradation. Even the final frontier, the sphere of the fixed stars, had to be removed to an unspecified height, since the apparent lack of annual parallax among the stars implied that it was impossible to remain at the relatively short distance of 20,000 e.r.

8 *Ibid.*, 7. Aristarchus' book *On the Sizes and Distances of the Sun and the Moon* determined the proportion between 18:1 and 20:1.

9 On the weight of the Earth in the work of many early modern authors, see Pietro D. Omodeo, *Copernicus in the Cultural Debates of the Renaissance: Reception, Legacy, Transformation* (Leiden: Brill, 2014).

3.3 The Distance of the Nova in Kepler's *De stella nova*

The problem of the least possible Copernican distance to the stars remained unresolved and rarely studied until the nova of 1604.[10] In Chapter 15 of *De stella nova* (DSN), in addition to defending the accuracy of the geometrical parallax technique,[11] Kepler demonstrated that it was impossible for the nova to be situated at a similar distance as the Moon, the Sun or Saturn. Kepler made clear from the beginning of his book that the nova could be seen in Prague on 11/21 October 1604 in the West through three stars "around 17.39', 17.4' or 17.42' Sagittarius, after sundown," and again on 2/12 February 1605 at sunrise "at 17°40⅓' Sagittarius, through the Eagle, through the Heart of Scorpio in 17°44⅓' corrected by Fabricius at 17°42' Sagittarius."[12] In fact, the nova could only be seen from Prague very close to the horizon, as the persistent poor weather made it difficult to make accurate observations, leaving a range of at least 1' or 2' of additional uncertainty.[13] However, this range of uncertainty was enough to determine that the nova was much farther away than the Moon. Thus, Kepler assumed a distance to the Moon of 60 e.r. and a parallax of 57', and displayed a figure [DSN: XV, 228] to show how two observers at different points on the Earth's surface should have seen the nova projected at different places among the fixed stars. Referring to a table he had included in the *Optics* (1604) to determine the apparent displacement of the star "if it had been within the confines of the course of the Moon," Kepler affirmed:

> Thus, the latitudinal parallax (assuming those things we have posited previously) will be 47'37", which produces a total difference in latitude (*a priori* 52'28") of almost five minutes ... Yet [the star] did not change location even slightly, certainly by nothing that could be compared with the amount given [i.e., 52 ½'].[14]

10 The importance of celestial developments in the decline of Ptolemy's system has been highlighted in several places. One example of note is the volume recently edited by Miguel Á. Granada, *Novas y cometas entre 1572 y 1618: Revolución cosmológica y renovación política y religiosa* (Barcelona: Universitat de Barcelona, 2012).

11 In his contribution to the volume, *Change and Continuity in Early Modern Cosmology* (Dordrecht: Springer, 2011), pp. 93–104, Patrick Boner presents a close analysis of the passage in question and highlights Kepler's criticism of Ambrosio Lorenzini's work on the nova as well.

12 *JKGW*, 1, 228.1–6.

13 *Ibid.*

14 *JKGW*, 1, 228.7–31: "[...] Parallaxis igitur latitudinis (positis, quae prius posueramus) 47'37", quae iam differentiam latitudinis facit (à priori 52'28") minuto rum ferè quinque ... Atqui locum nihil mutavit: certe nullo, quod cum hac quantitate comparari possit."

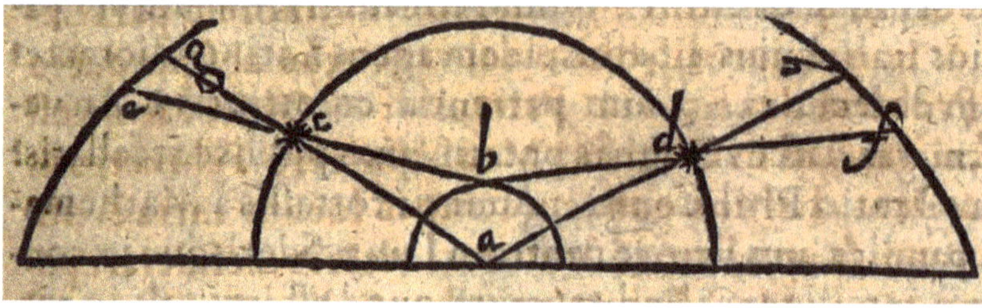

FIGURE 3.1 The parallax of the nova determined according to the distance of the Moon (*DSN*, p. 79).

As Kepler explains, the parallax is measured with a triangle whose base is the distance between the observers and the height the distance to the object. Accordingly, the vertex angle is proportional to the base and also inversely proportional to the height. Therefore, the scheme on page 79 is valid for representing any degree of parallax measurable with the Earth's radius. This measure can be expressed by the simple trigonometric formula D=1/Senθ.

After concluding that the nova was not sublunary, it was equally possible to exclude the nova as measuring a distance similar to that of the Sun. As his reference, Kepler assumed a distance of 1,200 e.r., which meant that the nova should exhibit a parallax close to 3' (the Sun's presumed parallax, as indicated on page 81). Kepler argued that even though the observations of the nova showed some discrepancies, these were not related to the position of the observer, but to the difficulty of calculating the exact location of the novelty from the surrounding stars. Nevertheless, Kepler affirmed: "Our star did not change location by 3 minutes."[15]

The next step was to demonstrate that the nova was located among the fixed stars. In order to do this, Kepler changed perspective, now applying the heliocentric model. We may recall that the distance to the stars according to Ptolemy was 20,000 e.r., and that Tycho Brahe later brought the Earth closer to "7,000 earth radii [to Saturn] and double that distance to the fixed stars."[16] In both cases, the absence of measurable parallax was explained by the lack of any way to measure more than the Earth's surface, since in these conceptions the Earth possessed no translational motion. For this reason, even within "the most accurate observation range" of 2', the maximum distance measurable considering only the earth's ratio is slightly less than 1,719—and this only under ideal conditions. However, if we consider the translational motion of the Earth, whose radius is more than 1,000 e.r. as Copernicus believed, we

15 *Ibid.*, 230.34–35.
16 *Ibid.*, 235.36–38.

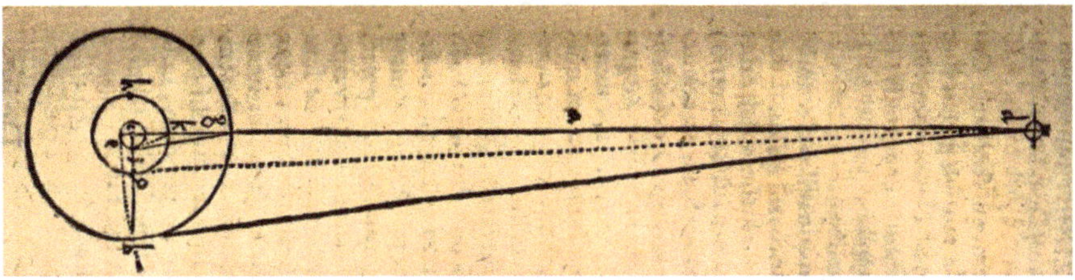

FIGURE 3.2 The parallax of the nova according to the heliocentric perspective (*DSN*, 82).

have a new parallax-triangle's base with a far greater range. Near the end of Chapter 15, Kepler states:

> In fact, as long as we continue to hold the common opinion of the immobility of the Earth, those distances that we have so far considered for the new star are truly childish [*puerilia*]. But if we open the immense Copernican abysses, my God! How high will this star rise?[17]

Kepler designed the above scheme (repeated five pages later) to calculate the parallax from a Copernican perspective. This calculation depends on the radius of the Earth's orbit, to which Kepler attributed a value of 1,200 e.r. in Chapter 15, changed in the following chapter to 1,432 e.r.,[18] thus matching a tenth of the estimated distance from the Sun to Saturn. Since the time between the two observations of the nova was almost four months, the Earth must have travelled through this maximum radius over the interval. Therefore, even with a distance similar to the one estimated to Saturn, a stationary object should measure a massive parallax of almost 6º:

> According to the Prutenic Tables, it appears that if Saturn (point D) remains motionless at the highest point of its orb (when it is ten times farther away from the Sun than the Earth) with the Earth departing from the line CD, which connects the Sun to Saturn, to the extreme side of its orb at O, then the parallax of Saturn or the angle CDO would be about 6º. However, since the star did not change its place even a tenth of a degree, and the angle CDO would not even occupy the sixtieth part of 6º, the star is necessarily higher than sixty times the distance to Saturn.[19]

17 *Ibid.*, 231.14–19: "Verùm puerilia sunt ista diastemata, quae pro novo sidere expendimus hactenus; dum in usitata sententia, de quiete telluris, manemus. At si Copernicanos immensitatis abyssos aperiamus; Deus bone, quantam in altitudinem sidus hoc elevabitur?"

18 *Ibid.*, 235.5.

19 *Ibid.*, 231.32–232.2. "Ex Prutenicis igitur videre est, si Saturnus altissimo sui orbis loco, qui sit D, (quando decuplo abest eius, quo abest Sol a Terra) staret immobilis; Terra ex CD

Because the nova did not even display a parallax of $^{1}/_{60}$ of these 6°, in other words, since it did not reach 6', its height must have been over sixty times the distance given for Saturn or 600 times the estimated distance between the Earth and the Sun. Following this play with proportions, Kepler concluded that even if the nova's height was as close as 720,000 e.r., it should have been possible to discern some degree of parallax. However, since no changes were observed, he concluded that the nova was three times the above figure—2,160,000 e.r. to be more precise, without exceeding the 2' range. In this way, Kepler determined that the nova was neither an atmospheric nor a planetary phenomenon, but rather a new star removed by a stunning distance.

3.4 The Pursuit of the Precise Distance of the Stars: The Method of Proportions

By the end of Chapter 15, Kepler had established the minimum distance to the stars according to the Copernican model and the accepted measures. However, even considering the magnitude of this achievement he did not regard these figures as more accurate. In Chapter 16, Kepler focuses on two complementary questions: to continue his calculations for a possible distance to the stars and to argue in favour of the Copernican model over the Ptolemaic alternative, following the path that Copernicus had opened by questioning the swiftness of the final sphere. Nevertheless, the most important innovation in this process was the move from the geocentric to the 'purely' heliocentric point of view. Kepler believed that the Copernican model was more harmonic, even with the huge gap between Saturn and the sphere of the fixed stars. In the Copernican model, the planets were organized by their periods; by contrast, the same order was absent from Ptolemy's model and his final sphere should follow an extremely swift motion. Kepler makes a "comparison of the incredible speed in Ptolemy with the incredible size in Copernicus" using geometrical and aesthetic arguments.[20] Those ideas had appeared for the first time in his correspondence with Herwart von Hohenburg, in the same place where Kepler ventured another calculation for a possible distance to the stars:

---

linea, quae Solem Saturno connectit, eiusque puncto K ad extremum orbitae suae latus O discedente; parallaxin Saturni, sive angulum CDO fore circiter 6. gradus. At cum ne decimam quidem unius gradus, locum sidus mutaverit, angulusque CDO occupaverit, hoc est ne sexagesimam quidem sex graduum; necesse est, altius esse sidus hoc, quam sexagecuplo distantiae Saturniae."

20 *Ibid.*, 234.10.

> Now it is time for me to turn to another objection. If the Earth moves and the Sun is still, it is necessary that the sphere of the fixed stars be similar to infinite. For the 1,200 radii of the Earth (the space between the Earth and the Sun) do not subtend one-third of a minute. And so, since 3,600 radii are less than one minute, both 216,000 are less than one degree and 77,760,000 are less than the ambit of the fixed stars. And 10,100,000 are smaller than the radius of the sphere of the fixed stars so the sphere of the Sun would scarcely be a ten-thousandth of the sphere of Saturn and this, in turn, a thousandth part of the sphere of the fixed stars. I respond that the objection is true, but it contains nothing more absurd than that immense swiftness found in Ptolemy.[21]

Although the calculation is not developed well, it provides a very interesting example of Kepler's work with geometrical proportions. Kepler postulated a tiny angle of 20", undetectable to the naked eye, and wanted to know the height of the parallax's triangle whose radial base coincided with the orbit of the Earth. He began with the commonly accepted solar parallax of 3' with 1AU or 1,200 e.r. of distance. Therefore, three times that distance is 1/3 of the vertex angle and the circumference of the circle with this radius is 3AU·2·3 (with π=3) or 216,000 e.r. Three times this figure is proportional to 1/3 of 1' (or 20"). But we do not use the radius of the Earth as the base, but rather the radius of the Earth's orbit; as a result, the entire calculation must be multiplied by 1,200. So, 3·216,000·1,200 = 77,760,000, which is the circumference of the great circle of fixed stars and 1/60 of this, or 10,100,000 (more precisely $12.96 \cdot 10^6$), the radius.

This geometrical procedure is similar to that in the DSN, where Kepler tries to find the real distance through a speculative argument. In the passage, Kepler ponders:

> Motion is the perfection of the world, which is like a certain life for it. Three things are required for motion: the mover, the mobile and place. The mover is the Sun; the mobile are [the planets] from Mercury to

21 Johannes Kepler to Herwart von Hohenburg, 28 March 1605, *JKGW*, 15, no. 340, ll. 246–255: "Commodus est mihi transitus ad objectionem aliam. Si Terra movetur, Sole stante, necesse est sphaeram fixarum esse infinitae similem. Nam mille ducentae semidiametri Terrae (·spacium inter Solem et Terram·) non subtendunt in fixis tertiam partem minuti. Itaque 3600 semidiametri minores sunt uno minuto, et 216000 minores uno gradu, et 77760000 minores ambitu fixarum. Et 10100000 minores semidiametro fixarum sphaerae, ita sphaera Solis vix esset decies millesima, Saturni millesima, sphaerae fixarum. Respondeo, objectio vera est, sed nihilo magis absurdi continet, quam immanis illa in Ptolemaeo celeritas."

> Saturn; and the place is [established by] the outermost sphere of the fixed stars. So if we may express something physical mathematically, the mobile [bodies] are the proportional means between the mover and place.[22]

Taken out of context, the above argument sounds arbitrary, since it does not explain the use of the proportional mean or any other proportion, nor even a random number to account for the parallax. However, if we consider Kepler's point of view from the *Mysterium cosmographicum* (1596), with its central idea of the harmonic proportions among the planets by means of the Platonic solids, it now seems logical to calculate the real distance to the stars by means of proportions between the supposedly known solar diameter and distance to Saturn. Accordingly, Kepler assigned 6 e.r. to the solar diameter, 1,432 e.r. to the Sun-Earth distance and ten times this distance to the height of Saturn. (We may express this as 6:14320::14320:R.) According to this procedure, the distance calculated to the stars is 34,077,066 $^2/_3$ [actually 34,177,066 $^2/_3$] and its vertex angle is 8 $^2/_3$".[23]

The figure is spectacular in comparison with the same distance given by Ptolemy or Brahe. In fact, it was one of the main objections raised by Tycho against the Copernican system.[24] However, Kepler attempts to show that it has a more reasonable proportion than the geocentric swiftness of the stars. He calculates that the speed of the sphere of fixed stars should be 2,257,500 mph, as opposed to the Copernican speed of Saturn, which is only 300 mph.[25] The former speed, in other words, is 7,500 times greater. Kepler then multiplied the figure by the geocentric distance to the sphere of fixed stars. The result is 150,000,000, a figure that defied physical interpretation. This number play was deployed by Kepler to convince his readers of the hidden disproportions in the geocentric theory. Meanwhile, Kepler took the opportunity to consider the Copernican planetary speeds caused by his hypothetical solar force. One relatively brief passage anticipates the fundamental theory of the *Astronomia nova* (1609):

---

22 *JKGW*, 1, 234.27–32. "Mundi perfectio est motus, quae eius quasi quaedam vita est. Ad motum tria requiruntur, Motor, Mobile, Locus. Motor est Sol: Mobilia sunt a Mercurio usque ad Saturnum: Locus est extima fixarum Sphaera. Quod si physicam rem licet enunciare mathematice; Mobilia sunt medium proportionale inter Motorem et Locum."

23 *Ibíd.*, 235–236.

24 J. D. Fernie, "The Historical Search for Stellar Parallax," *Journal of the Royal Astronomical Society of Canada*, 1975, 69: 153–161, on p. 155.

25 *JKGW*, 1, 234.4–6.

> For the fixed stars act as place and remain at rest. The sun acts as mover and remains motionless in the middle of the world, unless, according to what I proved to be consonant with physical causes in my commentaries on the motions of Mars, its body rotates in place, so that through this rotation it distributes an immaterial motive species [*species immateriata*] throughout the entire world. All the wandering stars may follow this species, in such a way that they move more swiftly the closer they are to the Sun.[26]

The physical image of the heliocentric system presented by Kepler suggests significant changes in the explanation of the concrete mechanism of the world. Up to this point, Kepler had presented a heliocentric system that provided a new minimal distance to the stars; it had also aimed to understand the gaps between the orbits, and begun laying the foundations for the solar force. All of these elements discredited the theoretical framework of geocentric cosmology, replacing it with a new and more consistent one through the heliocentric hypothesis. At the same time, the above changes shaped the Copernican hypothesis into a version that did not necessarily agree with other Copernican developments.

3.5 The Absolute Centrality of Kepler and the Homogeneous Universe of Bruno

Kepler expressed confidence in his cosmological principles. In his world vision, the absolute centrality of the Sun and the finitude of the universe fit almost perfectly with the hypothesis of the solar force, the proportion between the planets and even with the distance and (supposed) immobility of the stars. For Kepler, every element combined into a *symmetria*, the central idea behind his grandiose construction and the same topic already present in the work of

---

26 *Ibíd.*, 233.24–30. "Fixae enim pro loco sunt, et quiescunt: Sol pro Motore est, et tuetur mediam Mundi stationem immobilis; nisi quod physicis rationibus consentaneum effeci, in commentariis de motibus stellae Martis; corpus ipsius converti in suo spacio; ut hac conversione speciem immateriatam, motricem, per amplitudinem Mundi circumagitet, quam sidera errantia omnia sequantur, ut quodque propius, ita celerius." For the importance of the concept of solar force in Kepler's cosmology, see James R. Voelkel, *The Composition of Kepler's* Astronomia nova (Princeton: Princeton University Press, 2001), pp. 110–113.

Copernicus, only from a different approach. Echoing Goldstein and Hon,[27] Édouard Mehl explains that Copernicus did not consider *symmetria* essentially mathematical: "[for Copernicus] *symmetria* is not to be understood in a mathematical sense of commensurability (of the celestial motions); rather, *symmetria* is to be understood in the architectural and aesthetic sense of good proportion, in which each part is harmoniously related both to another and to the whole."[28] In contrast, *symmetria* for Kepler was in the first place mathematical and only secondarily aesthetic. Of course, Kepler believed that it was the cornerstone to the ancient Pythagorean philosophy and, as a consequence, a central part of what Patrick Boner calls "Kepler's Copernican campaign":

> A key part of Kepler's campaign was the proposal of a new understanding of cosmic proportion. In consideration of "the greatest gap between Saturn, the highest of the planets, and the sphere of the fixed stars," credited by Copernicus to the greatness of "the divine fabric [*fabrica*] of God Almighty," Kepler suggested another way of seeing greatness—in our relative smallness [...] he explained how the distance of the sphere of the fixed stars, "similar to infinity," suggested a sense of proportion that stressed, rather than negated, the nobility of human beings: God had granted mathematical minds to human beings, as small as 'specks of dust.'[29]

For Kepler, the Copernican cosmos should be determined *mathematically a priori*, moved by physical forces over a heterogeneous and anisotropic space. This approach to Copernican cosmology broke with the opinions of many others. For example, Galileo did not believe that *a priori* proof was possible and Kepler's teacher, Michael Maestlin, had little faith in the physical reality of the solar force. But an even bigger challenge for Kepler was the cosmology of Bruno, who had developed his own brand of Copernican heliocentrism. Bruno had argued that the universe must be infinite, homogeneous and isotropic. Since Kepler's entire thesis was based on the absolute centrality of the Sun,

27 Giora Hon and Bernard Goldstein, "Symmetry in Copernicus and Galileo," *Journal for the History of Astronomy*, 2004, 35: 273–292.

28 Édouard Mehl, "'Novum struam mundum': Kepler's Rebuilding of the Copernican 'symmetria mundi,'" in *Unifying Heaven and Earth: Essays in the History of Early Modern Cosmology.*, ed. Miguel Ángel Granada, Patrick J. Boner and Dario Tessicini (Barcelona: Universitat de Barcelona, 2016), 197–216, on p. 199.

29 Patrick J. Boner, "Kepler's Copernican Campaign and the New Star of 1604," in *Change and Continuity in Early Modern Cosmology*, ed. Patrick J. Boner (Dordrecht: Springer, 2011), 93–114, on p. 94.

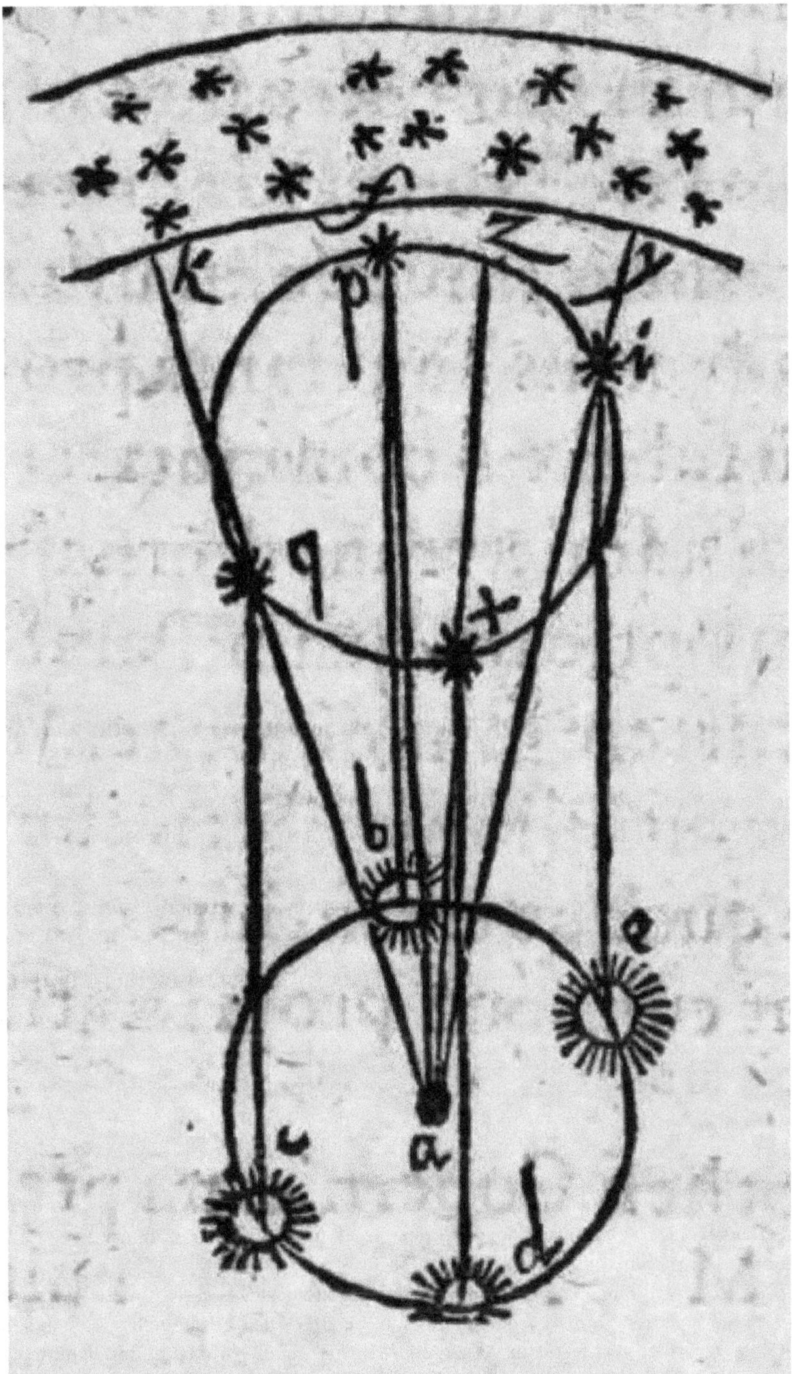

FIGURE 3.3 The circular descent of the new star according to Kepler (*DSN*, 91).

Bruno's cosmology was absolutely unacceptable to him. In *DSN* Chapter 21, Kepler seeks to refute the idea of infinite space with a homogeneous distribution of matter. Kepler's aim is to "deal with another sect of philosophers, the same ones who do not deduce the start of ratiocination from the senses (as Aristotle unjustly affirmed regarding the doctrine of the Pythagoreans recalled by Copernicus)."[30]

Despite the claim above, the first issue addressed by Kepler was not the problem of the infinite universe, but the idea that the nova had descended from an infinite height in circular motion, as if attached to a wheel. As Kepler introduced it, the idea made no mathematical sense. Any descent from the infinite is clearly impossible, but even a great finite circle would produce a very large parallax:

> For if on account of the convexity of the circle, the star completed its descent and began again to ascend, maintaining the same circular motion; then, when it was at the lowest point, it should move from one place to another before the fixed stars, for example, from the right side of the circle, from where it descended, to the left [side] from where, conversely, it made its ascent.[31]

The demonstration did not involve a complicated mathematical device, but merely pointed to the apparent change in place of the object.

A second, more serious problem was the possibility, suggested by Copernicus as well as Rheticus, that the elimination of motion extended the final sphere of the fixed stars infinitely outwards. In *DSN*, Kepler claims that such a concept of infinity "is taken up again from the ancient schools of pagan philosophers, especially through the following argument, namely that Aristotle proved from motion that the world is finite, but Copernicus, by removing motion from the sphere of the fixed stars, concedes that it can be infinite."[32] Kepler identifies

30 *JKGW*, 1, 251.30–32: "Sed est alia philosophantium secta, eorum, qui (ut Aristoteles de Pythagoreorum sententia, a Copernico repetita, immerito quidem pronunciat) non initium ratiocinationis ex sensibus deducunt [...]"

31 *Ibid.*, 252.14–18. "Etenim si propter circuli convexitatem, stella descensu peracto, coepit rursum ascendere, manente eodem circuli motu: ergo cum esset humilima, debuit sub fixis transferri de loco in locum, verbi gratia, a dextris circuli partibus, quibus demittebatur, ad sinistras, quibus rursum elevabatur."

32 *Ibid.*, 252.41–253.3.

William Gilbert and "*infelix ille Jordanus Brunus*" as the leading advocates of infinity.[33] He summarizes the bold proposal of Bruno as follows:

> But Bruno makes the world infinite, so that there are as many worlds as there are fixed stars, and this our mobile region may become one among countless worlds, with practically no difference from the other surrounding ones.[34]

The cosmic uniformity of Bruno implied that every star should be a sun, with its own court of planets orbiting it. This homogeneity removed any particular feature from the solar system, including the absolute centrality of the Sun (linked by Kepler with the solar force) and the existence of six unique planets (linked with the proportions). In this way, Bruno's cosmology directly undermined the foundations of Kepler's celestial physics, which was based on the existence of one unique and universal mover and one unique planetary system governed by the geometry of the Platonic solids and separated from the fixed stars by a proportional mean. Of course, in an infinite universe all of Kepler's efforts to uncover mathematical proportions between the parts and the whole as well as the aesthetic argument mentioned above fail to make any sense. It is one thing to say that the universe is "similar to infinite" and quite another to accept it as *truly unlimited*.[35]

Kepler's refutation of homogeneity is built entirely on a mathematical basis. Since it is not possible to prove or refute the infinite through observations, the only option is to address its implications. In a truly infinite universe that is governed purely by the principle of sufficient reason, matter should be uniformly distributed. However, the distribution of the stars across the sky does not appear uniform. Considering as a starting point that stars of the same magnitude are at a similar distance, it is not difficult to prove that the external appearance of the sky changes dramatically depending on the position of the observer. The well known example of Orion's Belt has been analyzed mathematically by

33 See Miguel Ángel Granada, "Kepler and Bruno on the Infinity of the Universe and of Solar Systems," *Journal for the History of Astronomy*, 2007, 39: 469–495.

34 *JKGW*, 1, 253.8–11: "Sed Brunus ita infinitum facit mundum, ut quot sunt stellae fixae, tot mundos, et hanc nostram regionem mobilium, unum ex innumerabilibus mundis faciat, nulla fere nota a caeteris circumpositis distinctam."

35 Cf. Miguel Ángel Granada, "Aristotle, Copernicus, Bruno: Centrality, the Principle of Movement and the Extension of the Universe," *Studies in History and Philosophy of Science*, 2004, 35: 91–114, on pp. 91–93.

Judith Field,[36] and Alexandre Koyré has famously portrayed Kepler's polemic with Bruno. As Kepler says:

> Thus, if someone were positioned on the Belt of Orion, having our sun and the center of the world above him, he would first see on the horizon a kind of perpetual sea of immense stars, as it were, appearing to touch one another; and from there, the more he raised his eyes, the fewer stars he would see and the less they would be touching, but they would gradually [appear to be] more and more sparse; however, if he looked straight upward, he would see the same [stars] as we do, but twice as small and two times closer together than we see them.[37]

Koyré is quite right when he says that presuppositions do not detract from the central argument of Kepler:

> It is a perfectly consistent and very well-reasoned argument. It is based on two premises, which, by the way, Kepler shared with his opponents. The first one is a direct consequence of the principle of sufficient reason and consists in admitting that, if the world has no limits and no particular, determined, structure, that is, if the world-space is infinite and uniform, then the distribution of the fixed stars in this universe must be uniform, too.[38]

Indeed, it is clear to Kepler that the external appearance of the sky changes dramatically from one place to another, just as the stars change for each observer depending on his or her position on the surface of the Earth. However, neither the combination of those presuppositions nor the deviation of the astronomical measurements allow Kepler to establish on a solid basis his version

36 Appendix 1 in Field (ref. 3), pp. 191–194. See also William H. Donahue, "Astronomy in the Service of Cosmology? Kepler's Arguments for a Finite Universe," in *Astronomy as a Model for the Sciences in Early Modern Times*, ed. Menso Folkerts and Andreas Kühne (Augsburg: Rauner, 2006), pp. 297–306.

37 *JKGW*, 1, 254.8–15: "Itaque si quis esset in illo Orionis cingulo constitutus, habens Solem nostrum, et mundi centrum, super suum verticem: videret ille primo perpetuum quoddam quasi mare ingentium stellarum, sese mutuo contingentium ad visum: inde, quo magis oculos levaret sursum, hoc minores videret stellas, hocque minus sese contingentes; sed iam paulatim magis atque magis sparsas; circa ipsum vero suum verticem, easdem nobiscum videret; sed duplo minores, tantoque invicem propiores, quam nos illas videmus."

38 Alexandre Koyré, *From the Closed World to the Infinite Universe* (Baltimore: Johns Hopkins University Press, 1957), pp. 61–62.

of the Copernican system. For his own part, Kepler aims to prove that a vast portion of the universe should be particularly different.

> For let it be admitted, among other principles, that the fixed stars are extended outward to infinity. Nevertheless, it is a fact that in this inner bosom there shall be an immense cavity, distinct from the spaces among the fixed stars and vastly different in proportion. Thus, if it occurred to somebody to examine only this cavity, even [if he were] ignorant of the eight small bodies flying around the centre of this space at a very small distance from it, and did not know what or how many they were, nevertheless, from the sole comparison of this void with the surrounding spherical region filled with stars, he would utterly conclude that this is a certain particular place and, in fact, the main cavity of the world.[39]

Even if it did exist, the above cavity does not prove the Keplerian model, since it may account for both the geocentric and geoheliocentric models as well. For all those cosmologies, the universe was finite, neither isotropic nor homogeneous. The supposed cavity would not be enough to prove which particular model was correct. For this reason, Galileo aptly focused his attention on the problem of the Earth's mobility as the best way to prove the heliocentric hypothesis. However, it was fundamental for Kepler that the geometry of the universe amount to a complete and physically fused astronomical landscape, as the distance and disposition of the stars was a widely recognised issue in astronomy that could reinforce the foundations of his cosmology. Since the Keplerian model was based on proportions, including the proportional means to measure the distance of the stars, his model would have been the only one that could correctly predict such a hypothetical cavity. In other words, if the cavity existed, then it would have been a significant piece of evidence to support the Keplerian model. Nevertheless, in DSN the trap or trick is that the same measurements used to propose the distance to the stars are employed to predict the cavity without any further proof.

39 *JKGW*, 1, 253.32–40. "Nam, esto inter principia, extendantur fixae in infinitum foras; verum tamen est, in hoc intimo sinu, cavitatem esse insignem, distinctamque ab intervallis, quae sunt inter fixas stellas, immani proportionis dissimilitudine: sic, ut si cui hanc solam cavitatem contingeret inspicere, ignoranti etiam de octo corpusculis, circa huius spacii centrum, angustissimo quidem intervallo, circumvolantibus, quae, qualia, quotve sint: tamen is ex sola hac vacuitate, comparata cum circumambeunte sphaerica regione, stellis refertissima, omnino esset conclusurus, peculiarem aliquem locum hic esse; adeoque praecipuum mundi sinum."

3.6 Conclusion

One of the major consequences of the hypothesis of the earth's mobility was the dramatic expansion of the universe. Not until *DSN* did a Copernican astronomer try, for the first time in full detail, to measure the physical limits of the cosmos according to the heliocentric hypothesis. In his account, Kepler dealt in detail with the enormous distance of the stars, a truly daunting interval for the era. The three possible solutions to this problem presented above were (1) the minimal distance according to the minimal range of error in the age of pre-telescopic astronomy; (2) a second proposal with a broader range; and (3) a final, further-reaching proposal, supported by a procedure following Kepler's own theory of proportions. The Keplerian model showed that the sun-centred system expressed *symmetria*, as far as it was understood mathematically. However, no matter how great the heliocentric universe could be, it should be finite for Kepler, neither homogeneous nor isotropic. The entire Keplerian proposal was based on the absolute centrality of the Sun, the uniqueness of the solar system, and a qualitative distinction between the Sun and the stars. These characteristics prompted a firm and fierce refutation of Bruno's infinite universe in *DSN*. It is perhaps ironic that Kepler used the problem of the distance of the stars—a disagreeable feature of the heliocentric model for many contemporary astronomers and philosophers—as an argument against Bruno. However, his refutation was made only in the field of mathematics. Almost in the same terms as Copernicus had expressed it before him, Kepler wrote: "It is thus certain that the inward world, toward the Sun and the planets, is finite and hollow in some way. The rest [of the world] remains for metaphysics to resolve."[40*]

Acknowledgements

I would like to express my gratitude to the Mexican National Council for Science and Technology (CONACYT) for financing my postdoctoral research at the University of Barcelona and to Miguel Ángel Granada, my supervisor during my stay. I also wish to thank Patrick Boner for overseeing the final stages of this paper. Finally, I thank Ernesto Schettino for introducing me to the fascinating relationship between Bruno and Kepler.

---

40 *JKGW*, 1, 256.15–17: "Est igitur certum, introrsum versus Solem et Planetas finitum esse mundum, et quodammodo excavatum. Quod reliquum est, Metaphysica superstruat."

CHAPTER 4

Celestial Novelty and the Science of the Stars: Kepler vs. Krabbe on Accuracy and Authority in Early Modern Germany

Patrick J. Boner

4.1 Introduction

As the scope of astronomy swelled over the sixteenth century, the science of the stars witnessed a wide variety of inferences about the world.* Wielding more precise instruments and mathematical principles, a growing number of astronomers arrived at novel ideas about nature. The problem of atmospheric refraction, an empirical phenomenon that prevented stargazers from calibrating and correcting their observations, compelled some to lay claim to the "most contentious issues" of the day.[1] In Kassel, Christoph Rothmann declared that the absence of refraction beyond a certain height above the horizon had "driven" him to eliminate the ether and extend the air to the stars.[2] The implication of his mathematical demonstration was not lost on Tycho Brahe, who refused to submit to such an "assault on natural philosophy."[3] Tycho regarded the correct evaluation of refraction as essential to restoring astronomy and affirming the philosophical foundation of his world system.[4] In a sweeping account

* The present study appears as part of the Research Project FFI 2015-64498-P, "Cosmología, teología y antropología en la primera fase de la Revolución Científica (1543–1633)," financed by the Spanish Ministry of Science and Innovation. I would like to thank Miguel Á. Granada and Robert A. Hatch for their many comments and suggestions.

1 Peter Barker, "Brahe and Rothmann on Atmospheric Refraction," in *Optics and Astronomy: Proceedings of the XXth International Congress of History of Science*, ed. Gérard Simon and Simone Débarbat (Turnhout: Brepols, 2001), pp. 43–50.

2 On the role of refraction in the world view of Rothmann, see Bruce T. Moran, "Christoph Rothmann, the Copernican Theory, and Institutional and Technical Influences on the Criticism of Aristotelian Cosmology," *Sixteenth Century Journal*, 1982, 13: 85–108, on pp. 101–107; W. G. L. Randles, *The Unmaking of the Medieval Christian Cosmos, 1500–1760: From Solid Heavens to Boundless Aether* (Aldershot: Ashgate, 1999), pp. 58–79.

3 *Ibid.*, p. 101. For a full account of his cometary theory and cosmology, see Miguel A. Granada, Adam Mosley, and Nicholas Jardine, *Christoph Rothmann's Discourse on the Comet of 1585: An Edition and Translation with Accompanying Essays* (Leiden: Brill, 2014).

4 Adam Mosley, *Bearing the Heavens: Tycho Brahe and the Astronomical Community of the Late Sixteenth Century* (Cambridge: Cambridge University Press, 2007), pp. 84–90. On the revolutionary consequences of refraction in the work of Tycho, see John Robert

 DOI:10.1163/9789004437272_006

of the subject, Johannes Kepler proclaimed the "true measure of refraction," tracing the behavior of light as it entered the atmosphere and reached the eyes of observers.[5] From cause to consequence, refraction urged many astronomers to deal in depth with the physical contours of the cosmos.[6]

Not every scholar was convinced by Kepler. From his place at the ducal court in Wolfenbüttel, Johannes Krabbe dismissed refraction as a source of doubt that was "still uncertain."[7] Rather than accept the recent work of Kepler, Krabbe denied the role of refraction to resolve the accuracy of his observations. When Krabbe turned his attention to the stunning luminary of 1604, he measured a motion that suggested the course of a comet beyond the sphere of Saturn.[8] The comet had moved just 30′ in a month, quietly coming to a halt and disappearing beneath the bright rays of the Sun during the day. While Krabbe promptly published a report claiming that the comet could still change in size and swiftness, he never considered it a star. His fierce resistance to Kepler, who furiously condemned the account, reveals the methods and motives of an early modern mathematician who fought to maintain his reputation. In the face of a leading figure, Krabbe urged Kepler to resolve their public conflict privately.[9] He sought to settle their differences more discreetly through the exchange of letters while continuing to promote his practice in print.

In the present study, I examine the career of Krabbe and how he strove to uphold it in response to the scathing attack of Kepler. Deemed a deceiver by Kepler, Krabbe defended his account in letters encouraging his famous critic "to forgive and forget" their personal differences.[10] To forge a friendship, Krabbe summoned mathematics and philosophy to focus their exchange and consolidate their common interests. My inquiry traces the lines of communication that linked Kepler and Krabbe, including their sources of influence across the Holy Roman Empire. I further consider the forces of power and prestige that

---

Christianson, *Tycho Brahe and the Measure of the Heavens* (London: Reaktion, 2020), pp. 125–126, 137.

5 *Astronomiae pars optica*, *JKGW*, 2, 88.

6 Together, these astronomers are portrayed as the spiritual children of Rheticus who "worked inductively" to reconsider physical reality; see Robert S. Westman, *The Copernican Question: Prognostication, Skepticism, and Celestial Order* (Berkeley: University of California Press, 2011), p. 294.

7 Johannes Krabbe to Johannes Kepler, 20 January 1610, *JKGW*, 16, no. 550, ll. 18–19.

8 When Rothmann perceived "no parallax" in the comet of 1585, he assigned it a place near "the sphere of Saturn." See Granada, Mosley, and Jardine (ref. 3), pp. 114–119.

9 While Krabbe described the exchange as 'private,' it is clear that others could have read their correspondence. On the larger nature of letters as "conversations that were often overheard," see Adam Mosley, Nicholas Jardine, and Karin Tybjerg, "Epistolary Culture, Editorial Practices, and the Propriety of Tycho's Astronomical Letters," *Journal for the History of Astronomy*, 2003, 34: 421–451, on pp. 422–425.

10 Krabbe (ref. 7), ll. 23–24.

shaped the conflict and encouraged the two scholars to arrive at a quiet resolution to their quarrel. Close ties between Prague and Wolfenbüttel suggest a similar context of patronage that valued clients according to the same form of currency. In my final analysis, I reflect on the nature of competence and convention as factors in the fraught fortitude of Krabbe. I also compare the encounter to other flashpoints in the career of Kepler that echo his claim to authority on the new star. Throughout the study, precision doubles as prestige in an early modern community of observers, patrons, and prophets.

4.2 The Nature of a Novelty

In December 1605, an imperial steward wrote to Kepler asking for his advice about the famous luminary that had first appeared in the sky more than a year ago. In his letter from Leipzig, Nikolaus Vicke called on Kepler to decide among the different ideas in circulation and determine the nature of the novelty. "I have listened to many opinions of learned men in conversation," Vicke wrote, "and I would be pleased to know your own verdict on the matter."[11] A central concern was the apparent motion of the star, which divided the discourse into a variety of views. Vicke noted that while some "assigned a planetary motion to the new star," others asserted it was entirely stationary.[12] Vicke was personally convinced by a local professor in Leipzig, Joachim Tanckius (1557–1609), that the bright object had never budged.[13] On the other hand, Vicke could not discard the work of the astrologer physician David Herlicius (1557–1636), who discerned a distant comet traversing the celestial ether beyond the sphere of Saturn. Herlicius had apparently relied on the observations of Johannes Krabbe (1553–1616) in Wolfenbüttel. Vicke requested that Kepler respond "without delay" and explain what he thought, "not only about the text of Herlicius, but also the observations of Krabbe."[14]

No doubt the purpose of the request was ultimately astrological. While the new star of 1604 inspired a surge of interest from every sphere of society, Vicke was mainly concerned with measuring the influence of the heavens more accurately. He asked the Imperial Mathematician to serve as a sort of 'quality

11 Nikolaus Vicke to Johannes Kepler, 11 December 1605, *JKGW*, 15, no. 362, ll. 1–10.

12 *Ibid.*, ll. 3–5.

13 On the role of Tanckius as a promoter of Paracelsian ideas, see Bruce T. Moran, *The Alchemical World of the German Court: Occult Philosophy and Chemical Medicine in the Circle of Morris of Hessen (1572–1632)* (Stuttgart: Franz Steiner Verlag, 1991).

14 Vicke to Kepler (ref. 11), ll. 10–13.

control' over the observations and predictions he encountered.[15] When a violent storm later struck the town of Halberstadt in January 1610, Vicke turned again to Kepler to consider the cause among the stars. Halberstadt had been severely hit "by thunder and lightning," and a fierce wind had "felled everything."[16] Vicke read in the planetary tables recently published by Krabbe that Mars and Jupiter had conjoined three days before the start of the storm. However, he expressed concern over his source. Why had Krabbe predicted "a wave of freezing cold weather," Vicke wondered, when the conjunction of Mars and Jupiter had occurred in a hot and humid sign?[17] Vicke was equally concerned about the quality of the tables. He asked Kepler to convey his opinion "on the observations and [planetary] tables" that Krabbe had shared with him during their time at court together in Wolfenbüttel.[18]

The new star had captured the court imagination. Across the Holy Roman Empire, accurate observations and persuasive predictions were sought as signs of privilege and sources of power. Reeling from rough weather and a raging witch craze,[19] Vicke wrote to Kepler during troubling times at the court of Duke Heinrich Julius (1564–1613). "So much for the severe weather," Vicke closed his letter to Kepler, "we have enough murder and hanging around here."[20] Yet astrology did not suddenly grow in demand and distinction during "times of crisis."[21] Fused in the very fabric of society, astrology served as a source of knowledge for patrons and practitioners who rarely "resorted" to it when they failed to determine the future every other way. Vicke regularly engaged in rituals that extended beyond the urgent concerns of administration. In a letter to David Fabricius (1564–1617), he described a book about astrology, chiromancy, and metoposcopy that he was preparing for publication. Vicke vowed to unveil "many mistakes" in the horoscopes of princes, "along with several other secrets

15 On the practice of patrons "testing the knowledge of the astrologers they employed," see Monica Azzolini, *The Duke and the Stars: Astrology and Politics in Renaissance Milan* (Cambridge, MA: Harvard University Press, 2013).

16 Nikolaus Vicke to Johannes Kepler, 18 January 1610, *JKGW*, 15, no. 549, ll. 9–11.

17 *Ibid.*, ll. 13–14.

18 *Ibid.*, ll. 4–6. It is clear that Vicke planned to share the responses of Kepler with others. On the public nature of private letters and the "invalid" notion that they provide the historian with 'raw' information, see Nicholas Jardine, "Kepler = Koestler: On Empathy and Genre in the History of the Sciences," *Journal for the History of Astronomy*, 2014, 45: 271–286, on p. 282.

19 On the association of the witch craze with poor harvests and rising crop prices, see Ulinka Rublack, *The Astronomer and the Witch: Johannes Kepler's Fight for His Mother* (Oxford: Oxford University Press, 2015), pp. 3–7.

20 Vicke to Kepler (ref. 16), ll. 14–15.

21 Azzolini (ref. 15), p. 10.

in astrological matters."[22] Fabricius eagerly anticipated the book and wrote favorably about it to Kepler.

Kepler apparently did not share his final verdict with Vicke until he finished his full study of the star in 1606. As he explained to Holy Roman Emperor Rudolph II (1552–1612), a thorough analysis could only come after the star had completely disappeared in February. An accurate account, in other words, required a complete record. Kepler described his distaste for astrology as a further cause for delay. "When the mind grows accustomed to geometrical demonstrations and considers the vile foundations [of astrology]," he remarked, "it long stands idle like a stubborn mule until, forced by blows and verbal abuse, it finally dips a foot into the pool."[23] Yet Kepler still could not avoid the subject of astrology in his sweeping study. He responded to the interests of his audience reluctantly, calling astrology "tangled and tedious, cultivated by many concerns and worries."[24] His principal purpose, as he expressed it, was to account for the motions of the heavens rather than their meaning. "I was not appointed by the emperor to be a public prophet," Kepler declared, "but to restore astronomy."[25] He claimed to follow the example of Tycho Brahe (1546–1601),[26] who had approached astrology with equal ambiguity.

4.3 False Ideas and Flawed Sources

Kepler had expressed the same sense of caution in a short report he wrote soon after the star appeared. Penned in German, the pamphlet appealed to patrons and contributed to "the great yield" that Kepler wryly predicted would be reaped by the presses following the appearance of the star.[27] Given the popularity of the naked-eye novelty, Kepler attempted to calm the clamor with accurate observations and somber words of warning. The circumstances surrounding the star were so stunning that he suggested humility and patience over any swift prediction. For Kepler, the meaning of the star was a matter of faith and fortitude. "The only thing certain about the star," he advised, "is

22 David Fabricius to Johannes Kepler, 20 January 1607, *JKGW*, 15, no. 408, ll. 387–388.

23 *JKGW*, 1, 151.33–36.

24 *Ibid.*, 151.32–33.

25 *Ibid.*, 354.12–14.

26 See Gábor Almási, "Tycho Brahe and the Separation of Astronomy from Astrology: The Making of a New Scientific Discourse," *Science in Context*, 2013, 26: 3–30. On the sense of shared destiny that Kepler expressed for Tycho, see Jürgen Hübner, *Die Theologie Johannes Keplers zwischen Orthodoxie und Naturwissenschaft* (Tübingen: J. C. B. Mohr, 1975), p. 263.

27 *JKGW*, 1, 398.18–19.

that it signifies either nothing at all for us or such highly significant things that they surpass the sense and reason of everyone."[28] While Kepler claimed that the common multitude could be swayed by any fool who posed as a prophet and pronounced "something credible,"[29] he still knew the stakes were high. No prediction about the new luminary could be taken lightly.[30] "Time will tell the true significance [of the star]," Kepler concluded, "so long as it pleases the Almighty."[31] Accuracy was the ambition of the astronomer and the elusive pursuit of the celestial prophet.

Kepler echoed his critical view of astrology in the final pages of *De stella nova*. He began with an apology explaining his focus on other factors and the disappointing absence of details about the future. "I beg my readers not to react angrily," he pleaded, "if my political and theological discourses about the disputes of our time are briefer than they expected."[32] Kepler excused his failure to forecast any further as a matter of priority and a consequence of his position. To forge a new science of the stars, nothing was more pressing than a fresh theoretical foundation and the revolutionary tables that his predecessor had promised the Holy Roman Emperor.[33] Kepler described his great effort as a gift for the public good rather than "spare ornaments for the sake of private individuals [*in privatorum gratiam parerga*]."[34] Why "turn others away," he argued, when his own theory could empower every careful observer to follow the motions of the heavens more faithfully? Kepler bravely envisioned an even broader role for his knowledge, which could one day contribute to universal peace and religious reconciliation. He witnessed the work of divine providence in the stars and wished to share it with the world.[35]

---

28 *Ibid.*, 396.3–5.

29 *Ibid.*, 398.24–28.

30 At a time when almanacs were widely known to go astray "concerning the weather," there is reason to believe that religious piety was not the only cause for restraint among astrologers. On the personal and practical motives of another stargazer, Sigfridus Aronus Forsius (1560–1624), see Martin Kjellgren, *Taming the Prophets: Astrology, Orthodoxy, and the Word of God in Early Modern Sweden* (Lund: Sekel, 2011).

31 *JKGW*, 1, 399.8–9.

32 *Ibid.*, 354.9–11.

33 On the conditions of his position as Imperial Mathematician and the obligation to publish the Rudolphine Tables, see James R. Voelkel, *The Composition of Kepler's* Astronomia nova (Princeton: Princeton University Press, 2001). According to Nicholas Jardine, the emergence of the world system as the object of astronomy "was a product of the competitive practices of courtly exchange." See *idem*, "The Places of Astronomy in Early-Modern Culture," *Journal for the History of Astronomy*, 1998, 29: 49–62, on p. 58.

34 *JKGW*, 1, 354.17.

35 On the "new kind of natural philosophy" that encouraged Lutheran scholars like Kepler to witness the work of God, see Sachiko Kusukawa, *The Transformation of Natural Philosophy:*

At a time when interpretation was paramount and in some cases even preceded "the correct identification of the object," Kepler eventually accepted the task of telling the future.[36] He proudly embraced the opinion espoused by Tycho about the new star of 1572. "No matter the cause," Kepler avowed, "I will follow in the footsteps of Tycho closely."[37] In essence, Kepler stated that the star had been sent to call to mind our sins and move us to repent more fervently. "Congruent with the Christian astrology taught by Philipp Melanchthon," Kepler endorsed a popular view among his peers.[38] Comparison with the Prince of Astronomers suggested an even stronger sense of precedent and prestige. "God exhibited a general sign to be seen by the eyes of everyone," Kepler affirmed, "to admonish individuals to examine themselves and their sins, and, once they understand their errors and offenses, to turn toward true penance."[39] Since astrology never followed a one-way course for Kepler, he asked his readers to react to the heavens with good humor and humility. Nothing certain could be found without first considering the individual circumstances of every subject under study.

Kepler brought his account to a close by urging his audience to consider the art of astrology more critically. He scorned those who had hastily sought the advice of others as soon as the star had shone. "When any portent divinely appears," he complained, "they immediately turn their eyes and ears away from it and toward the interpreter, regarding him rather than the new portent."[40] The problem for Kepler was that they missed the very details that were so vital to interpreting the star properly. Kepler conceived a causal astrology that focused primarily on "physical processes," from the precise positions of the planets and stars to the arrival of their light rays on earth.[41] He considered the physical nature of a novelty as the sole basis for sound study. At the heart of his criticism, Kepler noted a neglect of nature that had filled astrology with false practices

The Case of Philip Melanchthon (Cambridge: Cambridge University Press, 1995); cf. Gábor Almási, "Rethinking Sixteenth-Century 'Lutheran Astronomy,'" *Intellectual History Review*, 2013, 24: 1–16.

36 Charlotte Methuen, "'This Comet or New Star': Theology and the Interpretation of the Nova of 1572," *Perspectives on Science*, 1997, 5: 499–515, on p. 506.

37 *JKGW*, 1, 355.2–3.

38 Methuen (ref. 36), p. 506.

39 *JKGW*, 1, 354.39–355.1. On similar expressions of penance and punishment in the work of Maestlin, see Miguel A. Granada, "Michael Maestlin and His Unpublished Treatise on the Nova of 1604," *Journal for the History of Astronomy*, 2014, 45: 91–122, on pp. 108, 116–117.

40 *JKGW*, 1, 355.4–6.

41 On the physical astrology of Kepler as a continuation of Caspar Peucer (1525–1602), see Claudia Brosseder, *Im Bann der Sterne: Caspar Peucer, Philipp Melanchthon und andere Wittenberger Astrologen* (Berlin: Akademie Verlag, 2004), p. 302.

and principles. Accepting the unfounded opinion of an astrologer "who does not rely on reasons" had become the "perverse practice" of the general populace.[42] "They do not reckon there is any use for astronomy," he concluded, "except when they tie even our slightest actions to the stars" and determine everything in advance.[43]

Who did Kepler target in his assault on astrology? He was closely familiar with many other court astrologers, as well as a wide variety of celestial prophets who practiced in the surrounding provinces.[44] Despite his disapproval of common stargazers, Kepler knew the currency of credibility was set not only by the social elite but determined in the popular domain as well.[45] One astrologer had particularly roused his ire. At several points in *De stella nova*, Kepler accused Johannes Krabbe of playing the role of political advisor without paying due regard to the heavens. He proclaimed that Krabbe should never have published his prognostication, but shared it "only in private (or left it entirely silent),"[46] since Krabbe had said "nothing about the stars."[47] According to Kepler, Krabbe had committed a common crime among astrologers by failing to refer to the celestial causes. Krabbe had anticipated losses for the Holy Roman Empire "from apathy and indolence," a disturbing prediction he conveyed from the ducal court in Wolfenbüttel across the region.[48] Kepler could not make sense of it. "There may always be more signs of illnesses than doctors," he fumed in frustration.[49]

Krabbe struck a chord with Kepler. In the eyes of the Imperial Mathematician, Krabbe had failed to fulfill the principles of observational accuracy and causal

42 *JKGW*, 1, 355.4–10.

43 *Ibid.*, 355.24–28.

44 On his friendship with the astrologer physician Helisaeus Roeslin (1545–1616), see Miguel A. Granada, "Kepler v. Roeslin on the Interpretation of Kepler's Nova: (1) 1604–1606," *Journal for the History of Astronomy*, 2005, 36: 299–319.

45 Jardine (ref. 33), p. 52. On the astrological reform of Tycho as an attempt to preserve the art among private circles, see Steven Vanden Broecke, *The Limits of Influence: Pico, Louvain, and the Crisis of Renaissance Astrology* (Leiden: Brill, 2003).

46 *JKGW*, 1, 354.34–36.

47 *Ibid.* Kepler compared Krabbe to Fabricius, whose own prognostication he portrayed as a political commentary intended purely "to scare away enemies and encourage the Emperor." On Fabricius and his fascinating account of the star, see Miguel A. Granada, "Johannes Kepler and David Fabricius: Their Discussion on the Nova of 1604," in *Change and Continuity in Early Modern Cosmology*, ed. Patrick J. Boner (Dordrecht: Springer, 2011), pp. 67–92.

48 *JKGW*, 1, 354.32.

49 *Ibid.*, 354.36.

probability that prevailed in every branch of natural philosophy.[50] While Kepler would famously refer to astrology as the "foolish daughter" of astronomy, he saw no reason why she should not aspire to the ways of her "wise mother."[51] Many had "mixed the pepper with the mouse droppings," Kepler quipped, by mistakenly excluding astrology from the body of natural knowledge.[52] According to Kepler, Krabbe had further contributed to the neglect of nature in a relevant prediction regarding the Fiery Trigon. When he glimpsed "the rise of a third religion," Kepler remarked, Krabbe had only seen "through eyes turned away from heaven."[53] His attention had been focused entirely on the Holy Roman Empire and the rumors circulating at court. Contrary to Krabbe, Kepler embraced the bold endeavor of Giovanni Pico della Mirandola (1463–1494) to reduce astrology to a system of universal efficient causes. He joined Pico in refusing to predict any particular events and rejecting prophecies that snubbed the central role of the stars.[54]

Kepler took Krabbe to task on two further counts. First, he condemned Krabbe for feigning a natural cause for the new star. Several astrologers had assigned the origin to the spectacular conjunction of the superior planets in September 1604. "They now contend with complete confidence," Kepler carped, "that the great conjunction of the three superior planets was the progenitor cause [*causa progenitrix*] of the star."[55] Albin Moller (1541–1618), a pastor in Brandenburg, had identified a comet flickering in the sky like "a silver flame," deriving "the combined colors of Mars and Jupiter."[56] For Kepler, the explanation was about as satisfying as saying that a son would be born to him ten months after his marriage. Coincidence did not count as a cause. "While they simulate a natural act of generation with words," he objected, "in no way do they actually show how the cause led to the effect."[57] Kepler could find nothing about the physical role of nature in their rambling. In part, his accusation was a slight of hand that hid the fact that many astrologers explained things in roughly the same way as Aristotle. According to Moller, a dry exhalation had

50 See H. Darrel Rutkin, "Astrology," in *The Cambridge History of Science*, vol. 3, ed. Katharine Park and Lorraine Daston (Cambridge: Cambridge University Press, 2006), pp. 541–561, on p. 550.

51 *Tertius interveniens*, *JKGW*, 4, 161.10–17.

52 *Ibid.*, 230.31–36.

53 *JKGW*, 1, 333.8–10.

54 Rutkin (ref. 50), p. 548.

55 *JKGW*, 1, 275.24–26.

56 Albin Moller, *Gründlicher und warer Bericht von dem newen Cometstern* (Eisleben: Jakob Graubisch, 1605), 2v.

57 *JKGW*, 1, 275.29–30.

ascended to "the upper region of the air," where it had been set on fire by the motion of "the inflaming planets," Mars and Jupiter.[58]

In addition, Kepler charged Krabbe with fraud. Not only had Krabbe made "the most awful observations," but Kepler believed he had "arranged them in the wrong order."[59] Based on the double blunder, Krabbe argued for a comet that had moved the most by far on the first day it appeared. Frowning on any hint of motion, Kepler referred the matter to the more accurate observations of their mutual friend, Fabricius:

> Krabbe dared to attribute to our star the slightest motion, which over the total course of duration did not complete an arc as great as he wrongly observed right on the first day, 4/14 October, when he locates it at 17°0′ Sagittarius, while Fabricius discovered it the day before in East Frisia at 17°37′ or 17°40′ Sagittarius.[60]

According to Kepler, Krabbe had done far more than fumble the observations. His real sin lay with inflating their fidelity to found a false theory. "Since he commends his observations with so much certainty as to assign a slight motion to the star from their fidelity," Kepler seethed, "he will rightly be reproved as a deceiver."[61]

Kepler wrote poignantly to prevent the error from spreading to other scholars. In a community where false ideas could follow quickly from flawed sources, even an earnest calculation could sow the seed of a hollow harvest. "What will become of astronomy," Kepler gasped, "if, God forbid, poor posterity should assign the same faith to his observations of the planets as Krabbe pretends?"[62] To his dismay, Kepler learned that Herlicius had already identified the novelty as a comet upon consulting the account of Krabbe (Fig. 4.1). Krabbe had first observed the comet, Herlicius recalled, "right above the foot of Serpentarius."[63] According to Herlicius, it had followed a course high in the heavens between Saturn and the sphere of the fixed stars. "The comet is so high above the earth,"

58 Moller (ref. 56), 2v.

59 *JKGW*, 1, 161.37–38.

60 *Ibid.*, 161.38–162.1: "[…] ausus est stellae nostrae tribuere motum tardissimum, qui toto durationis tempore non tantum confecerit arcum, quantum Crabbus primo statim die 4/14. Octobris aberravit observando, dum eam in 17.0′ Sagitt. refert, quam Fabricius in Ostfrisia, die antecedente, invenit in 17.37′ vel 17.40′ Sagit."

61 *Ibid.*, 162.3–5.

62 *Ibid.*, 162.5–7.

63 David Herlicius, *Astronomische und historische Erklerung des newen Sterns oder ungeschwäntzten Cometen* (Szczecin: Joachim Rhete, 1604), 13r.

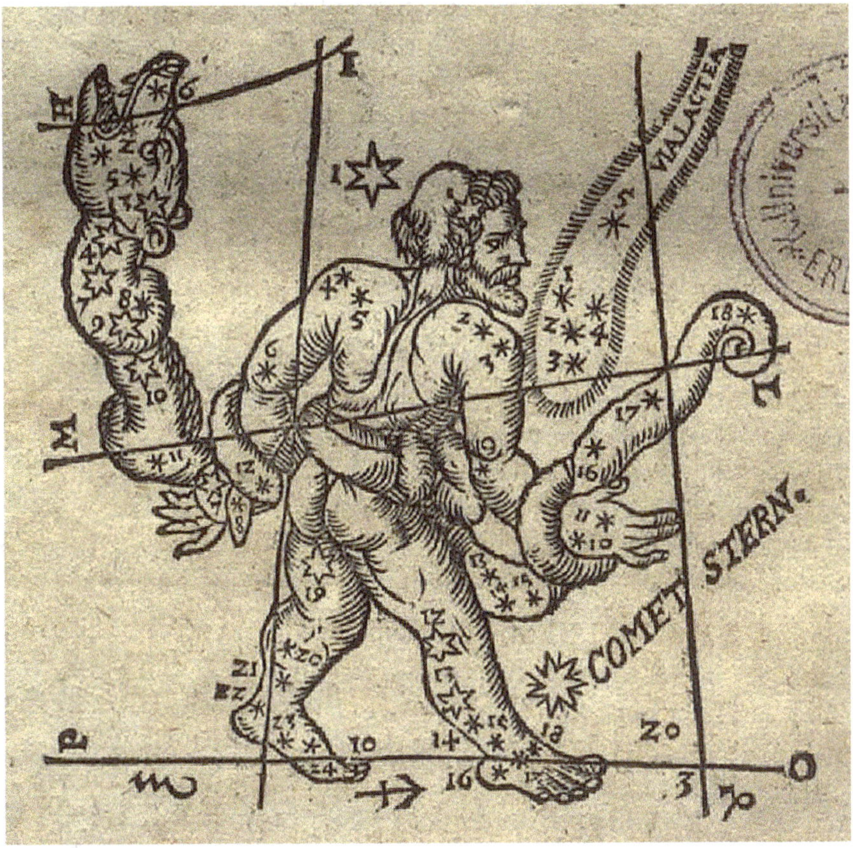

FIGURE 4.1 Original appearance of the comet according to David Herlicius, *Astronomische und historische Erklerung des newen Sterns oder ungeschwäntzten Cometen* (Szczecin, 1604), 1r.

Herlicius explained, “that if a bird could fly 24 miles every day, it would, according to my calculation, reach the comet in 4 years, 13 days, and 11 hours.”[64] Herlicius feared the astonishing circumstances of the comet could set a new standard for global misfortune. While “a similar star” had spelled disaster in 1572, the comet of 1604 had appeared even “bigger and brighter,” and “closer to the paths of the Sun and the planets.”[65]

---

64 *Ibid.*

65 *Ibid.* Herlicius had studied astrology at the University of Wittenberg and taught the subject for thirteen years at the University of Greifswald. On the personal autonomy of his private practice, in which he “determined how and when he would fulfill the request of a prince,” see Brosseder (ref. 41), pp. 48–49.

How did Herlicius account for the nature and substance of the comet in the celestial ether? The conflict with Aristotle was clear. Contrary to other comets that were born from the four elements "below the sphere of the Moon," the comet of 1604 was defined by Herlicius as a "supernatural accumulation of matter in the ethereal region."[66] Only God could produce such a spectacular miracle "beyond all understanding."[67] To support his view, Herlicius referred to "the scores of skilled astronomers" who had suggested the star of 1572 had stood "far higher in the firmament among the fixed stars."[68] Following their precedent, Herlicius professed that the comet of 1604 could have measured a similar distance. He further speculated that the comet had confirmed a popular theory among many metaphysicians. Referring to his place as a physician rather than a philosopher, Herlicius did not dispute that comets were "a congress or conjunction of two planetary light rays meeting together."[69] "That the heavens are free from generation and corruption," Herlicius recalled, had long been proven wrong "by other renowned astronomers."[70]

Herlicius was not the only target in the eyes of Kepler. In Linz, Balthasar Büttner had argued that the incredible physical properties of the comet conveyed a double meaning. On the one hand, Büttner suggested that such an "unspeakable" spectacle appeared as a sign of divine wrath "in response to our grave sins."[71] The blazing wonder had warned observers that God would one day reduce the world to flames in a fiery rapture. On the other hand, the safe removal of the comet from the surface of the earth signified a gift of divine grace. Seen as a sign of mercy, the comet expressed "a visible warning" urging the faithful to repent for their sins and reform their ways.[72] While Kepler arrived at a similar interpretation, he was primarily concerned with determining the physical details that could empower the right prophet to predict more precisely. There were physical factors, such as the fact that few comets were ever as radiant as the recent novelty, which had escaped the notice of the negligent.[73] Along with negligence came the diffusion of inaccurate observations that Kepler feared would prevent the possibility of prophecy in the first place. He roundly condemned Krabbe as a leading culprit.

66 Herlicius (ref. 63), 7v–8r.

67 *Ibid.*, 8r.

68 *Ibid.*

69 *Ibid.* On the growing trend among princes to request primarily medical advice rather than political counsel from court astrologers, see Brosseder (ref. 41), pp. 302–303.

70 Herlicius (ref. 63), 8r.

71 Balthasar Büttner, *Cometen Stundbüchlein* (Eisleben: Jakob Gaubisch, 1605), 26.

72 *Ibid.*

73 *JKGW*, 1, 274.15–21.

4.4 Johannes Krabbe and His Career in Astrology

Krabbe had prepared a short report on the comet for his patron, Duke Heinrich Julius of Braunschweig-Lüneburg. The report fulfilled one of his many responsibilities as court mathematician in Wolfenbüttel. As Krabbe later recalled, he had come to court "through an extraordinary act of God" in 1585 after a ducal visit to his native town of Münden.[74] Krabbe had been summoned to serve in every practical application, from servicing clocks to surveying the terrain of his sovereign. In a formal account of his appointment, Krabbe was described as a cartographer, instrument maker, and mathematician whose duties extended to calendars, tables, and tools spanning the science of the stars.[75] The son of a shipping family, Krabbe had studied briefly at the University of Helmstedt before befriending Jost Bürgi (1552–1632), who mentored Krabbe on mathematical and mechanical matters. Once in Wolfenbüttel, Krabbe soon became a close servant of the crippled sovereign, lifting him from bed in the morning and helping him lie down at night.[76] As power passed from father to son, Krabbe enjoyed greater interest from his new patron, Heinrich Julius, in the promotion of astronomy.[77]

At court, Krabbe was required to publish calendars and prognostications that allowed his readers to record their observations and reflect on the role of the stars in their lives. The task was typical of mathematicians and physicians across the Holy Roman Empire who were awarded by their ducal and municipal authorities monetarily.[78] In the preface to his prognostication for 1592, Krabbe described the divine nature of his work and praised the mathematical pillars of celestial prophecy. Echoing a common refrain among astrologers in Lutheran Germany, Krabbe scorned any skeptics who failed to appreciate his practice as divinely endorsed and mathematically inspired. "Should we,

74 Johannes Krabbe, *Newes Astrolabium, Sampt dessen Nutz und Gebrauch* (Frankfurt am Main: Matthias Becker, 1609), i–ii.

75 Karl Brethauer, "Johannes Krabbe Mundensis," in *Braunschweigisches Jahrbuch*, vol. 55, ed. J. König (Braunschweig: Braunschweigischer Geschichtsverein, 1974), pp. 72–89, on p. 79.

76 *Ibid.*, p. 78.

77 *Ibid.*, p. 80. For an intellectual profile of Duke Heinrich Julius, see Pietro D. Omodeo, "Astronomia, filosofia e teologia nel tardo Rinascimento tedesco: Heinrich Julius di Braunschweig e il soggiorno di Giordano Bruno in Germania," *Giornale critico della filosofia italiana*, 2011, 92: 307–326.

78 On the early calendars of Kepler and his professional predecessor in Graz, see Patrick J. Boner, "Finding Favour in the Heavens and Earth: Stadius, Kepler and Astrological Calendars in Early Modern Graz," in *Studia Copernicana*, vol. 42, ed. Richard L. Kremer and Jaroslaw Włodarczyk (Warsaw: Instytut Historii Nauki, 2009), pp. 159–177.

along with the godless Epicureans," Krabbe asked, "ridicule such a remarkable art and regard it as useless or nothing?"[79] His blanket claim against 'blind' Epicureans recalled the words of Philipp Melanchthon (1497–1560), who had freshly supplied astrology with a formal foundation in his sweeping reform of the arts and sciences.[80] A product of the new pedagogy, Krabbe discovered in mathematics an antidote to intellectual infirmity and a more certain path to predicting the future.

Krabbe proudly pointed to the accurate prediction of previous novelties in his prognostications. He conveyed an air of confidence in his account of the comet of 1596.[81] "According to several constellations and other causes considered," Krabbe wrote, "I announced in my prognostication for 1596 that a new comet would appear in the sky this year."[82] While he confessed that "no clear cause" had been prescribed for predicting the appearance of comets "according to a given month or year, much less an hour or day," several successes had encouraged Krabbe to continue practicing the art.[83] In particular, Krabbe referred to Peter Apian (1495–1552) and his chronicle of comets accurately predicted over the ages.[84] Apian once claimed to have anticipated the comet of 1532; only his doubt about "the supply of dry exhalation in the earth" had kept him from declaring the event with complete certainty.[85] Krabbe foresaw a bright future for "judicious astrologers" who drew on the insight of Apian and others to discern the course of the sky.[86] He reasoned that the record would allow the right scholars to consolidate the science of the stars, removing any reservation remaining among astronomers.

The evidence for determining the origin and effect of a comet on earth was necessarily historical.[87] In a common maneuver among astrologers, Krabbe compared the comet of 1596 with others that had appeared under some of

79 Johannes Krabbe, *Practica astrologica* (Berlin: Paul Donat, 1591), 1v.

80 Kusukawa (ref. 35), pp. 134–144.

81 On the popularity of prognostications in Lutheran Germany and the "confident claims" of their authors, see Robin B. Barnes, *Astrology and Reformation* (Oxford: Oxford University Press, 2016).

82 Johannes Krabbe, *Gründliche observation des newen Cometen lauff / höhe / und grösse, welcher in diesem 1596. Jahre / im Monat Julio am Himmel erschienen* (Magdeburg: Johann Francken, 1596), 1r.

83 *Ibid.*

84 *Ibid.*, 1v.

85 Peter Apian, *Ein kurtzer Bericht & Observation unnd Urtels des jüngst erschinnen Cometen* (Ingolstadt: Peter Apian, 1532), 11r.

86 Krabbe (ref. 82), 1v.

87 See Adam Mosley, "The History and Historiography of Early Modern Comets," in Granada, Mosley, and Jardine (ref. 3), pp. 282–325.

the same circumstances.[88] Krabbe noted that, soon after a similar comet had appeared in 1558, Holy Roman Emperor Charles V had "died in Spain from an acute fever."[89] Perhaps more pressing for his patron in Wolfenbüttel, Duke Franz Otto of Braunschweig-Lüneburg (1530–1559) had died less than a year later. Comets were widely thought to exercise their influence long after they vanished. Krabbe argued that the comet of 1596 would exert an impact "until the end of 1598."[90] While the nature of their influence was subject to interpretation, Krabbe warned his audience to weigh the matter warily. "No one can presume that the comet appeared in the sky without a purpose," he pronounced.[91] "Although many renowned astrologers will no doubt write their own interpretation," Krabbe declared, "I still wish to write my own modest account, even as a mere student of astrology."[92] In the face of complex calculations and critical contemporaries, a sense of humility never hurt.[93]

Krabbe strove to determine the course of the comet in the sky with greater precision than he situated it in history. Drawing a direct proportion between the distance and diurnal motion of the comet, Krabbe calculated that it had appeared more than 145,377 German miles "above the circumference of the earth."[94] In his calculation, Krabbe assumed the same "inviolable doctrine" of uniform circular motion that shaped the cometary theory of Michael Maestlin (1550–1631) and many other astronomers.[95] Conceiving the comet as part of "a sphere in the sky," Krabbe claimed that it had once covered 2°10″ in the course of a day.[96] Accordingly, he assigned it a supralunar location corresponding to the degree of diurnal displacement. "Seen from the circumference of the earth," Krabbe concluded, "the comet appeared higher in the heaven than the

88 On the role of comets structuring the course of history, see Tayra Lanuza Navarro, "La astrología como explicación científica de la historia: Los pronósticos españoles del siglo XVII," in *Synergia: Primer Encuentro de Jóvenes Investigadores en Historia de la Ciencia*, ed. Néstor Herrán, Josep Simon, Ximo Guillem-Llobat, Tayra Lanuza Navarro, Pedro Ruiz Castel, and Jaume Navarro (Madrid: CSIC, 2007), pp. 303–323.

89 Krabbe (ref. 82), 6v.

90 *Ibid.*, 7r.

91 *Ibid.*, 5v.

92 *Ibid.*

93 On the rhetorical strategies of Christoph Rothmann to win the sympathy of his audience in a contemporary account of the comet of 1585, see Nicholas Jardine, "How to Present a Copernican Comet: The Form and Tactics of Christoph Rothmann's *Dialexis* on the Comet of 1585," in Granada, Mosley, and Jardine (ref. 3), pp. 258–281.

94 Krabbe (ref. 82), 4v.

95 On the enduring role of circular regularity in early modern cometary theory, see Antonio Beltrán Marí, *Cometas, ciencia y religión: La polémica Galileo-Grassi* (Madrid: Tecnos, 2016), pp. xxix–xxxiv.

96 Krabbe (ref. 82), 4r.

Moon, namely between the sphere of the Moon and the sphere of Mercury."[97] Krabbe expressed concern that the comet appeared "much closer to Mercury than to the Moon."[98] As Krabbe well knew, moving in league "with the inconstant Mercury" was believed to bode poorly.[99]

How did Krabbe reconcile the supralunar course of comets with the traditional account of the heavens as constant and unchanging? There is no reason to believe that he aspired to break with the broadly held views of his time. In astrology, Krabbe saw many of the same signs and adhered to several of the same principles as his peers across Protestant Germany. He echoed a virtual chorus of voices that foresaw the frightening consequences of eclipses according to common opinion. In 1591, Krabbe warned that a lunar eclipse in the sign of Gemini, known as "the eighth house of death," would lead to "widespread mortality among children and young people from pestilence."[100] At another point, Krabbe advised against the same three 'evil aspects' among the planets that were commonly considered contrary to the positive configurations. When the Moon entered conjunction, opposition, or quadrature with Mars or Saturn, Krabbe counseled, "those days are infelicitous according to natural causes, and no one should start anything new, for it will obtain an evil beginning."[101] While Krabbe allowed for novelty in the heavens, he preserved a constant and conventional view of his practice.

On the other hand, Krabbe was among a growing number of astronomers who arrived at novel ideas about the nature of the world on the basis of mathematical devices and doctrines. Through the projection of parallax, Krabbe plotted the paths of comets according to the same principle he deployed to determine the displacement of distant objects on earth.[102] Krabbe understood the celestial application of surveying instruments that would similarly inspire Galileo to turn his spyglass to the stars. In a book about his new astrolabe, Krabbe declared that the device would serve "not only astronomers and physicians," but also builders, military officers, seafarers, surveyors, and every other

97 *Ibid.*, 5r.

98 *Ibid.*

99 *Ibid.*

100 Krabbe (ref. 82), 12r.

101 Johannes Krabbe, *Alter und Newer Schreibkalendar* (Erfurt: Martin Spangenberg, 1613), 1r.

102 On the generation of early modern astronomers and philosophers who envisioned "the unification of astronomy and philosophy," see Miguel A. Granada, "The Defence of the Movement of the Earth in Rothmann, Maestlin and Kepler: From Heavenly Geometry to Celestial Physics," in *Mechanics and Cosmology in the Medieval and Early Modern Period*, ed. Massimo Bucciantini, Michele Camerota, and Sophie Roux (Florence: Olschki, 2007), pp. 95–119.

profession that relied on the compass and ruler.[103] In part, Krabbe followed "well understood procedures" to position his case for the supralunar course of comets "within an existing astronomical practice."[104] In his report on the comet of 1604, Krabbe described a "commonly known" mathematical principle that proved the position of the marvel well beyond the Moon.[105] Drawing from general knowledge, Krabbe generated a distance as spectacular as the splendor of the celestial wonder itself.

4.5 The Brightly Burning Comet of 1604

A month after a bright light first blazed in the fall sky of 1604, Krabbe penned a short report on the radiant luminary. Responding to rampant concern across the Holy Roman Empire, the report was published in Augsburg, Erfurt, and Magdeburg. It appeared among a wave of works that flooded the provinces, as local presses fanned the flames of public intrigue. In his account, Krabbe described the course of a distant comet whose "brightly burning and twinkling light" evinced "a great marvel of divine omnipotence."[106] Similar in luminescence and luster to the luminary of 1572, the comet was considered by Krabbe to bear "no association with the course of the planets or stars," since it cut across their paths.[107] Krabbe did not mention the solid celestial spheres, abolished by other astronomers who had observed comets moving in opposition to the orbs or traversing their ethereal terrain.[108] Since Krabbe could not discern any sign of parallax "from which one may have customarily determined the distance," he resolved to measure it "according to the motion."[109] The result proved even greater than the distance assigned by Maestlin to the comet of 1580 in the sphere of Saturn.[110]

103 Krabbe (ref. 74), title page.

104 See Peter Dear, *Discipline and Experience: The Mathematical Way in the Scientific Revolution* (Chicago: University of Chicago Press, 1995), pp. 93–123.

105 Johannes Krabbe, *Cometa eo anno 1604. den 3. Tag Octobris / am Himmel erschienen* (Magdeburg: Johann Francken, 1605), 6v.

106 *Ibid.*, 5v.

107 *Ibid.*

108 On the dissolution of the celestial spheres in early modern cometology, see Miguel A. Granada, "Did Tycho Eliminate the Celestial Spheres Before 1586?," *Journal for the History of Astronomy*, 2006, 37: 125–145; Sara J. Schechner, *Comets, Popular Culture, and the Birth of Modern Cosmology* (Princeton: Princeton University Press, 1997), pp. 188–215.

109 Krabbe (ref. 105), 5v.

110 Granada (ref. 108), pp. 135–136. In his manuscript on the 'new star' of 1604, Maestlin referred to Krabbe and his cometary account; see Granada (ref. 39).

In the first chapter of his account, Krabbe explained how he had oriented his observations according to the fixed positions of Arcturus and Aquila, "as observed by Tycho Brahe."[111] Krabbe first observed the comet on 4 October at 17′0″ Sagittarius and spied it at the same location two days later. On 12 October, the comet appeared at 17′12″ Sagittarius; on 18 October, 17′20″ Sagittarius; and on 26 October, 17′28″ Sagittarius. Finally, Krabbe observed the comet on 1 November at 17′30″ Sagittarius, precisely where he perceived it nine days later. The final two observations convinced Krabbe that the comet had become "stationary or retrograde, turning back toward the South."[112] Without waiting for any further changes in magnitude or motion, Krabbe swiftly prepared his report for publication. We know that he completed his account before 29 November, when the comet would conjoin with Saturn and the Sun, unleashing "great changes in the air and on earth among men."[113] "Common experience," Krabbe wrote, had proven that comets preceded changes in local climates and cultures leading mainly to misfortune.[114]

At every turn, Krabbe endorsed his account with sources of evidence that were familiar and firmly established. With the same confidence he recounted the comet of 1596, Krabbe attested to several other successes. "In many other years," he maintained, "I showed that comets should appear in the sky according to sound causes."[115] The same applied to the comet of 1604, which Krabbe claimed to have predicted in his annual prognostication as well as his "observations on the great conjunction of Jupiter and Saturn [in 1603]."[116] Anticipating a clamor of critical voices, Krabbe called on Fabricius to confirm his correct prediction of comets.[117] Apparently, Fabricius had conferred with Krabbe at Wolfenbüttel, where he confirmed that Krabbe had successfully anticipated the appearance of comets in December 1599 and April 1601. "Fabricius told me in person," Krabbe pronounced, and "any diligent detractor" would say the same for the comet of 1604.[118] Equally clear was the fact that, contrary to some corners, the comet could not be confused with Venus. "As all astronomers

111 Krabbe (ref. 105), 3r.
112 *Ibid.*, 4v.
113 *Ibid.*
114 *Ibid.*, 7r.
115 *Ibid.*, 1v.
116 Fabricius enclosed the observations of Krabbe in a letter to Kepler. "I am sending you the observations made by Johannes Krabbe, a geometer at the court in Braunschweig," Fabricius told his friend in Prague. David Fabricius to Johannes Kepler, 27 October 1604, *JKGW*, 15, no. 297, ll. 24–28.
117 On Fabricius and his fame, see Menso Folkerts, "Der Astronom David Fabricius (1564–1617): Leben und Wirken," *Berichte zur Wissenschaftsgechichte*, 2000, 23: 127–142.
118 Krabbe (ref. 105), 1v–2r.

know," Krabbe explained, there was no way the morning star could have appeared in Sagittarius.[119]

Krabbe welcomed the work of other astronomers. "There is no doubt," he declared, "that other mathematicians have observed and described the comet with greater industry."[120] At no point, however, did Krabbe anticipate a dispute over the nature of the novelty. The motion of the comet was clear, no matter the measure of effort. "It is well known to every astronomer," Krabbe argued, "that the slower a star or planet proceeds in the sky, the higher or farther away it is from the Earth; and the faster and swifter it proceeds, the lower or nearer it is to the Earth."[121] According to the same steps he had followed for the comet of 1596, Krabbe drew a direct proportion between the motion of the comet and a distance that he determined more than 22,000,000 German miles away. Such a staggering figure required a removal beyond "the orb or sphere of Saturn," though "far closer to Saturn than the eighth sphere."[122] Krabbe arrived at the great distance with no apparent hint of sorrow or surprise. While he has been called "generally conservative,"[123] Krabbe grasped a growing trend among early modern astronomers to measure the physical contours of the cosmos through mathematical principles.

The location of the luminary in the far reaches of the heavens did not spell the end of celestial immutability. From his references to the comet, it is clear that Krabbe discerned a divine miracle that delivered a frightening message about the future. In the same spirit as Bartholomaeus Keckermann (ca. 1572–1609) and many other scholars,[124] Krabbe suggested that the comet was

119 *Ibid.*, 2r.

120 *Ibid.*, 7r.

121 *Ibid.*, 6v. In his own report on the comet, Fabricius described the same "well known" proportion among the planets. "The closer a planet is to earth," Fabricius wrote, "the more swiftly it moves," since the "circle or circuit" it traverses is smaller. On the other hand, "the higher a planet is elevated from the earth, or the further it is removed from it, the more slowly it moves." See David Fabricius, *Kurtzer und Gründtlicher Bericht von Erscheinung und Bedeutung deß grossen Newen WunderSterns* (Hamburg: Paul Langen, 1605), 7v. On the attempt by Tycho to determine the distance of Mars according to diurnal parallax, see Albert Van Helden, *Measuring the Universe: Cosmic Dimensions from Aristarchus to Ptolemy* (Chicago: University of Chicago Press, 1985), p. 49.

122 Krabbe (ref. 82), 7r. According to Van Helden, the disposition of the cosmos, including the distance of the fixed stars, "had not changed markedly since Ptolemy." See Van Helden (ref. 121), p. 53.

123 Barnes (ref. 81), 271.

124 On the views of Keckermann involving celestial novelties, see Miguel A. Granada, "Bartholomaeus Keckermann and Christoph Hunichius on Novas and Comets at the Beginning of the 17th Century: Two Opposing Views on the Relation Between Natural Philosophy and Mathematics," in *Natural Knowledge and Aristotelianism at Early Modern*

an extraordinary act of God surpassing the laws of nature. It had appeared in the sign of Sagittarius to announce "great disorder and division in all of Christianity," threatening members of the clergy "with death and defiance."[125] Krabbe painted a stark picture that expressed "the prevalent sense" of crisis sweeping across the Holy Roman Empire.[126] "A new prophet" would sway the masses with signs and wonders, Krabbe claimed, as "a new Reformation" gripped the empire "with great zeal."[127] With the dawn of a new religion and the rupture of society, several cities and countries would disappear entirely. Bracing for disaster, Krabbe contributed to "a discourse of dread" that spread through the very same endeavor to avoid misfortune by assessing the stars more accurately.[128]

4.6 Privately Resolving a Public Controversy

How did Krabbe respond to the scathing review? Kepler had charged Krabbe with deception and fraud, pouncing on his personal character. More than three years later, Krabbe wrote to Kepler to resolve their conflict. The long delay may reflect the circumstances of Duke Heinrich Julius, who had begun living in Prague intermittently and finally moved there permanently in 1610 as personal advisor to the Holy Roman Emperor.[129] Consequently, it is possible that the duke did not deliver a copy of *De stella nova* to Krabbe until long after receiving it. In his letter to Kepler, Krabbe explained that he had taken a while to read the book because he struggled with Latin and could only extend his "poor and simple powers of reason" so far.[130] Despite the delay, the passage of time had not healed all wounds. The Imperial Mathematician had questioned the very integrity of Krabbe, forcing him to salvage bonds he had forged and fostered for many years. "You attacked me so viciously," Krabbe complained, "that I was compelled to rescue my good name and reputation."[131] Krabbe suggested that

Protestant Universities, ed. Pietro D. Omodeo and Volkhard Wels (Wiesbaden: Otto Harrassowitz, 2019), 209–234.

125 Krabbe (ref. 105), 16r.
126 Barnes (ref. 81), 72.
127 Krabbe (ref. 105), 16r.
128 Barnes (ref. 81), 89.
129 Omodeo (ref. 77), 324.
130 Krabbe (ref. 7), ll. 2–8.
131 *Ibid.*, ll. 10–11.

Kepler, rather than scorn him as a swindler, "should have spared me from such scandalous words."[132]

Krabbe had good reason to mend his rift with Kepler. While Krabbe politely praised the Imperial Mathematician for "stooping so low" as to enter into discourse with him, the two scholars navigated the same general network of patronage and prestige.[133] Both belonged to a broader company of court astrologers and astronomers that based their reputation on the accuracy of their observations and the validity of their views. Krabbe had renewed hope of ending their dispute amicably with the recent arrival of a (lost) letter from Kepler. Regretting the public nature of their row, Krabbe encouraged Kepler to settle their differences confidentially. "Had you found a mistake in my observations," Krabbe insisted, "you should have written to me alone privately."[134] In his letter, Kepler had apparently shown a kinder side to Krabbe, his anger giving way to good will.[135] Once on the road to reconciliation, Krabbe repeated his request. Should Kepler ever spot a mistake, Krabbe asked him to write privately to explain where he had "strayed in the art [of astronomy]."[136] It is clear, however, that Krabbe never intended his invitation as an admission of error.

In fact, Krabbe stood by his account. In his reply to Kepler, Krabbe endorsed his observations with the same fidelity as the finest sources of the day. Accused of wrongly observing the great conjunction of 1603, Krabbe countered that Kepler, backed by Tycho, could have missed the mark as well. "In the end," he waxed vaguely, "we shall learn who observed rightly or wrongly."[137] Krabbe recalled the great care he had taken in the fall of 1604 to align his instrument with the meridian before locating the comet in the constellation of Sagittarius. Having designed and deployed his instrument diligently, Krabbe saw no reason to question the quality of his observations. "I would not know how to observe it any better," Krabbe declared.[138] In his eyes, the result remained the same. "There is no other way to say it," he sighed. "The star first possessed a motion that gradually diminished day by day until it could no longer be detected or discerned after three weeks of observation."[139] Krabbe admitted that an error

132 *Ibid.*, ll. 15–18.
133 *Ibid.*, ll. 3–8.
134 *Ibid.*, ll. 19–20.
135 On Kepler and his quest for harmony among the growing divisions of his day, see Aviva Rothman, *The Pursuit of Harmony: Kepler on Cosmos, Confession, and Community* (Chicago: University of Chicago Press, 2017). We may witness another "positive and hopeful" portrayal of Kepler in the refreshing work of Rublack (ref. 19).
136 Krabbe (ref. 7), ll. 25–26.
137 *Ibid.*, l. 30.
138 *Ibid.*, ll. 37–38.
139 *Ibid.*, ll. 38–41.

could "easily happen," but he saw no sign of false testimony or foul play in his faithful report.[140] The only doubt remaining was refraction, a subject Krabbe denounced as "still uncertain."[141]

Krabbe argued that refraction had played no more than a minor role in the record. Since it came from water "and not the air," Krabbe explained how the effect of moisture in the atmosphere above Wolfenbüttel had not altered the accuracy of his observations.[142] Why, he wondered, had Fabricius and Kepler scorned his account "so fiercely" in the first place?[143] Despite recent advances in the study of refraction, Krabbe claimed that it was still unclear and could not compromise his testimony. Kepler, on the contrary, voiced his disapproval keenly. He declared that Krabbe had not only defied the authority of Tycho,[144] but shown the world precisely where his error lay by denying the deflection of light rays. Krabbe calmly dismissed the objection with as much confidence as he predicted the appearance of another novelty. In the closing lines of his letter to Kepler, Krabbe announced that a comet would cross the sky in the summer of 1610. "I cannot resist telling you," he wrote, "that according to my astrology and what the constellations of the heavens [*constellationes coeli*] indicate, a new comet shall again appear to us in the summer of this year."[145]

In a second letter to Kepler, Krabbe measured the strength of his observations by a familiar scale. Replying to a (lost) letter from Kepler that had arrived in June, Krabbe returned to the same line of reasoning he had followed in his first letter. He began by allowing room for error in every observation. "For no man was found without fault or flaw," he argued, and "either vision or

140 *Ibid.*, ll. 50–55.

141 *Ibid.*, ll. 18–19. On the role of refraction in the revolutionary optics of Kepler, see Raz Chen-Morris, *Measuring Shadows: Kepler's Optics of Invisibility* (University Park: Pennsylvania State University Press, 2016).

142 Krabbe (ref. 7), l. 71.

143 *Ibid.*, l. 62. In his own report, Fabricius noted that "some astronomers" believed the star had moved "several minutes" soon after it appeared. Fabricius argued, however, that the angle of altitude had "always remained the same." He observed that the location of the star, slightly above the horizon and seen through "exhalations and vapors hovering in the air," had caused "a sensible measure" of refraction. "Such a refraction," Fabricius explained, altered the distances of the stars the more they approached the horizon. In the end, Fabricius deemed the notion of motion as nothing more than "an optical illusion caused by refraction." See Fabricius (ref. 121), 7r.

144 Tycho produced a posthumous table of refractions in the *Progymnasmata* (1602); see *TBOO*, 2, p. 64 (Sun), p. 136 (Moon). Essentially, he argued that objects appeared higher as they dipped below 30° and drew closer to the horizon. On his revealing exchange with Rothmann regarding the cause and nature of refraction, see Mosley (ref. 4), pp. 84–90.

145 Krabbe (ref. 7), ll. 140–143.

the instrument can deceive even the learned."[146] With Kepler in mind, Krabbe referred to the wide range of skills among the stargazers, recognizing that few, if any, could rival those of the Imperial Mathematician. "Just as a star surpasses others in brilliance," Krabbe quipped, "so may one man excel the others in skill and talent."[147] However, it was clear to Krabbe that no measure of intelligence could eclipse the reliability of regular and recurring observations. Krabbe had repeated every observation several times with the same result. "When an observation turns out the same three times," he reckoned, "it can and must be held for certain."[148] In his opinion, the only causes of error—a falsely aligned or faulty instrument, or perhaps the poor vision of the observer—did not bear directly on his own account.[149]

Kepler was not convinced by Krabbe. In his lost letters to Wolfenbüttel, he had advised Krabbe on the proper use of the astrolabe and observing the planets more precisely. In response, Krabbe thanked the Imperial Mathematician for his "faithful advice," humbly noting his "modest knowledge of the astrolabe and other mechanical devices."[150] No doubt their correspondence involved the book about the astrolabe that Krabbe had recently produced. Writing before reconciling with Kepler, Krabbe spoke harshly of him in the opening pages of the book. "I regard my critics, including Johannes Kepler and several other insolent defamers and detractors, as less than nothing."[151] Now at peace with Kepler, Krabbe expressed gratitude toward him through a sign of religious solemnity. Along with his debt to Kepler, Krabbe extended his appreciation to "God Almighty and Eternal."[152] Before closing, Krabbe explained why the comet he foresaw in the summer of 1610 had never appeared. "If it had not followed a slightly different course," he wrote, "we would have observed it."[153] There is no evidence, however, that Kepler ever held out hope of seeing it.

Near the end of his letter, Krabbe referred disapprovingly to the recent discoveries of an "Italian mathematician."[154] At a time when the telescope was widely viewed as a device of deception, Krabbe condemned it for preying on

146 Johannes Krabbe to Johannes Kepler, 5 October 1610, *JKGW*, 16, no. 595, ll. 22–24.

147 *Ibid.*, ll. 28–29.

148 *Ibid.*, ll. 30–32.

149 There is reason to believe that Krabbe implied the poor eyesight of Kepler as a cause of error. In his reference to the common blight of bad instruments and vision, Krabbe wrote: "just as you and every other observer have certainly experienced at some point." See *ibid.*, ll. 24–25.

150 *Ibid.*, ll. 109–112.

151 Krabbe (ref. 74), iv.

152 Krabbe (ref. 146), ll. 112–119.

153 *Ibid.*, ll. 120–122.

154 *Ibid.*, l. 123.

the innocent and proving the fancies of others. He described the four moons of Jupiter as "utter nonsense," opposing them so fiercely that he once fought with his chancellor over their existence.[155] Krabbe anticipated that Kepler, who had publicly endorsed the discoveries in April, would change his mind over one matter in particular. That "small stars become equal to those of the first magnitude," Krabbe complained, left the observer with "absolutely no sense of distance," but only "some flickering here and there."[156] Kepler had probed the optical properties of the stars to determine their distance in *De stella nova*.[157] Now with no sense of perspective in sight, Krabbe hoped that Kepler would "write to Galileo again" to discuss the disposition of the stars in further detail.[158] Apparently, no reply ever came from Kepler. But what had once been a bitter dispute was now an amicable exchange that allowed the two authors to air their differences away from the printed word.

4.7 Conclusion

Contrary to Kepler, Krabbe maintained the accuracy of his observations and assigned a motion to the marvel of 1604. While we cannot reduce their conflict merely to a matter of right and wrong, we may ask why Krabbe held so firmly to his account in the face of two leading figures. It was clear to Kepler why Krabbe resisted. Kepler flatly accused him of incompetency, claiming that Krabbe barely knew the first thing about observing. When Krabbe "kindly asked" Kepler to complete an empty chart of observations, including "the mean motion of the Sun and planets,"[159] Kepler replied with incredulity. Calculating the motion of a moving body was entirely different from observing it. "His chart only proves that the ignorant man still does not understand what an observation is," Kepler complained.[160] Even more damning, Kepler declared that the same incompetency had robbed Krabbe of realizing it. In other words, if the skills that Krabbe required to practice astronomy proficiently were the same

155 *Ibid.*, ll. 123–126.

156 *Ibid.*, ll. 123–126. Kepler confirmed the four satellites of Jupiter in September 1610 with a telescope he borrowed from the Elector of Cologne. Did Galileo fail to supply Kepler with a spyglass for fear of anticipating future discoveries? See Mario Biagioli, *Galileo's Instruments of Credit: Telescopes, Images, Secrecy* (Chicago: University of Chicago Press, 2006), pp. 89–90.

157 On his attempt to determine "the limits of the universe" according to the new cosmology, see the contribution by Javier Luna to this volume.

158 Krabbe (ref. 146), ll. 131–132.

159 Nikolaus Vicke to Kepler, 23 February 1610, *JKGW*, 16, no. 557, ll. 52–59.

160 Johannes Kepler to Nikolaus Vicke, 18 March 1610, *JKGW*, 16, no. 559, ll. 66–67.

ones he employed to evaluate his own abilities, there was no way that Krabbe could appraise his competence accurately. Known by modern scholars as "the failure of feedback," the problem often prevents us from identifying and improving on our mistakes.[161]

Another aspect of the conflict may shed further light on the fortune of Krabbe. We know that Kepler deployed a deep knowledge of optical demonstrations and was a keen spectator of court displays. During one demonstration in Dresden, Kepler identified an image in the air that not even the conductor of the experiment knew had been generated.[162] Tuned in to the techniques of others, Kepler relied on what we might regard as 'social comparison' to measure his own performance in relation to those of his peers.[163] In any given endeavor, Kepler calibrated his competence by observing the behavior of his contemporaries. From his place at the imperial court in Prague, Kepler culled a wide variety of observations and other sources and compared their quality. Krabbe, by contrast, did not appear willing to reconsider the accuracy of his own observations after enduring criticism from Fabricius and Kepler. By his own admission, Krabbe did not feel he could judge fairly between competing accounts. Comparing his tables to those of Kepler and Tycho, Krabbe simply suggested that the truth would "eventually come to light."[164] Faced with the opportunity to learn from the insight of others, Krabbe refused to contemplate any real form of revision.

There are, of course, many other reasons that may have prompted Krabbe to uphold his account. Several other stargazers, such as Herlicius, read and relied on it to arrive at similar interpretations. Encouraged by their approval, Krabbe knew that any alteration in the observations could spell the collapse of his account as a whole. Consistency was key. "What good would it have done to assign a motion to the comet," Krabbe asked rhetorically, "if I had not found it in my observations?"[165] In the end, Krabbe submitted his opinion to other scholars. "Let the learned consider my account and judge what they will," he forswore in his final letter to Kepler.[166] Krabbe could anticipate a positive response from

---

161 See Justin Kruger and David Dunning, "Unskilled and Unaware of It: How Difficulties in Recognizing One's Own Incompetence Lead to Inflated Self-Assessments," *Journal of Personality and Social Psychology*, 1999, 77: 1121–1134.

162 On Kepler and the experimental culture of optical 'wonder-working,' see Sven Dupré, "Wonder and Experiment in Kepler's *Optics* and *Dioptrics*," in *Studia Copernicana* (ref. 78), pp. 223–237.

163 Kruger and Dunning (ref. 161), 1131.

164 Krabbe (ref. 7), ll. 27–31.

165 Krabbe (ref. 146), ll. 45–47.

166 *Ibid.*, ll. 49–50. Contrary to Krabbe, Fabricius argued that the "unbelievable" nature of his interpretation only served to prove the point. On the return of "the Golden Age" of the

many of his peers, since eliminating any remnant of motion would remove the star to the final sphere of the cosmos. At no point did Krabbe suggest change in the sphere of the fixed stars. Even Fabricius, who argued that stars newly appeared in the sky, concluded that they were, in fact, created by God from the beginning and "ignited or illuminated at certain times" to serve as signs "through their visible appearance."[167] When it came to the stars, there was nothing new under the Sun.

The interrelated factors of patronage and power elucidate further features of the dispute. Before ever touching a telescope, Kepler delivered his verdict on the discoveries of Galileo as their first and foremost judge.[168] Through his renown and title, the Imperial Mathematician conferred "high credibility" on the discoveries at a point when the Paduan sorely needed the support.[169] Kepler similarly asserted his authority on the new star. Despite the difficulties of observational discontinuity, he stood at the center of a broad network of observers that extended far beyond the Holy Roman Empire.[170] Closely connected to Prague, Krabbe knew the word of Kepler carried great weight in Wolfenbüttel. A piercing review from the Imperial Mathematician tainted his practice and tarnished his reputation, forcing Krabbe to save face among patrons and peers. Krabbe preferred to settle their differences privately while promoting in print his instruments, observations, and predictions. He soon published a popular book about his astrolabe and continued to produce his annual prognostications. Bearing witness to the quality of his work, Krabbe strove to build and broaden trust in a sprawling community of stargazers and their sponsors.

Holy Roman Empire according to Fabricius, see Granada (ref. 47), p. 74.

167 Fabricius (ref. 121), 5r.

168 On his assumption of "a complementary, yet superior role to Galileo," see Westman (ref. 6), pp. 460–465.

169 Biagioli (ref. 156), p. 34.

170 On the emergence of 'learned commerce' as "a defining feature of the early modern world," see Paula Findlen, *Empires of Knowledge: Scientific Networks in the Early Modern World* (London: Routledge, 2019), pp. 1–22.

CHAPTER 5

Stars, Crystals and Courts: Johannes Kepler and Anselmus Boëtius de Boodt

Jonathan Regier

5.1 Introduction

The new star of 1604 twinkled marvelously. Kepler devoted a chapter to its scintillation in *De stella nova* (1606). Two centuries later, François Arago reviewed the opinions of his predecessors in the course of his own exposé on stellar scintillation. He did not mince words about Kepler's account:

> In several passages, he compares the stars to carved diamonds, in which the slightest movement gives birth to the colors of the rainbow. He imagines that the stars can have angular sections, regions that are unequally luminous, and he explains in this way how they do not need to make one complete revolution for each scintillation. Analyzing the explanation of Scaliger, Kepler places the action of the air among the secondary causes without importance. All this, let's frankly admit, hardly seems worthy of Kepler's genius.[1]

Arago does not even bother mentioning that Kepler had these crystalline stars pulsing with heat and life. What makes Kepler's view even more fascinating was his presence at the court of one of Europe's most passionate aficionados of gems and precious stones, Rudolf II, whose collection was built on pillars of beauty, rarity, exoticism, and medical utility. Rudolf's personal physician, Anselmus Boëtius de Boodt (1550–1632), wrote a major early modern treatise on mineralogy, the *Gemmarum et lapidum historia* (Hanau, 1609). This work,

1 François Arago, *Oeuvres complètes*, ed. J. A. Barral, vol. 4 (Paris: Gide, 1858), p. 67: "Dans plusieurs passages, il compare les étoiles à des diamants taillés à facettes, dans lesquels le moindre mouvement fait naitre les couleurs de l'arc-en-ciel. Il imagine que les astres peuvent avoir des parties anguleuses, des régions inégalement lumineuses, et explique ainsi comment il n'est pas nécessaire qu'elles fassent une révolution totale à chaque scintillation. En analysant l'explication de Scaliger, Kepler range l'action de l'air au nombre des causes secondaire et sans importance du phénomène. Tout cela, avouons-le franchement, semble peu digne du génie de Kepler."

 | DOI:10.1163/9789004437272_007

which happened to appear the same year as the *Astronomia nova*, harbors a vision of nature with a striking resemblance to that of Kepler. Their natural philosophies seem to share a root system, as if they both participated in a common project or institution—*which indeed they did*. This is important because it suggests how Kepler, who spent an intellectually fertile period in Rudolf's Prague, was shaped by Rudolfine culture. In what follows, I will look at a few tantalizing similarities. The point is not to claim any direct transfer of ideas between the two scholars. As far as I know, they never mention one another by name, although Robert Halleux has suggested that Kepler might allude to de Boodt in his New Year's gift on the six-cornered snowflake.[2] At the moment, it suffices to examine how the two echo one another. Kepler's activity was about prediction, calculation and explanation, his material experimentation limited to lenses and light. Understandably, matter theory is one of the sketchiest aspects of his natural philosophy. On the other hand, de Boodt made his career studying, evaluating, and experimenting on, earthly matter. He marshaled that knowledge toward the accumulation of wealth, the restoration of health and the protection against disease. In short, reading de Boodt may provide us additional clues about how Kepler's philosophy meshed with the wider intellectual enterprise of Rudolf's court, beyond astrology and astronomy.

5.2 Kepler and Gems

Kepler opens his discussion on stellar scintillation by referencing Julius Caesar Scaliger's *Exotericarum exercitationum liber XV* (Paris, 1557), a work that he had esteemed highly in his youth. Scaliger embraced a variety of causes.[3] Kepler lists them as "magnitude, brightness, sidereal motion, the medium of air, and the motion of light in the star."[4] Kepler is especially convinced that scintillation should be attributed to changes within the star itself. The inner force that causes light must have its own proper motion.[5] Kepler admits to difficulties in

2 Johann Kepler, *L'étrenne ou la neige sexangulaire*, trans. Robert Halleux (Paris: Vrin, 1975), pp. 56 and 91, n. 31. On the role of courtly "ethos" and culture in the astronomy of Kepler, see Nicholas Jardine, "The Places of Astronomy in Early Modern Culture," *Journal for the History of Astronomy*, 1998, 29: 49–62.

3 Julius Caesar Scaliger, *Exotericarum exercitationum liber xv* (Frankfurt: Andreas Wechel, 1582), Exercise LXIII, especially pp. 240–242.

4 *JKGW*, 1, 241.9–11: "J. C. Scaliger *Exotericarum Exercitationum* 63. causas affert ad scintillationem concurrentes quinque: Magnitudinem, Claritatem, Motum sideris, Medium aeris, et Motum luminis in sidere."

5 *Ibid.*, 241.24–25: "Placet Scaligero et, ipse vehementer approbo, naturalem esse in sideris corpore vim luminis sese spargentis non sine motu."

knowing exactly how this is so. First, he draws a comparison with the beating of the heart and arteries, where the heart is like the Sun and the light is like life. But it cannot be, he concedes, that the star squeezes light throughout its body like *spiritus*. Light is so pure that it needs no impulsion. The motion inside the stellar body must not be a change of place—a local motion—but a qualitative change rendering the body luminous.[6] Here, Kepler hits on a better example, that of "live coals" (Kepler adds that they acquired the name of "live" from this alternating vigor). If we could speed up the wavering of hot coals, he continues, they would look a lot like the twinkling of Venus.[7] On just a visual level, Kepler says, the most apt comparison with stars is the flickering of torches, as the poets tell us. But the wavering of firelight is caused by the conflict of flame and air: air saps the fire; fire is renewed by its fuel and dissipates the surrounding vapor. Where are any of these effects to be found in the super-rarified environment of the fixed stars? So, he concludes, twinkling must be internal to the bodies, which are eternal and transparent.[8]

Kepler then introduces another cause of scintillation.[9] He suggests that the stars turn on their axes and shimmer diamond-like with the passage of their facets and angles. This would be a secondary cause, explaining differences in flashing and color. It would require some eternal faculty in the stars (presumably an animal faculty). Kepler prefers this reason to others because it is clean and simple: differences in twinkling and color are reduced to the optical characteristics of the body. Kepler preempts a question from some of his more Aristotelian readers. If the stellar body has surface differences, as well as differences in interior density, such as we see in diaphanous rocks, this may suggest stars to be impermanent. Kepler recalls that gems in the

6 *Ibid.*, 241.33–35: "Sed tamen analogus aliquis intelligatur in stella motus, non localis, sed alterationis, secundum eam qualitatem, qua stella lucidum est corpus."

7 *Ibid.*, 241.35–242.3: "Imo vero vivum propono huius alterationis exemplum, in vivis carbonibus; qui ab hoc ipso motu, et quasi vigore alternante, nomen vivorum sunt adepti. Etsi hoc exemplum, si et celeritatem ei maiorem affingas, Veneris scintillationi magis respondet."

8 *Ibid.*, 242.27–32: "Itaque sint ista exempla non substantiae, sed luminis: de fixis vero maneat sententia Scaligeri; vim inesse in corpore perenni, adde et pellucido, luminis reciprocantis, analogon deflagrationi nostrarum flammarum. Nostro vero sideri, vel idem corpus, vel eandem vim luminis tribuemus, prout postea videbitur."

9 Scaliger had himself criticized Cardano for attributing scintillation uniquely to the extreme speed of the stars, whose whirling carries them a vast distance around the Earth in one day. While Scaliger seems to accept that this movement results in some twinkling, it cannot be the only cause. Kepler rejects any translational motion among stars, unnecessary in Copernican astronomy.

Earth—though sometimes spotted, fissured, cloudy—are for all intents and purposes eternal, too.[10]

In the chapters that follow his account of scintillation, Kepler covers the conception of the star. These chapters are notable for their view of nature as generative and suffused with animal faculties.[11] The new star was created by a vital faculty in the aether operating much like the Earth's vital faculty when the latter fashions new stones, insects, or plants. Kepler speaks here about bodily heat, humidity, fats and humors, and he makes little distinction between the stony and fleshy:

> It is clear enough that the Earth operates through such an inner faculty that engenders some metals in other parts, that fosters heat in the interior, that absorbs sea water, digests it and makes it evaporate at the surface, provides matter for rivers and rains, and, relying on none other than some hidden impression responding to the animal imagination, introduces incredible geometric shapes in fossils [...]. For this is that faculty which, whenever it finds some superfluous matter, converts it into a small animal that serves the nature of things [...] Vapor appears to be a certain instrument of both [the Earth and the human body] that is generated within the body. The exhalation envelops the exterior, filling with its humidity the hidden cavities of bodies where its heat can persist. Once this faculty has entered the outside body (attracted, in kind, to heat or the effusion of its vapor), it begins to fashion those marvelous things through a divine impulse, received at creation and retained down to the present time, charged with ordering every part to an end goal.[12]

---

10 *JKGW*, 1, 243.18–23: "Quare non metuo, ut perpetua esse non possint corpora stellarum, si angulosa, aut si intus inaequaliter densa sunt, ut solent Uniones, partibus aliis aliter pellucidi. Nam et hostis illinc abest, qui vim inferat; et sunt nobis in hac Tellure gemmae, si vim auferas, aeternae; et eae tamen partim maculosae, partim fissae, partim nubilae."

11 For an overview and extended analysis of Kepler's vitalism, see Patrick J. Boner, *Kepler's Cosmological Synthesis: Astrology, Mechanism and the Soul* (Leiden: Brill, 2013). On Kepler's vitalism in a medical context, see Jonathan Regier, "Kepler's Theory of Force and His Medical Sources," *Early Science and Medicine*, 2014, 19: 1–27.

12 *JKGW*, 1, 268.11–269.7: "Satis constat de Terra, pollere illam intus huiusmodi facultate; quae aliis partibus alia gignit metalla; quae calorem intus fovet; quae marinas aquas absorbet, digerit, sublimes evaporare facit, fluviis et pluviis materiam sufficit; quae mirabiles figuras geometricas fossilibus indit, non alia re, nisi quadam impressione occulta, quae respondet imaginationi animalium, subnixa [...] Haec nempe est illa, quae quoties invenit superfluam aliquam materiam; convertit eam in animalculum tale, quod rerum naturae serviat [...] Utriusque organon quoddam vapor esse videtur, qui in corpore genitus, id forinsecus circumdat exspiratus, implens madore suo sinus corporum occultiores, ubi calor inclusus durare potest. Hoc enim facto statim in eum egressa facultas haec, corporis

To summarize, the vital faculty of the Earth was installed at creation and tasked with conservation and perfection. It is similar to the vitality of the human body, or of the celestial reaches. It manifests as a creative and industrious heat acting on oily and fatty humors, matter ripe for life and transformation. Regarding this latter point, it bears mentioning that life, as Kepler understands it, generally requires heat and humidity, which the celestial bodies produce to varying degrees (more on this point below). Kepler also describes in more detail how this vital faculty behaves on a daily basis: wherever it finds some kind of unused or degenerated matter, it transforms the matter into a higher form of being, whether a small animal, insect, or plant.[13] Finally, it is worth noting that Kepler was aware of the concept of seminal reasons—the formal seeds (*logoi spermatikoi*) of Stoic origin present in a variety of sixteenth-century contexts, including Paracelsian. Earlier in the *De stella nova*, noting the patterns of color in flowers, Kepler had suggested that the vital faculty of all living things was equivalent to the "*rationes seminariae*."[14]

In the above, we encounter a complex of ideas well represented throughout Kepler's corpus. One is that nature operates via souls and their faculties or instruments. The other is that gems and metals are produced by life forces and sometimes contain those forces within themselves. There are many possible sources for Kepler's ideas here. Gems, minerals and precious metals were frequently associated with vegetative and animal bodies in the Renaissance. They could be thought to grow in the Earth like plants, expanding in branch-like or root-like formations, and they often showed up in the human body, in the kidney and bladder, with extremely painful and life-threatening implications.[15] Cicero writes of the formative powers of the Earth's womb in Book II of the *De natura deorum*.[16] In the *Naturales quaestiones*, Seneca compares the fluids of the human body to the many fluids of the earth, all these fluids serving different purposes (metals, according to Seneca, are a kind of terrestrial fluid

incola (utique prolecta [prolectata] calore, et effluvio sui vaporis), ista mira machinari incipit: instinctu divino per creationem accepto, hactenusque retento, de singulis membris ad finem ordinandis admonita."

13 Cf. Boner (ref. 11), p. 31.

14 *JKGW*, 1, 185.35–37. For more on this passage and the chapter in which it appears in *De stella nova*, see Patrick J. Boner, "Kepler on the New Star: *De stella nova*, Chapters 7–9," *Culture and Cosmos*, 2010, 14: 209–234.

15 There is an extensive, if somewhat scattered, secondary literature on early modern body stones. See, for example, Margaret Healy, "A Most Troublesome and Dangerous Ailment: Encounters with the Stone in Early Modern Europe," *Journal de La Renaissance*, 2005, 3: 207–216.

16 Cicero, *De natura deorum / Academica*, trans. H. Rackham, Loeb Classical Library (Cambridge, MA: Harvard University Press, 1933), 2.33, pp. 203–205.

that hardens).[17] Cardano writes that gemstones are not simply alive but also undergo a lifecycle of disease, old age and death.[18] Kepler, familiar with all the above authors, would have had to look no further than Gilbert's *De magnete* (London, 1600) for a vital theory of metal formation.[19]

A link between stones and celestial bodies was also widely accepted. Ficino has celestial bodies and constellations presiding over gemstones and metals.[20] Agrippa systematically lists metals and stones under the planets and signs presiding over them.[21] Yet in Kepler we find something altogether different—planets and stars are said to have a composition materially comparable to stones. Kepler discusses stones in his *De fundamentis astrologiae certioribus* (Prague, 1602), a first systematic account of his astrology written early in his Prague period while he was engaged in securing his position at Rudolf's court.[22] Here, he defends astrology as a natural philosophical enterprise, a study of the effects of celestial bodies, via their light, on earthly bodies. The chief powers of celestial light are to heat and moisten. Ptolemy's *Tetrabiblos* remains the example of how to think of planetary influence, yet

17 Seneca, *Natural Questions*, trans. Harry M. Hine (Chicago: University of Chicago Press, 2010), 3.15, p. 34.

18 Girolamo Cardano, *The* De Subtilitate *of Girolamo Cardano*, trans. John M. Forrester (Tempe, Arizona: Arizona Center for Medieval and Renaissance Studies, 2013), 1.7, pp. 366–367. Cardano devotes Book VII of *De subtilitate* to gemstones.

19 Gilbert accepts only one true element in the universe, true earth, which possesses a pure magnetic capability and an innate tendency to rotate around a fixed axis. The rest of the matter we see around us—non-magnetic, heterogeneous, perpetually changing—is "a kind of veil," a shallow and corrupted covering to the essential sphere of the earth. William Gilbert, *On the Magnet* (*De Magnete*), trans. Silvanus Phillips Thompson (New York: Basic Books, 1958), p. 120. Hence, Gilbert must explain how everything else comes to be: air, water and other metals. The answer is that condensed vapors travel upward from the center of the Earth, becoming trapped just below the surface or inside mountains. Here, as in a womb, they cool and congeal into humors that are the proximate causes of metals, "like the blood and semen in the generation of animals." *Ibid.*, pp. 20–21.

20 See, for example, the first two chapters of the *Liber de vita coelitus comparanda* in Marsilio Ficino, *Three Books on Life*, ed. and trans. Carol V. Kaske and John R. Clark (Binghamton, NY: Renaissance Society of America, 1989), pp. 243–255.

21 See Book I, Chapters 22–32, in Henry Cornelius Agrippa of Nettesheim, *Three Books of Occult Philosophy*, ed. Donald Tyson, trans. James Freake, Llewellyn's Sourcebook Series (St. Paul, MN: Llewellyn Worldwide, 1993).

22 See Judith V. Field, "A Lutheran Astrologer: Johannes Kepler," *Archive for History of Exact Sciences*, 1984, 31: 189–272, especially p. 195: "*De Fundamentis Astrologiae Certioribus* was written after the death of Tycho Brahe. Kepler's desire to obtain the now vacant post of Imperial Mathematician no doubt contributed to his decision to write the work at this particular time. One can, nonetheless, trace other, older and more intellectually respectable, origins for the work."

whereas Ptolemy assigned the planets a range of influence corresponding to the four elements, Kepler reduces the causal palette to heating and moistening. Every planet provides either a deficiency or an excess of each of these two powers. When Kepler imagines the material constitution of planets, he thinks of gems and rocks. So Saturn, which "[...] has an excess of humidity and is deficient in heat [...]," is like ice. Jupiter is like a ruby, "[...] so that it may be very transparent, to account for its abundant intrinsic light, and even and red to account for its brightness and colour." Mars is dry and hot like coal. Venus is like amber because of its brightness, and like amber, Venus must also have a soft body since it humidifies more than it warms. Mercury is like sapphire, given its transparency and heat.[23] The above shows that Kepler considered both planets and stones to imbue light with powers deriving from their physical constitution and to operate on living bodies in a similar, possibly identical, manner.[24] Moreover, although heating and moistening are the main astrological effects of planetary bodies, a range of powers might also be transmitted from planet and stone alike: "[...] for any power a Planet shows in its effect there will be some analogous disposition in the body of the Planet."[25]

Stones and metals were also, of course, socially operative as embodiments of wealth and class. Kepler exploited their multivalent functionality when as an ambitious young mathematician, he sought to create a precious astronomical sculpture in order to further his standing in Württemberg. In early 1596, he was finishing his first book, the *Mysterium cosmographicum* (Tubingen, 1596), which he believed would unveil a discovery of historic importance. An advantage of Copernican astronomy was that it neatly established planetary distances as ratios of the Earth-Sun distance. Kepler, having exhausted the summer of 1595 in search of what lay behind those distances, had found what he believed to be the divine motive. God, it seems, had established the universe based on the Platonic solids, five figures that are mathematically unique (maximally symmetrical) and textually unique (they are the subject of Euclid's *Elements* Book XIII). In a word, Kepler thought he had hit on the divine idea behind the world's architecture. He also viewed the *Mysterium* as an opportunity to elevate his professional status[26] and as an occasion to make inroads

23 *Ibid.*, p. 247.

24 To illustrate how light can transmit chemical properties of objects, Kepler uses the example of sunlight reflected off a cement wall. When the light is strong, he says, it receives and transmits the corrosive effects of lime in the cement and poses a danger to people standing nearby. *Ibid.*, p. 246.

25 *Ibid.*

26 Kepler would send copies to leading mathematicians, including Tycho Brahe, Nicolaus Reimers Ursus, and Galileo.

with local nobility. Hence, in February 1596, he wrote to Duke Friedrich I of Württemberg announcing his great *inventum*. His letter, hardly touching on astronomical and philosophical details, was brief and to the point, as would be expected from a relatively unknown supplicant. Writing in German with a sprinkling of Latin technical terms, Kepler proposed a large, decorative silverwork that would "fittingly and charmingly" (*füeglich und zierlich*) capture his new cosmography. Kepler's project was a lavish liquor bowl.[27] The bowl, he wrote, would be the Duke's very own *scyphus Nestoris*, a reference to King Nestor's cup in the *Iliad*, a cup so heavily decked-out that few men but old Nestor could easily lift it. Kepler's bowl would feature the planetary orbs set at their polyhedral distances, with each orb serving as a reservoir for a particular kind of alcohol, which would flow via hidden pipes to spigots on the outer rim of the bowl. The strength and cost of the alcohol would vary with distance from the sun, presumably to represent the solar virtue diminishing with distance.[28] The body of the sun would hold a precious *aqua vitae*, to be delivered to the solar spigot; the Sun-Mercury orb would hold *branntwein*, and so on until the "old bad wine or beer" contained in the Jupiter-Saturn orb. Meanwhile, Kepler wanted the planets to be depicted by precious stones: Saturn as diamond, Jupiter as hyacinth, Mars as ruby or spindle, Earth as turquoise or lodestone, Venus as amber, Mercury as crystal, the Sun as carbuncle, and the Moon as pearl.[29] Kepler does not offer reasons for this selection, appealing only to the work's importance and to the Duke's taste. Yet the latter was a major benefactor of alchemy and would have immediately recognized the general relationship between mineral, alcohol and the heavens, well-established as this relationship was in a wide swath of natural philosophy. The distillation of alcohol was probably the most important of all chemical-medical activities, and the distillation process itself

27 Johannes Kepler to Duke Friedrich of Württemberg, 17 February 1596, *JKGW*, 13, no. 28. An account of this episode can be found in Walther von Dyck, *Nova Kepleriana: Wiederaufgefundene Drucke und Handschriften von Johannes Kepler. Vol. 8. Die Keplerbriefe auf der Braunschweigischen Landesbibliothek in Wolfenbüttel. II. Teil. Zusammen mit zugehörigen Aktenstücken der Landesbibliothek in Stuttgart*, (Munich: Bayerische Akademie der Wissenschaften, 1934), pp. 4–6. A slightly expanded account can be found in Arthur Koestler, *The Sleepwalkers* (New York: Macmillan, 1959), pp. 268–271. Adam Mosley examines the liquor bowl and the later clockwork design in "Objects of Knowledge: Mathematics and Models in Sixteenth-Century Cosmology and Astronomy," in *Transmitting Knowledge: Words, Images, and Instruments in Early Modern Europe*, ed. Sachiko Kusukawa and Ian Maclean (Oxford: Oxford University Press, 2006), pp. 193–216, on pp. 201–209.

28 An exception is the shell of the Earth (including the moon), which holds water. Mosley sees the bad wine or beer in the Saturn orb as a reminder of the decay that Saturn represented.

29 Kepler (ref. 27), ll. 74–79.

was imbued with profound philosophical significance. Highly refined elixirs, produced by instruments pioneered during the late-medieval period and early Renaissance, were frequently likened to celestial aether.[30] (Years later, Kepler's peppery exchange with Thomas Harriot led him to consider liquor under a different light: as a counterexample to atomism. In their discussion on optics, Harriot, an atomist, maintained that transparency required vacuum spaces.[31] Kepler asked how liquor, eminently transparent, could maintain within itself pockets of vacuum when compressed. How would these pockets, filled with nothing, support themselves against the pressure? If liquor cannot support vacuum spaces, Kepler concluded, no diaphanous material can: "All diaphanous things derive from liquor."[32])

Kepler could not have chosen a better target for his budding courtly ambitions than Duke Friedrich, a passionate collector and patron. Friedrich sought to elevate his lands through discovery and cutting-edge industry: mining, alchemy, textiles. In 1608, the year of his death, his alchemical laboratory "accommodated a team of twenty-one people operating fourteen portable copper furnaces and four ovens."[33] He also had a propensity for extravagance and symbol.[34] Once the Duke received Kepler's letter, he immediately asked for a copper model, which Kepler fashioned in paper because he could afford no better. The Duke then asked for a report on the proposal from Michael Maestlin, Kepler's mentor and former professor of mathematics at the University of Tübingen. Maestlin enthused over the project. The Duke accepted. But this was only the beginning of a long ending. The Duke asked that the model take the form of a

30 For the range of chemical and medical meaning attached to distillation, see Sergius Kodera, "The Art of the Distillation of 'Spirits' as a Technological Model for Human Physiology: The Cases of Marsilio Ficino, Joseph Duchesne and Francis Bacon," in *Blood, Sweat and Tears: The Changing Concepts of Physiology from Antiquity into Early Modern Europe*, ed. Manfred Horstmanshoff, Helen King, and Claus Zittel (Leiden: Brill, 2012), pp. 139–170. For Renaissance "distillation manuals," see Allen G. Debus, *The Chemical Philosophy: Paracelsian Science and Medicine in the Sixteenth and Seventeenth Centuries* (New York: Science History Publications, 1977), pp. 21–22.

31 *JKGW*, 16, no. 497, ll. 28–30: "Fateor meam sententiam fundari in doctrina de vacuo vt recte collegisti. Nullam Dyaphanitatem agnosco nisi ratione vacui."

32 *JKGW*, 16, no. 536, ll. 41–42: "Jam si liquor non recipit vacua, nihil quod sit diaphanum, recipit. Ex liquore enim omnia diaphana existunt."

33 Ulinka Rublack, *The Astronomer and the Witch: Johannes Kepler's Fight for His Mother* (Oxford: Oxford University Press, 2015), p. 49.

34 Duke Friedrich famously waged an embarrassing five-year campaign of charm on Queen Elizabeth to be named a Knight of the Garter, and so was immortalized as a punch line in Shakespeare's *The Merry Wives of Windsor*. In 1599, he oversaw a sumptuous parade filled with New World decor and had himself decked out as Lady America. *Ibid.*, p. 46.

globe rather than a liquor bowl.[35] Kepler oversaw the initial stages in Stuttgart but soon after had to leave for home and work in Graz. Well over a year later, around October 1597, Maestlin anxiously asked Kepler for news about the *opus argenteum*. Maestlin admitted to Kepler that he had entertained worries at the outset but had thought enough about Kepler's competence and honor, and of the project, to remain silent. Now his fears had come to pass. "I dread terribly," he wrote, "that the illustrious Duke will be offended."[36] The main problem, as he understood it, was rising cost, and he wished to communicate the best course of action to the court, once Kepler had provided a best course of action. His and Kepler's reputations were at stake. Kepler responded in January 1598 with a long and at times angry letter.[37] Most of the blame, by his account, belonged to the goldsmith, whom he portrayed as uniquely incompetent. To be fair, it seems that the goldsmith from the outset had never really understood what was expected. In any case, the work was being done in Stuttgart, while Kepler lived in Graz, a distance requiring days of travel. Kepler had expected his voice to be heeded among the Duke's counselors, and his orders to be passed along to the goldsmith, but the councilors had ignored his frequent complaints. They were now, he thought, trying to drop all the blame on his shoulders:

> I have not renounced my home. Even if I am absent, I am still under the protection of the Prince, and I have a small patrimony in Württemberg. And what man of any standing would endure having his reputation torn to pieces in that illustrious city [Stuttgart]? I will not bear having my good name destroyed in the Duke's eyes by unjust accusations. If I cannot at the same time maintain the grace of the Duke *and* his counselors, then I will thus demonstrate my innocence most clearly to him concerning those other indignant [counselors].[38]

Kepler's foray into the realm of precious materials—in this case, gems and silver—was also a foray into the courtly sphere of power. Intellectually, he could instrumentalize the nobility of these precious materials by integrating

---

35 Johannes Kepler to Duke Friedrich of Wurttemberg, 28 May 1596, *JKGW*, 13, no. 42.

36 Michael Maestlin to Johannes Kepler, 30 October 1597, *JKGW*, 13, no. 80, ll. 80–81: "Etenim Illustrissimum principem offensum iri vehementer metuo."

37 Johannes Kepler to Michael Maestlin, 6 January 1598, *JKGW*, 13, no. 85.

38 *Ibid.*, ll. 305–311: "Ego vero patriam meam non abjuravi: et quid tum, si absens sum, qui nihilominus in fide sum principis, qui patrimoniolum habeo in Wirtembergia. Et quis ingenuus ferat lancinari famam suam in illustri urbe. Itaque mihi ferendum non est, ut iniqua criminatione principi deferar. Si non possum simul retinere gratiam et principis et consiliariorum: ergo iratis his innocentiam meam illi demonstrabo luculentissime."

them into his natural philosophy. Yet he lacked the leverage—social and material—required to organize them into an actual object. Kepler's letter to Maestlin amounted to an admission of failure and a request to abandon the project entirely in favor of something new, a planetarium built from cheaper material, either copper or brass, that would display movement around the sun along with planetary distances. As Walther von Dyck notes, this second project would soon dissolve. Kepler would be exiled from Graz, begin working with Tycho Brahe, and enter into another, much greater, sphere of power at the court of Rudolf II in Prague. Nevertheless, the episode demonstrates Kepler's eagerness to connect his celestial research to the chemical projects popular at European courts.

5.3 Kepler and de Boodt

Kepler had had a number of personal contacts who could have influenced his thinking about stones and metals. His first employer in Prague, Tycho Brahe, was an experienced chemical practitioner, oriented largely toward medical application. Tycho's *pyronomic* or *spagyric* activities[39] have been well documented, from his laboratory at Uraniborg, to the remedies and elixirs that he produced,[40] to his personal relationship with Petrus Severinus, the Danish royal physician and one of the most influential Paracelsian philosophers in all of Europe.[41] Brahe expected to continue his chemical activities in Prague when he left Denmark amid scandal and took up the post of Imperial Mathematician. At the time of his death, a laboratory was being built for him at the castle in

39 For Tycho's aversion to the "alchemical" and preference for the "pyronomic" or "spagyric," see John Robert Christianson, *On Tycho's Island: Tycho Brahe, Science, and Culture in the Sixteenth Century* (Cambridge: Cambridge University Press, 2000), p. 91.

40 On Tycho's alleged attempt to cure Christoph Rothmann of syphilis, see Peter Barker, "How Rothmann Changed His Mind," *Centaurus*, 2004, 46, 41–57.

41 For the laboratory, see Jole Shackelford, "Tycho Brahe, Laboratory Design, and the Aim of Science: Reading Plans in Context," *Isis* 84, 1993, 84: 211–30. On Tycho and Severinus, see Jole Shackelford, "Providence, Power, and Cosmic Causality in Early Modern Astronomy: The Case of Tycho Brahe and Petrus Severinus," in *Tycho Brahe and Prague: Crossroads of European Science*, ed. John Christianson et al. (Frankfurt: Verlag Harri Deutsch, 2002), pp. 46–69. For an overview of Tycho's chemical activities, see Vladimír Karpenko and Ivo Purš, "Tycho Brahe: Between Astronomy and Alchemy," in *Rudolf II and Alchemy: Exploring the Secrets of Nature in Central Europe in the 16th and 17th Centuries*, ed. Ivo Purš and Vladimír Karpenko (Prague: Artefactum, 2016), pp. 459–88. Various information on Tycho's chemical practices can also be found throughout Victor E. Thoren, *The Lord of Uraniborg: A Biography of Tycho Brahe* (Cambridge: Cambridge University Press, 1990).

Benátky nad Jizerou that Rudolf had given him for his new residence.[42] It is even possible that Kepler witnessed some of Brahe's experimentations in Prague or at the Benátky castle.[43]

After Tycho's death, Rudolf's court at Prague likely remained for Kepler a source of information about material productions of the Earth and heavens. Rudolf is remembered by historians for his passionate interest in alchemy. His castle was the site of several laboratories, where he was personally involved in experimentation. Over large stretches of time, he seems to have been more available to his alchemists and philosophers of nature than he was to certain functionaries and dignitaries of state.[44] Rudolf's fascination with stones and gems was not only alchemical in the strict sense. He was one of Europe's great collectors of precious stones and a major patron of glyptic works—decorations including vases and rich inlays. Following the example of Ferdinando I de' Medici, Rudolf commissioned works of *pietre dure* from the best craftsmen in Milan, sending them Bohemian stones to be cut and set, and managed to entice a Milanese master craftsman to join his court in 1588.[45] We should also note that the expansion of trade and voyage in the sixteenth century brought a surge of exotic stones from the Americas and Asia, especially India and China.[46]

At Rudolf's court, the chief authority on precious stones was the Bruges-born Anselmus Boëtius de Boodt, who hailed from an aristocratic, Catholic family.[47] After training in law, de Boodt studied with the physician-philosopher

42 Shackelford, "Tycho Brahe" (ref. 41), p. 220.

43 Kepler describes in his *Tertius interveniens* (1610) how he tasted a strong brandy that Brahe had distilled from roses. Karpenko and Purš (ref. 41), p. 488.

44 The classic work on Rudolf and his occult interests is R. J. W. Evans, *Rudolf II and His World: A Study in Intellectual History, 1576–1612* (Oxford: Oxford University Press, 1984), especially Chapter 6, pp. 196–242. See also Ivo Purš, "Rudolf II's Patronage of Alchemy and the Natural Sciences," in *Alchemy and Rudolf II* (ref. 40), pp. 139–204.

45 Anna Maria Giusti, *Pietre Dure: The Art of Semiprecious Stonework*, trans. Fabio Barry (Los Angeles: Getty Publications, 2006), pp. 109–16.

46 Donald F. Lach, *Asia in the Making of Europe*, Volume 2, *A Century of Wonder* (Chicago: University of Chicago Press, 1970), pp. 113–122.

47 For a recent overview of de Boodt's education and career, see Ivo Purš, "Anselmus Boëtius de Boodt: Physician, Mineralogist and Alchemist," in *Alchemy and Rudolf II* (ref. 41), pp. 535–537. For an analysis of de Boodt's philosophy of nature, specifically his use of *spiritus* and seminal reasons, see Hiro Hirai, *Le concept de semence dans les théories de la matière à la Renaissance: de Marsile Ficin à Pierre Gassendi* (Turnhout: Brepols Publishers, 2005), 375–399. Also, see Robert Halleux, "L'œuvre minérologique d'Anselme Boèce de Boodt (1550–1632)," *Histoire et Nature*, 1979, 14: 63–78; and Lynn Thorndike, *A History of Magic and Experimental Science*, vol. 6 (New York: Columbia University Press, 1941), pp. 318–324.

Thomas Erastus in Heidelberg, then took a doctoral degree in medicine at the University of Padua. From 1588 to 1604 de Boodt served as court physician in Prague. The year 1604 saw him promoted to Rudolf's personal physician, a post he retained until the latter's death in 1612. His main function at court was mineralogical work, much of it devoted to classification and verification, and some of it of an alchemical nature. De Boodt's personal views on alchemy and the possibilities of transmutation remain ambiguous, although Nicolas Barnaud reported that he had had some luck transmuting mercury into gold.[48] De Boodt's role as physician was not ancillary to his mineralogical work. On the contrary, the two activities were closely connected, as can be seen from his *Gemmarum et lapidum historia*, first published in 1609. The *Historia* is an encyclopedic treatise on gems and stones, and one of the most influential early modern books on mineralogy.[49] Not only does it treat all manner of mineralogical topics—from the generation of stones, to their identification, decorative use, classification, falsification, and valuation—, it is an exhaustive compendium of known stones, with chapters devoted to their faculties and powers. These chapters focus on how a given stone can protect and heal the body, to such an extent that a contemporary reader might come to see gemstones as a kind of wearable technology of the Renaissance. So we learn that, "as the authors tell us," carbuncle or true ruby strongly resists venom, protects against plague (a leitmotif of stone powers), chases away melancholy and terror, and generally keeps the body safe from sickness. Rubies can also darken in the presence of misfortune and return to their original color once the danger has passed. However, they shorten sleep and agitate the blood, so that wearers will be quick to anger.[50] In de Boodt's chapter on the faculties and powers of pearl, he writes that pearls can strengthen the heart and vital spirits. Hence, they resist venoms, plague and corruption, and cure cardiac problems, melancholy and fainting. De Boodt continues with a list of afflictions that can be soothed with pearl, then insists on a complex manner of preparation (a kind of cleaning and purifying) to be completed before mixing the pearl into medicines that achieve a variety of effects. De Boodt assures us that the best remedy, far and away, against venoms and animal bites is a precise mixture of powdered pearl,

48 *Ibid.*, pp. 549–552.

49 The only scholar of whom I am aware to note similarities between Kepler and de Boodt is Robert Halleux in his 1975 French translation of *Strena seu de nive sexangula*. In particular, he underlines the similar appeals to a vital faculty. See Kepler, *L'étrenne* (ref. 2), pp. 21–22.

50 Anselmus Boëtius de Boodt, *Gemmarum et lapidum historia* (Hanau: Claudius Marnius, 1609), 2.14, p. 73.

powdered intestinal stones, powdered unicorn horn, and the powdered horn or heart bone of a stag.[51]

All of this might seem distant from Kepler, and here we might signal a major difference between the two: de Boodt is an atomist, whereas Kepler is a dedicated anti-Epicurean.[52] But de Boodt and Kepler have much in common, especially when we turn to de Boodt's understanding of natural cause and generation, as well as a few other assorted themes. Early in the *Historia*, de Boodt grounds generation—animal, plant and mineral—in a terrestrial spirit acting via heat and light.[53] When God created the universe from nothing, says de Boodt, he established earth and water as the fundamental matter. At first, the Earth was deserted and sterile, with neither ornament nor seed. On the third day of creation, God made the Earth participate in a formative and seminal faculty, so that it could produce trees, herbs and everything benefiting from a vegetative soul. This faculty is described as the terrestrial vehicle of the creative spirit that, as *Genesis* recounts, first dwelt above the waters. Aided by celestial light and heat, this spirit assures the generation, alteration and change of all things. The spirit is hot in potential, and its heat is made active through the effect of light.

> [...] it thus seems to be the author of all movement and the efficient cause of all things. Reposing on things and setting in them a seminal and formative faculty, it is like the architect that fashions trees and plants, that paints them and multiplies them to infinity. God, good and great, created this same spirit in animals, so that they could fulfill their functions, conserve their life and propagate their species. He also infused this same spirit in man, as the nearest instrument of the soul, by which he could conserve his life and posterity. This spirit is therefore author of all things after God and is the first instrument of God present in the entrails of the Earth and in water and air. It is never idle. But by its innate heat,

51 *Historia* 2.38, p. 87. For more on unicorn horns as a remedy for poison, see Joan Evans, *Magical Jewels of the Middle Ages and the Renaissance, Particularly in England* (Oxford: Clarendon Press, 1922), pp. 176–77. Thorndike reports that de Boodt elsewhere expresses skepticism about the unicorn's existence; Thorndike (ref. 47), pp. 323–324.

52 Although de Boodt's atomism is unnoticeable through most of the *Historia*. However, he does invoke atoms to account for the transparency of stones. Transparency is caused, he says, by the tight assemblage of very small atoms. *Historia*, 1.14, p. 21. Transparency, in turn, is one of the most important features of stones. His system of classification is based on a hierarchy where the most transparent occupy the highest rung, the least transparent the lowest. See *ibid.*, *Ad lectorem*.

53 Also see Purš, "Anselmus Boëtius de Boodt" (ref. 47), pp. 540–541; and Hirai Hirai (ref. 47), pp. 378–382.

> and roused by the external heat of the heavens, it stirs up and forms the matter with which it meets, and transforms this matter into various species according to the variety of the seed joined to it or the matter mixed within it. This variety is the reason why, in addition to animals, insects and vegetables, there is such a variety and multitude of mixed things that are generated. [...] For when it encounters a good amount of matter that has already undergone multiple alterations [already is sufficiently mixed], it shapes it into nobler mixtures, such as imperfect animals, and insects, herbs, trees, and vegetables. When it encounters rude matter with little alteration, it produces metals, minerals, rocks or gems, as if it were their proximate, efficient cause. Either the matter or the spirit itself must contain the seed of future form. Were this not so, nothing would be produced or generated.[54]

In short, de Boodt's formative spirit is more or less identical to Kepler's terrestrial spirit. Both are instruments of God in nature and were established at creation. Both are specifically purifying, with a similar transformative activity. Both are identified as a kind of heat that provides form. Both are said to be equivalent to seminal reasons.[55] And both are celestial and animal, such that the Earth's generative faculty is of the same kind as those of the celestial region

54 *Historia* 1.8, p. 8: "[...] ita motus omnis autor, & efficiens rerum omnium causa videatur. Rebus enim insidens rerumque seminalem, & formatricem facultatem in se ferens, tanquam architectus quidam, plantam seu arborem fabricat, pingit, ac tandem in infinitum multiplicat. Similem Spiritum ubi animantia crearet. D. Opt. Max. illis ad varia officia obeunda, vitamque custodiendam, ac speciem multiplicandam dedit. Homini etiam eundem inspirauit tanquam animae proximum instrumentum quo vitam & posteritatem conseruare posset. Hic ut rerum omnium fere autor post Deum est, ac proximum quasi eius instrumentum terrae visceribus, aquae & aeri insidens, ociosus nunquam manet: Sed perpetuo materiam quam nanciscitur calore, qui illi innatus est, ac ab externo coelesti excitatur, exagitat, format, ac in varias species, pro seminis sibi adiuncti aut materiae immixti varietate mutat. Haec varietas efficit ut praeter animantia, insecta, & vegetabilia; tam varia, & multa mixtorum sint genera. Nam ut pictores ex albo, nigro, rubro, ceruleo, & flauo colore omne colorum genus facere possunt, ita ex rebus a Deo prius formatis & diuersis, licet paucioribus infinitas res componere, & efficere potest hic Spiritus formatrice facultate instructus. Dum enim materiam multas alterationes passam nanciscitur, ac ipse copiosus est, nobiliora mixta ut animantia imperfecta, insecta, herbas, arbores, & vegetabilia efformat. Dum rudem ac parum passam, metalla, mineralia, lapides, aut gemmas tanquam proxima earum efficiens causa, producit. Verum vel materia Seminarium futurae formae, vel ipse Spiritus continere debet. Id si non fuerit nihil generatur, aut producitur."

55 For de Boodt's account of generation and the activity of the vital spirit, see *Historia* 1.8, pp. 8–9.

and human body. And, as seen in the above quotes, Kepler and de Boodt describe them in very similar ways.

There is another point of similarity relating to the Earth's vital faculty, specifically the question why crystals are hexagonal. Kepler actually asks this question about snowflakes, assuming an analogous mechanism at work between ice and crystals. Kepler tackles the snowflake question in his *Strena, seu de nive sexangula* (Frankfurt, 1611), a gift of charm and erudition to his friend and loyal patron in Prague, the councilor and diplomat Johann Matthäus Wacker von Wackenfels (1550–1619). As the title indicates, the work was a New Year's gift for 1609–1610 and published in 1611. That year, Rudolf would abdicate the throne to his brother Matthias. Kepler, amid political uncertainty and terrible personal loss, would begin plans to relocate to Linz, where he lived from early 1612. The *Strena*, then, marks the end of Kepler's decade in Prague. Robert Halleux, in his French translation of the work, has pointed out a compelling similarity between Kepler and de Boodt: both attempt a material explanation for the manifestation of hexagons before throwing up their hands and appealing to their architectural world spirits. "De Boodt's process is exactly parallel to Kepler's," writes Halleux, "with the same hesitation between a purely mechanical explanation and an explanation that is animistic and Neoplatonic."[56] Here would be an opportune moment to compare their respective arguments more closely. De Boodt and Kepler both begin their respective examinations by considering what Andrea Cesalpino (1519–1603) had to say about hexagonal crystals in his *De metallicis libri tres* (Rome, 1596). Cesalpino's argument runs roughly as follows: crystals form together in a mass before they begin to separate into individuals; due to the purity of their sap, they form identical figures that fill the space without gaps; there are only three such figures available: the triangle, square and hexagon; of these, the hexagon approaches closest to the circle, particularly as it can be divided into six equilateral triangles sharing the same center; hence, crystals are hexagonal.[57] Both Kepler and de Boodt disagree with the above account. Crystals do not form side-by-side in a consistent mass, says de Boodt: they grow freely in clusters, as if from a common root. Kepler, for his part, writes that snowflakes are not connected during their formation, so no material reason exists why they would adopt a plane-filling shape. De

56 Kepler, *L'étrenne* (ref. 2), pp. 21–22. "La démarche de Boèce de Boodt est exactement parallèle à celle de Kepler, avec la même hésitation entre une explication purement mécaniste et une explication animiste et néoplatonicienne."

57 Andrea Cesalpino, *De metallicis libri tres* (Rome: Aloysius Zanetti, 1596), 2.19, pp. 96–99. Halleux provides a summary of Cesalpino's account in Kepler, *L'étrenne* (ref. 2), pp. 19–20. De Boodt accurately summarizes Cesalpino's account in *Historia*, 1.13, p. 18. Kepler responds to Cesalpino in *L'étrenne*, pp. 79–80. Cf. *JKGW*, 4, 278–280.

Boodt explains the production of angles chemically, but this does not help him to answer for the number of angles.[58] Why six? "I esteem," he writes, "that this is because nature never does more when it can more easily do less. Nature acts more easily [with the hexagon]: because the hexagon holds and brings together the circular with more order, proportion, and equality."[59] In fact, de Boodt agrees with Cesalpino that the hexagon is the most noble of polygons. But his demonstration of this nobility takes a different turn. The hexagon, writes de Boodt, "divides the circle at its center by three lines such that there form six equilateral triangles."[60] Pentagons and heptagons, he continues, cannot divide the circle equitably. Squares and octagons can, but their constituent triangles are not equilateral. De Boodt's demonstration is rudimentary, and somewhat awkward, but the point is clear. The hexagon's simplicity originates from how it divides the circle via inscription. To put it another way, the nobility of a polygon can be ascertained by inscription. Certainly, such an idea does not originate with Kepler, although it is one of the most prevalent of all Keplerian themes.[61] In any case, de Boodt concludes by noting his displeasure with his own purely material-mathematical account:

> The hexagonal figure is therefore the most perfect of polygons, and most apt for the circle to be contracted into it, and so the crystal hardens into a hexagon. But if I may confess sincerely, I am not entirely satisfied with this. I think that nature has given the hexagonal shape [to the crystal], so that by this outward mark [the crystal] is differentiated, not otherwise than how nature gives to the leaves of trees and the flowers of herbs their

58 If salt is not present in high quantity, the stone will take on a round form; cf. *Historia*, 1.13, pp. 19–20.

59 *Ibid.*, p. 20: "[…] hexagonam figuram oriri, quae rotundae quam proxima est, ac aliis omnibus figuris angularibus perfectior, quia ex sex triangulis rectilineis aequalium laterum constat, habetur. Fieri a. potius sex angulos quam plures puto ob id quod natura per plura nunquam faciat quod per pauciora commodius facere potest. Facit autem commodius, quia rotundum citius, ordinatius & aequalius contrahit."

60 *Ibid.*: "Dividit enim circulum per centrum tribus lineis tali modo, ut sex inde fiant trianguli aequilateri, ac quilibet centrum habeat ab omni angulo & a cuiusque lateris medio aequidistans; quod in nulla alia figura plurium angulorum vel pauciorum contingit."

61 See Jonathan Regier, "An Unfolding Geometry: Appropriating Proclus in the *Harmonice Mundi* (1619)," in *Unifying Heaven and Earth: Essays in the History of Modern Cosmology*, ed. Miguel A. Granada, Patrick J. Boner, and Dario Tessicini (Barcelona: Universitat de Barcelona, 2016), pp. 217–237.

> particular figures, which are fashioned by this artisan spirit and formative faculty in an unknown manner.[62]

Thus de Boodt. Kepler, having established that snowflakes are not hexagonal because of material necessity, also settles on their creation via the Earth's vital faculty. So, he asks, is the hexagon ingrained as a privileged figure in the Earth soul? No, because the Earth knows and exercises all of geometry, meaning that it puts diverse figures into rocks and plants.[63] Kepler submits that the faculty may act differently according to the humor that it encounters.[64] And with that suggestion, he addresses his patron directly and concludes his New Year's treatise:

> Let the chemists, then, say whether snow contains any salt, and what kind of salt, and what shape it assumes at other times. As for me, after knocking on the doors of chemistry, when I see how much remains to be said in order to capture the cause of the thing, I prefer to listen to what you think, most ingenious man, than to tire myself out through further discussion.[65]

But what is the sense in rejecting a purely mechanical or chemical explanation—based entirely on the movement of salts or other materials—in preference for a terrestrial spirit? De Boodt and Kepler's skepticism begins with acknowledging the complexity of the chemical phenomenon. This complexity signals to both that a simpler cause can be found by focusing on the agent.[66] The simi-

---

62 *Historia*, 1.13, p. 20: "Est itaque hexagona figura omnium figurarum perfectissima & aptissima, ut circulus in eam contrahatur, & per consequens ut Crystallus in eam coaguletur. Sed, ut ingenuè fatear, mihi non plane satisfacio, & naturam ut Crystalus hac nota ab aliis gemmis distinguatur, hexagonam figuram dedisse autumo, non secus quam arborum frondibus & herbarum floribus peculiares suas figuras dat, quae ab architectonico spiritus & formatrice facultate ignoto nobis modo fabricantur."

63 For Kepler's most thorough account of the Earth's soul and vital faculty, see Johannes Kepler, *The Harmony of the World*, trans. E. J. Aiton, A. M. Duncan, and J. V. Field (Philadelphia: American Philosophical Society, 1997), pp. 358–385.

64 *JKGW*, 4, 279.40–280.1: "Itaque verisimile est hanc facultatem formatricem pro diuerso humore, diuersam fieri."

65 *JKGW*, 4, 280.2–6: "Dicant igitur Chymici, an in Niue sit aliquid salis, et quod nam salis genus, et quam illud alias induat figuram. Ego namque, pulsatis Chymiae foribus, cum videam, quantum restet dicendum, vt causa rei habeatur: malo abs te, Vir solertissime, quid sentias, audire, quam disserendo amplius fatigari." Cf. Johannes Kepler, *The Six-Cornered Snowflake: A New Year's Gift*, trans. Jacques Bromberg (Philadelphia: Paul Dry Books, 2010), pp. 112–113.

66 As Halleux notes in ref. 2.

larity in the appeal to an agent, and the similarity in the nature of said agent, point us to the theoretical grounding shared by de Boodt and Kepler. The sixteenth century saw many "*spiritus mundi*" schemes, where a world soul was said to act via a corporeal spirit, its first instrument. I would argue that a hallmark of these schemes, from Ficino onward, is that they support a medicalized view of nature: the human body forms a causal continuum with the rest of nature, meaning that natural forces can be channeled, *détournées*, and the like, in order to protect or conserve the body. Robert Westman, noting the incredible cast of artisans and intellectuals at Rudolf's court, has observed that, "although they shared many common concerns and competences, the court figures who contributed under the emperor's aegis to the theory of the heavens were not the emperor's mechanicians but [...] his physicians."[67] This seems to me an accurate conclusion to draw. We might take it a step further. Considering Evans' classic account of Rudolf's favorites among the alchemists and chemists at his court, it is clear that de Boodt was not the exception: Rudolf's most trusted practitioners in these arts were also physicians, by and large.[68]

Another way to examine the similarity between de Boodt and Kepler is through the prism of cause or force. After all, what is the vital faculty if not a universal ensemble of formal and efficient cause? In turn, we might refine this question by asking about celestial forces in particular. To begin with, I am unaware of a moment in the *Historia* where de Boodt compares planetary bodies with metals or gemstones. De Boodt would have considered unfounded the sympathies seen in Ficino and Agrippa, a point we will touch upon momentarily. As for whether there might be an actual material likeness between planets and certain stones, such as Kepler believes there to be, de Boodt says nothing. In the *Historia* 1.26, he covers in detail the kinds of powers that stones can have. The brief answer is that they can only possess powers based on their form, essence or matter (i.e., mixture of elements). In short, they can only act as what de Boodt calls "natural causes." They cannot produce supernatural effects, the domain of supernatural beings. Moreover, de Boodt does not think that these forces are placed in stones by the celestial bodies; God was Himself more than capable of ascribing to stones their respective powers for the use

67 Robert S. Westman, *The Copernican Question: Prognostication, Skepticism, and Celestial Order* (Berkeley and Los Angeles: University of California Press, 2011), p. 238.

68 "Behind the legends and against the background of imposters and simpletons there were at Rudolf's court significant representatives of alchemy in its widest sense, and several of them lived as members of the Emperor's permanent establishment in the capacity of court physician (*Leibartz*)." Evans (ref. 44), p. 203. See *ibid.*, pp. 203–208, for the roll call of physician-alchemists at the service of Rudolf.

of mankind.[69] Ivan Purš and Hiro Hirai have highlighted a wonderful passage later in *Historia* 1.26, where de Boodt criticizes the belief that stones, by their shapes or by characters engraved on them, can be in "sympathy" with stars or astrological aspects. The meaning of such shapes and characters is human convention only, he writes, just like astrological signs, which were invented by astrologers to aid them in teaching.[70] Purš indicates that de Boodt seems to embrace Pico della Mirandola's disputations against astrology. If so, it is a qualified embrace. De Boodt does not reject astrology outright; he asks for natural causes. He is open to the possibility that celestial aspects and states can influence terrestrial matter, via real forces acting upon entities materially disposed to receive them.[71] All this signals another correspondence between him and Kepler. In Chapters 8 and 9 of the *De stella nova*, Kepler undertakes a discussion with Pico over what is and is not natural in astrology, and he accepts the lion's share of Pico's criticism.[72] The difference is that Kepler does not reject everything, especially not the efficacy of astrological aspects, which he holds to be empirically irrefutable and explicable via natural causes.[73] Hence, both Kepler and de Boodt are open to some form of astrology—what we might term a natural-philosophical astrology. And unlike Pico, who acknowledged only three types of celestial cause—heat, motion and light—, de Boodt and Kepler both admit that forces other than these can be transmitted. As de Boodt argues:

69 *Historia*, 1. 26, p. 47.

70 Purš, "Anselmus Boëtius de Boodt" (ref. 47), pp. 548–549. Hirai (ref. 47), pp. 397–398. Thorndike has also drawn attention to this passage: Thorndike (ref. 47), pp. 320–321.

71 *Historia*, 1.26, pp. 53–54.

72 *JKGW*, 1, 184–194. See, for example, *ibid.*, 184.8–16, where Kepler thoroughly agrees with Pico that there is nothing natural about the zodiacal signs and triplicities, nothing about the association of signs with animals, nor triplicities with elements: "Dixi hactenus, neque naturalem esse distributionem zodiaci in signa duodecim, neque perpetuam vel accuratam, hanc, quae a motibus superiorum fit, in triplicitates quatuor distinctionem, triumque distantium connexionem: Dictum etiam, neque signis proprietates ullas eorum animalium, neque signorum Triplicitatibus, qualitates eorum elementorum inesse, a quibus denominantur. Quae omnia cum centum amplius annis ante me docuerit Joannes ille Picus, Mirandulae comes: iure videri possim totus in eius sententiam de Astrologiae vanitate concedere." See also Sheila J. Rabin, "Kepler's Attitude Toward Pico and the Anti-Astrology Polemic," *Renaissance Quarterly*, 1997, 50: 750–770. On the role of "natural" elements in the astrology of Kepler, see the following contribution by Aviva Rothman.

73 For his most articulate account of the astrological aspects in Book 4 of the *Harmonice mundi* (1619), see Judith V. Field, *Kepler's Geometrical Cosmology* (London: Athlone, 1988), pp. 127–142.

> The sky and the celestial bodies as common and universal causes govern this inferior world and transmit forces by their rays and by their heat: but the effects that follow always result from this, and are neither stimulated nor impeded by figure or character [inscribed on rocks]. Such effects are easily recognized by experience and offend neither nature nor intellect, even if the reason perhaps cannot be understood. Thus the northern zone of the sky always turns the piece of iron rubbed by the magnet and placed in equilibrium toward it, even though the reason why may not be known [...][74]

5.4 Conclusion

By way of the above quotation, we come to a surprising difference between our two subjects, and one that leads us to our concluding remarks. As an example of celestial force, de Boodt adopts a magnetic theory common from the medieval period until the seventeenth century: that the lodestone is attracted by the polestar or, rather in this case, by the "northern zone of the sky." Yet nearly a decade before the *Historia*'s publication, William Gilbert had offered profound evidence that the Earth was itself a magnetic body, thereby explaining the behavior of compasses. Indeed, I can find no mention of Gilbert in the *Historia*. If we assume some form of dialogue or discourse between Kepler and de Boodt, this lacuna is particularly difficult to explain. Kepler made Gilbert's magnetic philosophy a centerpiece of his physical astronomy, to be laid out in the *Astronomia nova*. Why would de Boodt not have entertained Gilbert's opinions even if he disagreed, as he seemed ready to do with other contemporaries or near contemporaries?[75] Yet it is perhaps less important to ascertain whether de Boodt and Kepler discussed their work with one another. The fact is that they share a theoretical grounding or hard core, even if their research and output cover different subjects. And we are not discussing two secondary figures at Rudolf's court. One was a personal physician to Rudolf and his

74 *Historia*, 1.26, p. 56: "Coelum & astra tanquam causae communes & universales regunt quidem ista inferiora ac vires per radios & calorem rebus immittunt: sed effectus qui inde proueniunt semper inde proueniunt, nec propter figuram aut charaterem promouentur aut impediuntur. Qui autem inde proueniunt facile experientia deprehenduntur, nec contra naturam aut rationem esse deprehenduntur, etiamsi ratio fortassis intelligi non possit. Sic coeli plaga septentrionalis ferrum magnete tactum & in aequilibrio positum ad se respicere cogit semper, etiamsi ratio cur id fiat ignota [...]"

75 For de Boodt's citation of authors, see Thorndike (ref. 47), pp. 318–319.

chief mineralogist. The other was his chief mathematician and astrologer. Between the two, heaven and Earth were nicely covered—and heaven and Earth meshed together coherently. The similarities we find between de Boodt and Kepler strongly suggest that Kepler successfully adapted his work to the wider Rudolfine philosophical institution. His celestial physics and astrology lent themselves to a matter theory specifically directed toward the use and renewal of terrestrial substances for the preservation of health and well-being.

Acknowledgements

Many thanks to Patrick Boner for this invitation and for the scholarly generosity that he has shown me over the years. Thanks are also due to Patrick and Miguel Á. Granada for organizing the workshop at Universitat de Barcelona that gave rise to this volume. I am also grateful to workshop participants for their questions and comments. A finished version of this paper was presented at the 2018 meeting of the International Commission on the History of Geological Sciences in Mexico City. I would like to thank the organizers of that conference, as well as Tina Asmussen and Pietro Daniel Omodeo for putting together the panel.

CHAPTER 6

Kepler's Astrological Play

Aviva Rothman

6.1 Introduction: Nature vs. Culture

Kepler saw himself as both a practitioner and a reformer of astrology. In his capacity as a reformer of the discipline, he devoted himself to "throwing out the chaff and keeping the grain,"[1] much as Luther had done with theology. "No one," he insisted, "should deem it unbelievable that out of astrological foolishness and godlessness a useful sense and holiness could not also be found."[2] Luther, Kepler believed, had eliminated the idols of Roman Catholicism and retained only Christianity's original, godly doctrines and rites.[3] Kepler cast his astrological pursuits in a similar light. Yet just what was Kepler himself actually separating and discarding as astrological chaff, or idols, and what was he retaining? Typically, Kepler's astrological reforms have been described as a separation of the natural—the grain—from the cultural—the chaff. I quote, as just one example, Gérard Simon, who argues that in his astrological work Kepler "is rigorous in distinguishing the natural from what we might call the cultural."[4]

This formulation of Kepler's astrological reforms—nature in, culture out—is problematic for many reasons. Most obviously, it's not clear just what culture means here in the first place; as Raymond Williams notes, "culture is one of the two or three most complicated words in the English language."[5] In its early modern usage, it was typically understood as a process: the tending of something, either literal or metaphorical.[6] But even if we use it in its (still ad-

1 Johannes Kepler to Michael Maestlin, 15 March 1598, *JKGW*, 13, no. 89, ll. 177–178.

2 *Tertius interveniens*, *JKGW*, 4, 161.32–38.

3 While it is true that Kepler largely understood Luther's reforms in this way, I am greatly oversimplifying Kepler's attitude toward the Catholicism of his day, one which was far less antagonistic than many of his Protestant contemporaries, as I've described elsewhere. See Aviva Rothman, *The Pursuit of Harmony: Kepler on Cosmos, Confession, and Community* (Chicago: University of Chicago Press, 2017), in particular Chapter 3, "Kepler and the Catholic Church," pp. 109–144.

4 See "Kepler's Astrology: The Direction of a Reform," *Vistas in Astronomy*, 1975, 18: 439–448, on p. 444.

5 Raymond Williams, *Keywords: A Vocabulary of Culture and Society* (Oxford: Oxford University Press, 1985), p. 87.

6 *Ibid.*

 | DOI:10.1163/9789004437272_008

mittedly multivalent) modern sense, in which culture has something to do with human production and historical development, or with interpretation, symbolic systems, and meaning-making, a clear nature-culture boundary simply didn't exist in the sixteenth and seventeenth centuries. For a world in which macrocosm and microcosm mirrored one another, where nature, history, metaphor, and myth were often different forms of expressing the same truths, and where webs of signification linked all spheres of life, the natural often *was* cultural, and vice versa. As just one example, we might think of the realm of music, whose tripartite division by Boethius was embraced by many in the sixteenth and seventeenth centuries, including Kepler himself. That division, of *musica mundana*, *musica humana*, and *musica instrumentalis* linked the celestial realm to the realm of the human soul and finally to the very instrumental music that human artists created.[7] All were intertwined and reflected the same divine origins, and this entanglement made it impossible to neatly divide the products of nature from the products of man.

The popularity *of Wunderkammern*—cabinets of curiosities—likewise showed just how porous the boundary between nature and culture was in the sixteenth and seventeenth centuries, as Lorraine Daston and Katherine Park have demonstrated.[8] Crammed with both natural and artificial objects whose mimicking of each other was as much a source of wonder as the strangeness of each one alone, the *Wunderkammer* was a place in which, as the Paduan physician Fortunio Liceti wrote in 1616, "I see the convergence of both Nature and Art."[9] Daston and Park have likewise highlighted the ways that monsters in early modernity had "a kind of twilight status," not neatly fitting into any one category. Monsters could be both products of nature and signs imbued with divine meaning, for God used nature, and in particular natural wonders, to convey his messages.[10] If this is puzzling to us, it is because of the "comparative coarseness of the modern distinctions" as compared to their early modern counterparts.[11] Just as "the more nuanced ontology of the early modern period did not treat natural and supernatural causes as mutually exclusive,"[12] so too "in the late sixteenth and early seventeenth centuries, the ancient opposition between art and nature first blurred and then dissolved."[13]

---

7 See Rothman (ref. 3), pp. 9–12.

8 See Lorraine Daston and Katherine Park, *Wonders and the Order of Nature, 1150–1750* (New York: Zone Books, 1998), Chapter 7, "Wonders of Art, Wonders of Nature," pp. 255–301.

9 *Ibid.*, p. 200.

10 *Ibid.*, p. 183.

11 *Ibid.*, p. 192.

12 *Ibid.*

13 *Ibid.*, p. 260.

We might recall, too, the ways in which both nature and scripture were understood to be books, whose meanings needed to be read and interpreted. Hermeneutics was an art that linked the priest and the poet to the philosopher of nature, for all sought to reveal hidden, layered meanings in their texts. The interpretation of both the Book of Scripture and the Book of Nature were likewise understood to inform and illuminate each other.[14] To be sure, this hermeneutical approach to nature ultimately gave way to a "new, non-symbolic conception of the nature of things," one which Peter Harrison, for example, has linked to Protestant readings of God's two books.[15] Kepler himself, however, does not fit easily into Harrison's story, as Kepler's worldview was still clearly sacramental and symbolic. In the most obvious example, he still understood the very structure of the cosmos to represent and illuminate the doctrine of the Trinity (as had Cusa before him)—and vice versa. As Kepler wrote to his teacher, Michael Maestlin, "in the orderly world [there are] the fixed stars, the sun, the aura or intermediate ether: and in the trinity, the son, the father, and the spirit."[16] Likewise, the materialism which Harrison argues arose from this new perspective would have been anathema to Kepler. As Patrick Boner has so clearly illuminated, far from someone whose clockwork universe worked regularly and mechanistically, Kepler's cosmos was vital, generative, and often unpredictable.[17] Boner has highlighted in particular how Kepler relied on metaphor and analogy in order to explore the connections between the celestial and the terrestrial, and the human and non-human realms.

Nature and culture are thus not as easily separable as they may appear, especially at this moment in time. Further, it is particularly problematic to speak of a reform of astrology in light of a separation of nature from culture because astrology itself was so thoroughly a cultural art. It was, to put it differently, the quintessential *interpretative* art, a "universal hermeneutics" that connected the Book of Nature directly to the Book of Scripture, and to the scriptural prophecies in particular.[18] As astrologers hoped to use this hermeneutics to interpret the biblical and historical past, the uncertain present, and the (usually

14 See, e.g., Kenneth Howell, *God's Two Books: Copernican Cosmology and Biblical Interpretation in Early Modern Science* (Notre Dame: University of Notre Dame Press, 2002).

15 Peter Harrison, *The Bible, Protestantism, and the Rise of Natural Science* (Cambridge: Cambridge University Press, 2001), p.114.

16 Johannes Kepler to Michael Maestlin, 3 October 1595, *JKGW*, 13, no. 23, ll. 73–74.

17 See Patrick J. Boner, *Kepler's Cosmological Synthesis: Astrology, Mechanism and the Soul* (Leiden: Brill, 2013).

18 See Claudia Brosseder, "The Writing in the Wittenberg Sky: Astrology in Sixteenth-Century Germany," *Journal of the History of Ideas*, 2005, 66: 557–576.

apocalyptic) future simultaneously, we might conceive of astrology as history, poetry, and prophecy combined.[19] What would it mean for an interpretive art that bore on the meaning of human life, past, present, and future, to be anything *but* cultural?

6.2 Kepler's Astrology

Let us look to what Kepler himself explicitly rejected, and what he did not. Most clearly, he rejected the twelve signs of the zodiac. "This division," he wrote in the *De stella nova* (1606), "is not taken from the nature of things and does not express a natural division, but is merely geometrical or arithmetical. Astrologers, go and find other things!"[20] The names of the zodiacal signs in particular he labeled a "human fantasy."[21] His own discoveries he credited not to the influence of the heavens, but rather to the influence of his predecessors and patrons: "My stars were not Mercury as morning star in the angle of the seventh house, in quartile with Mars, but they were Copernicus, they were Tycho Brahe ... not Saturn the overlord of Mercury, but Rudolph and Matthias, each a Caesar Augustus."[22] Kepler himself, then, emphasized that astrology must rely "on the nature of things," rather than on geometry alone—a distinction that itself is still unclear, as geometry, for Kepler, was the very archetype of the nature of things. He likewise emphasized that there is no direct correlation between the things in the heavens and the specific achievements of people on earth.

Writing retrospectively of his *De stella nova*, Kepler explained that in that book he had fully explained his rejection of the zodiac's "division into twelve equal parts, the houses, the rulerships, the triplicities, etc. I reject all these things, retaining only the aspects and converting astrology to harmonic doctrine."[23] Yet the *De stella nova* was not merely a refutation or reform of traditional astrology, but also a defense of certain parts of the discipline against Pico and others who sought to demote it entirely. Alongside his

19 See, for example, Håkan Håkansson, "Tycho the Prophet: History, Astrology, and the Apocalypse in Early Modern Science," in *The Word and the World: Biblical Exegesis and Early Modern Science*, ed. Peter J. Forshaw and Kevin Killeen (Dordrecht: Springer, 2007), pp. 137–156.

20 *De stella nova*, *JKGW*, 1, 172.20–22.

21 *Ibid.*, 176.7–8.

22 Johannes Kepler, *The Harmony of the World*, trans. E. J. Aiton, A. M. Duncan, and J. V. Field (Philadelphia: American Philosophical Society, 1997), p. 377.

23 Johannes Kepler to Thomas Harriot, 2 October 1606, *JKGW*, 15, no. 394, ll. 74–77.

defense of astrological aspects,[24] "practically the only thing in astrology worth preserving,"[25] Kepler sought to defend the theory of great conjunctions, attacked by Pico in his own work as a form of Chaldean idolatry.[26] In so doing, he also seemed to accept some version of the very theory of triplicities, or trigons, he had in other places claimed to reject. He explained:

> Saturn and Jupiter ... arrange themselves such that two of their mutual consecutive meetings are approximately one third of the zodiac apart from one another. In this way it is brought about that at any time, three signs of the zodiac, combined by the authorities into one trigon, obtain special power from the conjunctions of the superior [planets] in moving (I do not say forcing) the nature of sublunary affairs. Giovanni Pico Mirandola has not yet taken this part of astrology away from me.[27]

Struggling with the fact that Kepler, throughout his work, appeared both to reject trigons *and* to accept them, Sheila Rabin writes that with the trigons, Kepler here "granted astrological validity to an artificial construction based on physical fact."[28] Yet this description too does little to clarify the divisions that Kepler has drawn, and merely adds extra categories to the mix—we now have nature and culture, artifice, the physical, and the geometrical, and it is not yet fully clear how these divisions relate to astrological reform more generally—particularly if this is an instance wherein the "artificial" has astrological validity.

Let us look further at just what Kepler did with the theory of trigons and great conjunctions in the *De stella nova*. There, Kepler relied on—and seemed himself to continue—a long tradition of conjunctionist astrology that based itself on the rare or "great" conjunctions of Saturn and Jupiter, with Jupiter's twelve-year cycle around the zodiac combined with Saturn's thirty-year cycle causing the two planets to meet every twenty years as they moved across the zodiac. The zodiac itself was broken down into four "trigons" or "triplicities," based on the four elements. Now Kepler was quite clear in the *De stella nova* that the association between the trigons and these specific elements was arbitrary and inconsistent. Yet the trigons themselves he took to be astrologically meaningful, precisely because of the succession of great conjunctions that

24 For an explanation of Kepler's theory of astrological aspects, see Boner (ref. 17), pp. 50–53.

25 *De stella nova*, *JKGW* 1, p. 166 l. 41- p. 167, l. 1.

26 For a discussion of Kepler's engagement with Pico's work, see Sheila J. Rabin, "Kepler's Attitude Toward Pico and the Anti-Astrology Polemic," *Renaissance Quarterly*, 1997, 50: 750–70.

27 *De stella nova*, *JKGW* 1, p. 181, ll. 7–13.

28 Rabin (ref. 26), p. 755.

passed through them. Indeed, it was by tracing the conjunctions of Jupiter and Saturn that he had arrived at his polyhedral hypothesis (wherein the planetary orbits could be explained via the Platonic solids) in the first place, a theory which was to form the centerpiece of his first book, the *Mysterium cosmographicum* (1596), and was to inspire his future cosmological pursuits.

Those twenty-year great conjunctions of Saturn and Jupiter in particular recurred in the same trigon for about 200 years, and thus moved through all four trigons (and back to the one where it had begun) in a period of 800 years. Conjunctionist astrology used this 800-year cycle to create "an astrology of history itself,"[29] and Kepler was no exception. The new star of 1604 was particularly exciting to him because of its occurrence at the moment of a great conjunction in the fiery trigon, the first in 800 years. In his own reconstruction of history in the *De stella nova,* Kepler used this 800-year cycle to chart out some major historical figures and transformations, beginning with Adam and the creation of the world in 4000 BC, moving through the birth of Christ in approximately the year 0 (though, not coincidentally, his own reconstruction of the chronology put the specific date four years earlier[30]) and arriving at the current conjunction around the year 1600, which must too, he believed, signify great things. Kepler ended, finally, with a nod toward the future, where he wondered what was in store for his native Germany in the year 2400.[31]

On the one hand, Kepler cautioned that the events he described in his historical analysis could not be taken as *effects* of the great conjunctions in the fiery trigon, and he labeled the 800-year divisions "a most useful memory aid" for the division of time.[32] On the other, he took the conjunctions themselves (and, in this case, the new star that appeared alongside the current great conjunction) to be astrologically significant, and to point directly (not merely coincidentally) to noteworthy historical moments. God, in Kepler's view, had chosen the particular place of each planet at the moment of creation, and such a choice must have been significant, both for that moment and for the future plan of the cosmos. As Kepler had argued in his *Mysterium cosmographicum,* "certainly God did not start the motions blindly, but from some single definite beginning, some illustrious conjunction of stars, and at the beginning of the

29 Kennerly M. Woodey, "Dante and the Doctrine of Great Conjunctions," *Dante Studies,* 1997, 95: 119–134, on p. 120.

30 See Anthony Grafton, *Worlds Made by Words: Scholarship and Community in the Modern West* (Cambridge: Harvard University Press, 2009), Chapter 6, "Chronology, Controversy, and Community in the Republic of Letters: The Case of Kepler," pp. 114–136.

31 *De stella nova, JKGW,* 1, 183.1–24.

32 *Ibid.,* 182.24–25.

zodiac, which the Creator fashioned according to the inclination of the Earth, our home, because everything is for the sake of Man."[33]

The conjunction of planets, their location along the zodiac, new stars—while these things might not always directly change the course of human life, they still might convey messages *about* those lives, sent by a Creator whose pen was celestial as much as it was scriptural. God knew that conjunctions were significant to astrologers, and in fact likely intended them to be so; he made sure to send his new star at precisely the time of just such a conjunction, because he knew people would be watching.[34] As Kepler wrote in his German report about the new star, "since all Mathematicians would have been certain to turn their diligent sight to the time and place of this conjunction, the star's appearance ... resembled a public spectacle, a Triumph, or the entry of a mighty potentate."[35] He continued to insist that the star was connected to the conjunction of planets to the extent that such things formed part of God's message to his people: "God himself ... intended to indicate something notable for the human race, and therefore marked the time and place of this conjunction ... to be remembered forever."[36] The combination of a new star and a great conjunction 1600 years earlier had marked the birth of Christ, and in the *De stella nova* Kepler called on his readers to "imitate the example of the Magi, who had the prophecies of Daniel for the coming of the Messiah, when a marvelous star appeared at the very same time of the great conjunction ... and were excited to study books and ... seek He whom the stars had indicated ... the Savior of the world."[37] Kepler clearly accepted the astrological significance of both the conjunction in the fiery trigon and the new star itself, though he hesitated to interpret the precise meaning of that significance, and implied that the astrological events were signs, rather than causes, of great events on earth. Still, his astrological gaze looked back at history, forward toward prophecy, and squarely at the present, for divine meaning, in ways that seem to hearken directly back to the hermeneutical astrology of the past.

---

33 *Ibid.*, 78.6–9.

34 Indeed, Kepler took God's celestial pen so seriously that he extended the traditional theory of accommodation from God's words in scripture to God's astronomical messages: see Rothman (ref. 3), pp. 116–125.

35 *Gründtlicher Bericht von einem ungewöhnlichen newen Stern*, JKGW, 1, 395.21–25. This report is also translated in "A Thorough Description of an Extraordinary New Star Which First Appeared in October of This Year, 1604," J. V. Field and Anton Postl, *Vistas in Astronomy*, 1977, 20: 333–339.

36 *Gründtlicher Bericht von einem ungewöhnlichen newen Stern*, JKGW, 1, 396.35–40.

37 *De stella nova*, JKGW, 1, 347.17–22.

6.3 The Category of Play

How, then, are we to understand Kepler's larger conception of his own reforms and his status as astrologer? I refer not to the specifics of what he included and excluded (aspects in, zodiac out), which have been illuminated so helpfully in recent work on Kepler's astrology.[38] I refer, rather, to the larger categories (nature, culture, artifice, real, physical, geometrical) and the ways that Kepler's astrological worldview fits within them. Here, I want to suggest a category that may be better suited to capturing the complexity of Kepler's astrological endeavors, and that may help us better understand his own astrological goals: the category of *lusus*, or play, as illuminated particularly in the work of Paula Findlen. Findlen herself invokes the notion of play in her work as a way to avoid historically problematic dichotomies, specifically that of the rational and irrational. In so doing, she praises recent scholars of the occult tradition of science for recognizing that such a tradition was flourishing in early modernity, but chastises them at the same time for the ways in which they have categorized it: for their insistence on "the persistence of the 'irrational' as an oppositional structure, at odds with the process of modern science."[39] Rather than assume this and other traditional dichotomies, Findlen hopes to investigate "the gradations of these categories" using the idea of *lusus*, a "construct that drew on a wide range of discursive practices ...[and thus provides] a way of unraveling the tangled webs of meaning that constituted science from the Renaissance to the Enlightenment."[40]

Play, Findlen thus makes clear, was a feature of early modern knowledge-making practices that should complicate our canonical divisions and categories. It was something that characterized at one and the same time the acts of God, the acts of man, and the acts of nature. The last is particularly significant to Findlen, for if nature was playful, then nature itself was poetic, even imaginative. It could create, transform, and subvert our traditional assumptions and expectations. Like human poets, it relied on analogy and metaphor to make its points. Play thus complicated not merely the rational/irrational dichotomy but also the natural/artificial one. *Lusus* characterized the acts of both nature and man simultaneously, and it was never clear which one preceded the other; playful nature imitated man as much as playful man imitated nature, and both relied, at times, on illusion to do so. "The scientific jokes, both of nature and of

38 See Boner (ref. 17); Simon (ref. 4).

39 Paula Findlen, "Jokes of Nature and Jokes of Knowledge: The Playfulness of Scientific Discourse in Early Modern Europe," *Renaissance Quarterly*, 1990, 43: 292–331, on p. 295.

40 *Ibid.*, p. 296.

knowledge," writes Findlen, "were constructed to parallel each other in their attempts to keep the boundaries between natural and artificial flexible, if not dissolve them entirely."[41] As the creative and destructive upheavals of both the natural and the socio-political world revealed, "the only thing that remained clear was that no boundaries were absolute and no categories transparent."[42] Relying on a theory of correspondences and archetypes and a vision of a God who communicated *via* nature, *lusus* also testified to an animated nature that communicated for itself.

Kepler himself identified with the playful side of nature and with the lure of play more generally. As he wrote in his own self-horoscope, he was someone who enjoyed riddles and also tended toward pretense, disguise, and deceit; all of this, he explained, had roots "in common with jokes and jest."[43] As a person characterized by *lusus* of all kinds, it is no surprise that Kepler felt drawn not just to mathematics but also to poetry, as he also described in his self-horoscope. He moved through comedies, acrostics, anagrams, and various forms of lyric poetry, and was particularly fond of allegories, "which he worked out to the most minute detail, dragging in far-fetched comparisons."[44] Pico had argued earlier for a sharp contrast between poetry and philosophy. "For what does the rhetorician do but seek to lie, to deceive, to be circuitous, and to insinuate?" he wrote to Ermolao Barbaro.

> ... It is your claim that with the near magical lure of eloquence, you can transform at will things themselves in appearance and character ... Can such a one have any affinity with the philosopher, whose sole purpose is to know, and to demonstrate the truth to others? No philosopher could serve the truth who sought merely to create impressions and to seduce with language.[45]

Kepler, by contrast, accepted no such distinction between the two realms. "Grant me license," he wrote in the margins of his *Harmony of the World* (1619), "to imitate both Plato's *Atlantis* in story-telling and Cicero's *Scipio* in dreaming." To successfully produce natural knowledge often required, according to

41 *Ibid.*, p. 324.

42 *Ibid.*, p. 323.

43 See Kepler's self-horoscope, *JKGW*, 19, pp. 328–337.

44 *Ibid.*, p. 328.

45 Quoted in Louis Dupré, *Passage to Modernity: An Essay in the Hermeneutics of Nature and Culture* (New Haven: Yale University Press, 1995), p. 109.

Kepler, "follow[ing] the thread of analogy and pass[ing] through the labyrinths of the mysteries of nature."[46]

As Nicholas Jardine has shown, play and *ioco-seria*, serious jokes, were an essential part of Kepler's approach to the natural world: "what is remarkable in Kepler," Jardine writes, "is the sheer range of playfulness he finds and celebrates in nature."[47] Particularly noteworthy is how for Kepler the notion of play linked the products of God, the products of nature, and the products of man. As Kepler explained in his *Tertius interveniens* (1610), "as God the Creator played, so He also taught nature, as His image, to play the very game which He played before her."[48] Nature's play was like God's play, and God's play was like man's play, and in fact it was sometimes not clear who modeled whom: "God, in the image of one of our architects," Kepler wrote in the *Mysterium cosmographicum*, "approached the shaping of the world with order and pattern ... as if it were not art that imitated Nature, but God himself who had considered the custom of building of the man of the future."[49] On one level, Kepler is here suggesting that God created both man and the game that man would later play. On another, he is playing with the very idea of God's play, suggesting that God looked ahead in time and used human art to model nature, rather than the reverse. And while the nature he made was typically ordered, like human art, it could also be disordered; sometimes it could seem capricious indeed. "For the closer I approach her," Kepler wrote in the *New Astronomy* (1609) about his struggles with the orbit of Mars, "the more petulant her games become, and the more she again and again sneaks out of the seeker's grasp just when he is about to seize her through more twists. Nevertheless, she does not cease to invite me to seize her, as though delighting in my mistakes."[50]

Play was revelatory precisely because of these linkages between the divine, the natural, and the human. Man, created in the image of God, played like him, and was uniquely equipped to decode his play and ascertain its meaning. It follows that the things that man created were not arbitrary or meaningless; even human constructs could be illuminating, insofar as all forms of creation could, and often did, mimic each other, and humans created in much the same way as their own Creator did. When writing about how his astrology, astronomy, and

---

46 *Harmony of the World* (ref. 22), pp. 495–496.

47 See Nicholas Jardine, "God's 'Ideal Reader': Kepler and His Serious Jokes," in *Johannes Kepler: From Tübingen to Żagan*, ed. R. L. Kremer and J. Wlodarczyk, *Studia Copernicana*, 2009, 42: 41–51, on p. 46.

48 *Tertius interveniens*, *JKGW*, 4, 246.23–24.

49 *Mysterium cosmographicum*, *JKGW*, 1, 6.7–10.

50 Johannes Kepler, *New Astronomy*, trans. William H. Donahue (Cambridge, Cambridge University Press, 1993), p. 428.

musical theory were all linked, Kepler explained to Herwart von Hohenburg that "although these things are the inventions of men, nevertheless man is the image of the creator."[51] In other words, culture was natural too. "Not every hunch is wrong," he wrote soon afterward in another letter to Herwart, for if man was the image of God, then "it is possible that he thinks the same way as God in matters which concern the adornment of the world."[52] Play thus not only revealed the pattern of the cosmos, but also the mind of God, and was a way for man to emulate his Creator and come to understand him, from whatever angles possible.

6.4 Astrological Play and Its Dangers

I want to suggest that Kepler's astrology, in particular, is best understood through this lens. Kepler opened his *Optics* by talking about cosmic phenomena more generally as a form of play, but in his discussion he focused specifically on eclipses—cosmic events with long-standing astrological relevance. He wrote,

> For this theater of the world is so ordered that there exist in it suitable signs by which human minds, like likenesses of God, are not only invited to study the divine works ... but are also assisted in inquiring more deeply. For I implore you: what is the cause, if not this, for nature's playing such games in the sun's and moon's bodies, by which ... humans, as history bears witness, are turned to wonder and stupefaction ... ?[53]

Play invited wonder, which invited inquiry, which invited astronomical discovery. When it came to astrology, cosmic play invited inquiry not merely into natural causes, but also into human lives, past, present, and future. In his discussion of the new star of 1604, Kepler therefore explained that God, as architect, took such "delight" in his interactions with his human creations that he used astrological signs as "public significations of his concern for us," and did

51 Johannes Kepler to Herwart von Hohenburg, 6 August 1599, *JKGW*, 14, no. 130, ll. 650–651.

52 Johannes Kepler to Herwart von Hohenburg, 14 September 1599, *JKGW*, 14, no. 134, ll. 438–441.

53 Johannes Kepler, *Optics*, trans. William H. Donahue (Santa Fe: Green Lion Press, 2000), p. 15. For an account of Kepler's *Optics* that emphasizes the role of play, see Raz Chen-Morris, *Measuring Shadows: Kepler's Optics of Invisibility* (University Park: Penn State University Press, 2016).

so in ways that were so distinctive and noteworthy that they "could not remain hidden or fail to excite us exceedingly."[54]

Let us also recall the precise way that Kepler described his astrological reforms, which I quoted right at the start. "No one should deem it unbelievable," Kepler wrote, "that out of astrological foolishness and godlessness a useful sense and holiness could not also be found."[55] Though "useful sense" comes very close to the contemporary meaning of Kepler's German, it's worth noting that the actual German—"eine nützliche Witz"—would translate still more literally as "a useful wit." Wit in Kepler's day, that is, was a kind of sense, or knowledge ('Witz' derives etymologically from the German wissen). Today, of course, the translation would more accurately be "a useful joke." This linguistic shift—from wit as sense to wit as joke—is indicative of the fact that the boundaries between traditional knowledge and jokes, or play, has always been fluid. In the eighteenth and early nineteenth centuries, in fact, the German Romantics specifically embraced wit as a form of knowledge that linked the aesthetic to the scientific, and that relied upon analogies and correspondences. In the views of those who embraced it, wit revealed that "the perception of beauty could not be divorced from the understanding of truth"; further, "Witz, as both knowledge and humor," was able "to take minor moments of similarity and turn them into essential relations."[56] The "nützliche Witz" that Kepler hoped to find in astrology was a kind of play that comes remarkably close to this later understanding of the term.

The interpretation of such cosmic play might be as poetic as scriptural interpretation, for both were creative, often playful ways to elicit human responses. "I once saw a poetic description of a comet," Kepler noted in his discussion of the comet of 1607, "in which the comet with its long tail was extremely elegantly compared with a young heretic who would seduce the people with fanatical speeches. I certainly do not reject this interpretation."[57] God was both a poet and "an architect, by whose counsel the star comes into conjunction with the planets"[58]—in both roles, much like humans, he sought to delight, enlighten, and offer something at once beautiful and useful. Astrology was a hermeneutics of play, wherein interpretation, illumination, and redemption

54 *De stella nova*, JKGW, 1, 280.18–24.

55 *Tertius interveniens*, JKGW, 4, 161.33–39: " … niemandt für ungläublich halten / daß auß der Astrologischen Narrheit und Gottlosigkeit / nicht auch eine nützliche Witz und Heyligthumb / … gefunden werden köndte."

56 See Elena Russo, *Styles of Enlightenment: Taste, Politics, and Authorship in Eighteenth-Century France* (Baltimore: Johns Hopkins University Press, 2007), p. 152.

57 *De cometis libelli tres*, JKGW, 8, 245.16–20.

58 *De stella nova*, JKGW, 1, 290.2–3.

all went hand in hand. And indeed, as there might be no one correct interpretation of a poem, so too there might be no one correct interpretation of an astrological phenomenon, as Kepler insisted time and time again; the key was to seek inspiration when possible, and find personal meaning in universal messages.

This is not to say that *anything* at all was possible, and we should again remind ourselves that Kepler saw himself as a *reformer* of astrology, rather than simply a proponent of all of its standard claims. With astrology understood as play, it is clear the opposition between nature and culture does not quite capture the character of Kepler's reforms. Yet the category of play does allow us to reframe and better understand Kepler's concerns about traditional astrology and his attempts to reform it. His reforms centered on a danger which we can approach by means of a question: just what are we playing *with* in our astrological games? Here, Kepler saw two possibilities: either we are playing with things that exist in the world *physically*—natural objects which could then function as cultural signs, forms of divine or natural play to be interpreted and imbued with meaning. Or we are playing with our own *inventions*, things with no physical existence at all, and then interpreting the things that we ourselves have created. He is distinguishing, that is, not between nature and culture per se, but rather between the attribution of (cultural or symbolic) meaning to "the nature of things" physically present in the world and the attribution of such meaning to the "inventions of men," present only in the imagination.

As Judith Field notes, Kepler's emphasis on the physical rather than the imaginary in astrology is highly similar to his emphasis on the very same distinction in astronomy, when he rejects planetary orbits based on the "mean Sun," a mathematical construct, in favor of orbits centered on the actual position of the sun itself.[59] His rejection of zodiacal signs is likewise based on the fact that he perceived them to be random divisions, not linked to anything physical in the cosmos—just imaginary lines drawn through physical space. By contrast, the great conjunctions were based on the actual physical bodies of the planets, and even the trigons could be understood as physically created and demarcated by the successive positions of those conjunctions. The figures that Kepler drew on the blackboard in his days as a teacher in Graz, when he first thought up his theory of the Platonic solids, revealed shapes created by the movements of Saturn and Jupiter, rather than shapes imagined in the human mind and projected outward. The nature of things needed to be based on things themselves, rather than figments of the imagination.

59 Judith V. Field, "A Lutheran Astrologer: Johannes Kepler," *Archive for History of Exact Sciences*, 1984, 31: 189–272, on p. 201.

This did not mean that the imagination could not then *interpret* those things—this was precisely what astrology was all about. The interpretations themselves were, in certain key ways, always human constructs or cultural products, even if their production was intended by the Creator who had inspired them via his signs and modeled on his own both rational and playful nature. The real question was what their referents were. If those referents were *only* the imaginary inventions of men, then instead of divinely sanctioned play were two possibilities, both problematic. The first was idolatry, itself just another name for the worshipping of one's own creations. Kepler made this point directly, when he compared his own astrological reforms to Luther's reforms of Catholic errors, as we saw at the start. In fact, Kepler likened certain Catholic customs that he believed tended toward idolatry to the astrological practices that he hoped to reform, like the worship of saints to the belief in zodiacal houses, and he referred to the "stupid" and "foolish" uses of astrology as "impious."[60] The zodiacal houses themselves were mistaken human constructs, wrongly imbued with power; they were further granted planets who "ruled" over them much like lesser deities, thus compounding the taint of idolatry present in traditional astrological doctrine and practice. In reforming these doctrines and removing those practices, Kepler would be smashing astrology's idols and stripping its altars, paving the way for a purified, God-centered practice—in his words, removing "godlessness" and replacing it with "holiness." Rather than making his own gods, he would help people interpret the messages sent by the real one.

The wrong kind of astrological play, then, could become idolatry, wherein people worshipped their own creations. A related problematic possibility was that the wrong kind of play became not just the worship of one's creation, but the worship of oneself, or one's image—that is, the problem of narcissism. Mythic Narcissus, ignorant of himself and misled by his senses, mistook the image of himself for the real thing, and erred still further by falling in love with it. Narcissus thus represented the dangers of mistaking imaginary constructs for real things of value, and this was a danger Kepler felt acutely when it came to knowledge-making practices of his day, be they cosmological or astrological. He himself had, at times, fallen for this trap, he explained in his footnotes to the *Mysterium cosmographicum*, when he noted that some of his early cosmological forays had proceeded "by the wandering steps of ambiguous words,"[61] which had led him to be "confused by [an] imaginary coincidence, like someone who touches his left hand with his right hand unawares in the dark and is

60 *De stella nova*, JKGW, 1, 336.27–30. See also Boner (ref 17), p. 45.

61 *Mysterium cosmographicum*, JKGW, 8, 114.11.

scared."[62] Copernicus himself had done this, not with his own image but with that of his predecessor, Ptolemy, insofar as he "took it upon himself to express Ptolemy, not the nature of things."[63] Apian likewise had followed Ptolemy's image rather than the things in the world, and had thus "spent so many hours and wasted so many ingenious meditations trying to express, by means of spirals and corolla and helices and volutes, a vast labyrinth of the most intricate curves, a human figment which the nature of things clearly disowns."[64] In so doing, he, like Narcissus, had "entertained his mind with … tricks, in which he rivaled nature herself," but which were only "empty artistry."[65]

Kepler's later dispute with Robert Fludd over the nature of Fludd's numerical symbolism likewise hinged on this issue. According to Kepler, Fludd's harmonies were "mere symbols … poetic or rhetorical rather than philosophical or mathematical," and his analogies were often created by "drag[ging] things into the air." In other words, Fludd's problem was that he worshipped "his own picture of the world" rather than "the world itself."[66] Fludd was a "numerical prophet" who "takes over false things," while Kepler positioned himself as someone who played, like Fludd, but played with *real* things, rather than his own image projected outward onto the world. "I too play with symbols," he wrote, not of Fludd this time but of Andreas Reinhard's work on the monochord, "but I play such that I do not forget that I am playing. For nothing is proved by symbols … unless by sure reasons it can be demonstrated that they are not merely symbolic, but are descriptions of the ways in which two things are connected and of the causes of these connections."[67] It was not symbolism, poetry, or games that were problematic, in Kepler's view, but rather the nature of the things with which one played and that one interpreted. His games, he wrote in the *Tertius interveniens*, avoided the problems of past astrological narcissism because they were based on real things and produced real knowledge. "As God and nature play," he wrote, "so must men play after them: not a silly child's game, but a natural instinct implanted by God,"[68] then turned toward understanding the world.

62 *Ibid.*, 115.9–10.

63 *New Astronomy* (ref. 50), p. 172.

64 *Ibid.*

65 *Ibid.*

66 Appendix to Book v of the *Harmonice mundi*, JKGW, 6, 377.1–3; cf. Robert S. Westman, "Nature, Art, and Psyche: Jung, Pauli, and the Kepler-Fludd Polemic," in *Occult & Scientific Mentalities in the Renaissance*, ed. Brian Vickers (Cambridge: Cambridge University Press, 1984), pp. 177–229, on p. 180; Judith V. Field, *Kepler's Geometrical Cosmology* (London: Athlone, 1988), pp. 179–187.

67 Johannes Kepler to Joachim Tanckius, 12 May 1608, JKGW, 16, no. 493, ll. 150–156.

68 *Tertius interveniens*, JKGW, 4, 246.32–34.

I do not mean to suggest that Kepler believed truth to always be like a game, nor that he rejected certain knowledge in favor of playful uncertainty. As Raz Chen-Morris demonstrates, there were *certain* games that Kepler believed could reveal a truth that was always certain and univocal, so long as one had the right interpretive tools—these were mathematical games.[69] Of course, even mathematics itself was a flexible art that could be used to imagine multiple possible outcomes and calculate accordingly.[70] When it came to astronomy, mathematics revealed the true structure and motions of the cosmos. But when it came to uncertain disciplines, and in particular astrology, Kepler understood himself to be playing a different kind of game entirely. This was a game that "signified things through allegories, not directly, clearly, or distinctly,"[71] a game involving "labyrinths of meaning."[72] And we need to keep these games in mind as well when we consider his larger worldview. It was not that the world was no longer a text, nor that metaphor and analogy were no longer significant or legitimate tools to turn upon the world. Astrology was still cultural; it was still metaphorical and analogical, and astrological signs were still texts to be weighed and interpreted, with multiple possible meanings, all somewhat uncertain. The trick in playing the game was to be sure that one focused on allusion, rather than illusion; God, rather than idols; and the world itself, rather than a narcissistic outward projection of human imagination and desire. This was the core of Kepler's astrological reform—a division not between nature and culture, but between real games and idolatrous or narcissistic ones.

Acknowledgements

Many thanks to all the participants of the 2017 workshop on "Kepler's Star: Context and Controversy" at the University of Barcelona for their helpful comments and suggestions. Special thanks to Patrick Boner for his detailed comments afterward, all of which greatly improved this article.

69 Chen-Morris (ref. 53), pp. 147–158.
70 See Rothman (ref. 3), pp. 246–253.
71 *De Stella Nova*, JKGW 1, p. 337, ll. 12–13.
72 *Ibid.*, p. 337, ll. 31–32.

CHAPTER 7

The Nova of 1600 in Cygnus and the Christianization of the Constellations

Miguel Á. Granada

7.1 Introduction

Kepler's *De stella nova* (1606) contains two separate chapters,* each with its own frontispiece and dedicatory letter: the second and final chapter is a "Consideration of the most recent opinion of the Pole Lawrence Suslyga on the true year of the birth of Christ our Saviour"; the first, following Chapter 27 and preceding the second part dealing with the signification and natural effects of the nova of 1604, is an "Astronomical report on the star of the third magnitude in the constellation of Cygnus, which remained unknown until 1600 and has still not extinguished" (*De stella tertii honoris in Cygno, quae usque ad annum M.DC. fuit incognita, necdum extinguitur, narratio astronomica*).[1] In the following study, I will focus on the *Report on the Star in Cygnus* and discuss similar reactions to this star from later German astronomers and other scholars that relate in various ways to Kepler's *Narratio*.

The nova of 1600 was very different from the previous one in Cassiopeia (1572) and the subsequent one in Serpentarius (1604), the subject of Kepler's *De stella nova*. These were stars of the first magnitude, stemming from the explosion of galactic supernovas (no such star has appeared since); they were observed in Europe, for various reasons,[2] from the moment of their explosion

* This article is the result of research conducted by the project "Cosmología, teología y antropología en la primera fase de la Revolución Científica (1543–1633)," funded by the Spanish Government (Ministerio de Economía y Competitividad, Project FFI2015–64498-P (AEI/FEDER UE) for the triennium 2016–2018). I wish to thank Patrick J. Boner for his careful reading and improvements on the original version of this article.

1 *JKGW*, 1 ("De stella ... in Cygno"), 293–311. The account is translated into French in J. Képler, *L'étoile nouvelle dans le pied du Serpentaire suivi de L'étoile inconnue du Cygne*, trans. Jean Peyroux (Paris: Blanchard, 1988), pp. 217–232, and into German in J. Kepler, *Über den neuen Stern im Fuss des Schlangenträgers*, epilogue by Eberhard Knobloch, trans. Otto and Eva Schönberger and Eberhard Knobloch (Würzburg: Königshausen & Neumann, 2006), pp. 173–193.

2 The nova of 1572 appeared in Cassiopeia, a northern constellation very close to the zenith, the nova of 1604 at the foot of Serpentarius, close to the ecliptic and at the precise moment

 | DOI:10.1163/9789004437272_009

FIGURE 7.1 Nova of 1600 in Blaeu's Globe of 1602. Peter van der Krogt, *Globi Neerlandici* (1993), p. 198.

and were soon subject to changes of magnitude and colour before disappearing in 1574 and 1605 respectively. By contrast, the nova in Cygnus (now known as P Cygni) initially exhibited a magnitude of the third degree and remained practically without change until 1606 (it was still visible in 1623 with the aid of the *perspicillum* or telescope). In accordance with successive appearances, the nova in Cygnus is now recognised as a variable star, like *Mira Ceti,* first observed by David Fabricius (1564–1617) in 1596 and again in 1609. It is no surprise that its smaller magnitude and greater unaltered permanence begged the question whether it really was a nova or rather an original star, present in the sky from the first creation but omitted from the stellar catalogues, both

(near the end of September 1604) when astrologers and astronomers were viewing the close conjunction of Mars with Saturn and Jupiter in Sagittarius (beneath the foot of Serpentarius), at the beginning of a new period of the 'fiery trigon.'

ancient and modern, until being discovered in 1600 by Willem Janszoon Blaeu (1571–1638) and included in his celestial globe of 1602.[3]

It is also no surprise that Kepler's *Narratio* aims to rigorously address the issue of the novelty or antiquity of the star (positioned near the star in the breast of the Swan [*pectus Cygni*; now γ Cygni]) and to resolve the matter definitively. Nevertheless, there is more to the account than the age of the luminary. At one moment, Kepler recalls an idea that he had in November 1601 about the star (without even posing the question of its possible novelty, but assuming that it was an old star), namely a new configuration of the constellation; this opened the way for later authors to address, in connection with other novelties in the heavens, the question of the 'transformation' or 'new configuration' of the stellar vault. It is our purpose to present, even if only cursorily, the most salient moments in this development, focusing on the work of Helisaeus Roeslin, Wilhelm Schickard, Kepler's son-in-law Jacob Bartsch, and finally Julius Schiller.

We may first recall the Christian interpretation of the new star of 1572 in Cassiopeia by Cornelius Gemma (1535–1578) in his widely diffused *De naturae divinis characterismis* (1575). Here, Gemma indicated that the nova formed with the other stars in the constellation the figure of the Crucifixion, as an omen of the portentous meaning of the miraculous star as well as the imminent changes in human society it heralded as a *prodromus* of the second coming of Christ.[4] However, this suggestion was vehemently denounced and bitterly rejected by Tycho Brahe (1546–1601) in his critical review of the

---

3 *JKGW*, 1, 307.1–3, where 1601 is given as the date of publication of the globe. On Blaeu, see John R. Christianson, *On Tycho's Island: Tycho Brahe and His Assistants, 1570–1601* (Cambridge: Cambridge University Press, 2000), pp. 254–256, where no mention is made of the nova of 1600. For Blaeu's observations of the novas and their representation in the stellar globes he manufactured, see Rob van Gent, "De nieuwe sternen van 1572, 1600 en 1604 op de hemelglobes van Willem Jansz. Blaeu," *Caert-Thresoor: Tijdschrift voor de Geschiedenis van de Kartografie in Nederland*, 1993, 12: 40–46. See also Peter van der Krogt, *Globi Neerlandici: The Production of Globes in the Low Countries* (Utrecht: HES, 1993), pp. 165–168, 503–505, with the reproduction on p. 167 of the description of the nova of 1600 in the globe of 1602: "Novam illam Stellam (quae anno 1600. primum in pectore Cygni apparuit atque etiamnunc immota manet) ex diligenti nostra ad caudae [α Cygni] Lyraeque [α Lyrae, Vega] lucidas observatione Longit: ♒ 16 P.15'. Latit. B 55 P. 30' habere comperimus" (see Fig. 7.1 for a reproduction of the description in the celestial globe of 1602). The globe of 1616 gives 18.8.1600 (surely old style) as the date of the first observation of the nova; see *Globi Neerlandici*, p. 517.

4 Cornelius Gemma, *De naturae divinis characterismis seu raris et admirandis spectaculis, indiciis, proprietatibus rerum in partibus singulis universi* (Antwerp: Christophe Plantin, 1575), vol. 2, p. 130. See also Dario Tessicini, "'Vere gemmeum est?': Cornelio Gemma e la stella nuova del 1572," in *Cornelius Gemma: Cosmology, Medicine and Natural Philosophy in Renaissance*

literature on the nova carried out in the posthumously published *Astronomiae instauratae progymnasmata* (1602). There, Brahe denounced Gemma for his blasphemous attempt to crucify Christ a second time in the heavens.[5]

No doubt this controversial passage left a lasting impression on Kepler. After the sudden death of Tycho in 1601, Kepler had become the editor of his work. Kepler also knew quite well Gemma's *De naturae divinae characterismis*, a treatise he greatly appreciated, at least for Gemma's ideas on the "*spiritus mundi*."[6]

A second intragalactic nova appeared in 1604, this time in the foot of the Serpent Bearer, close to the ecliptic, at the end of September, when Mars was approaching Jupiter and Saturn, whose assembly had brought to fulfilment the great conjunction in 1603 that inaugurated a new period of the fiery trigon, the seventh since the creation of the world according to biblical chronology. This second nova, in addition to its wider 'context,' occasioned the publication of many treatises of manifold kinds, among them Kepler's *De stella nova in pede Serpentarii* in 1606.

As indicated above, Kepler's massive work contained as a conclusion to the first part the "Astronomical report on the star of the third magnitude in the constellation of Cygnus." The full title did not affirm that the star was a nova; it only indicated that it had first been identified in 1600 and was still present in the heavens six years later. Kepler recalled in the report that he had seen the star in November 1601, without any surprise and assuming it was therefore an old star. As testimony of this, Kepler described his "fancy" or idea that a Christian could "most aptly imagine" in the constellation the figure of Christ crucified, with the star in the breast of the swan, representing the head of the

---

Louvain, ed. Hiro Hirai (Pisa-Rome: Fabrizio Serra, 2008), pp. 51–64, with further bibliographical references.

5 Tycho Brahe, *Astronomiae instauratae progymnasmata*, in *TBOO*, 3, 80: "Eo tandem in his decretorum Stellae decisionibus [Gemma] prolabitur, ut Christi crucifixi imaginem stellis assuere, et denuo quasi inter sidera crucifigere non exhorrescat, idque nimis superstitiose, ne quid amplius dicam."

6 Cf. Kepler's letter of 21 February 1605 to Wolfgang Wilhelm von Neuburg, in *JKGW*, 15, no. 332, ll. 111–117: "Derowegen ich hie *Cornelii Gemmae philosophiam* der warheit am ähnlichisten sein vermeine, wölicher *in libris cosmocriticis* fürgibet, das die ganze weite welt *unum aliquod animal* und drinnen ein ybermächtiger *spiritus mundi* seye, und wie im menschlichen leibe die *spiritus animales quovis fulmine citius* aus dem hirn in alle glider gehen, also gehe diser *spiritus mundi* durch die ganze welt und vereinige himmel und erden mit einander, das ein mitleiden zwischen ihnen seye," cit. in Miguel A. Granada, "'A quo moventur planetae?': Kepler et la question de l'agent du mouvement planétaire après la disparition des orbes solides," *Galilaeana*, 2010, 7: 111–141, on p. 129. For a full translation of this letter, see the appendix to Patrick J. Boner, "Kepler's Imprecise Astrology," in *Kepler: La physique céleste*, ed. Édouard Mehl (Paris: Les Belles Lettres, 2011), pp. 115–132.

crucified Christ leaning to the left, and the star under discussion, very close to this one and called in the dedicatory letter a "miracle" of God, representing the breast of Christ ("pectus Crucifixi"), so that "this fancy kept thus far the memory of this star."[7] Kepler's work on the nova in Serpentarius linked the two stars through a Christological dimension; for, in addition to his 'fancy,' the nova of 1604 was widely interpreted as a sign of the second coming of Christ in the "fiery trigon" that had begun in September 1603.[8] Arguably, then, Kepler's *De stella nova* could have been the source, or at least one of the sources, of the adoption of both stars by the Rosicrucian manifesto *Confessio fratertinatis* (1615), as the two extraordinary phenomena in the heavens heralding the new age of global reformation announced and called for by the secret fraternity founded by Christian Rosenkreutz:

> De sua quidem voluntate jam praemisit nuncios Deus, Stellas in Serpentario atque Cygno exortas, quae magna profectò magni Consilii signacula illud docere possunt, quàm junctis iis, quae humanum ingenium adinvenit, suae occultae scripturae inservire faciat, ut Liber Naturae in omnium quidem oculis expansus adapertus sit; pauci tamen vel legere omnino, vel intelligere possint.[9]

7 *JKGW*, 1, 297.20–30: "[...] hanc ipsam [stellam], de qua agitur, eo ipso tempore, Novembri puta anni 1601. viderim quod minime haesitans memoria mihi dictat. Nam cum eo tempore post occasum Solis versaretur sidus in parte coeli occidua, porrigeret rostrum versus Horizontem, caudam sursum, alas ad dextram et sinistram: Equidem, cogitabam, in stellis Cygni, si Christianus aliquis de novo inciperet fingere imagines; aptissimam figuram inveniret crucifixi, cum inclinato capite. Stella enim, de qua dubitamus, tunc mihi ad dextram paulo erat, pectus crucifixi repraesentans; quae vero in pectore Cygni, deflectebat ad sinistram, in crucifixi caput concedens. Haec imaginatio mihi memoriam huius stellae hactenus conservaverat."

8 On the eschatological interpretation of the nova of 1604, see Miguel A. Granada, "Kepler v. Roeslin on the Interpretation of Kepler's Nova: (1) 1604–1606," *Journal for the History of Astronomy*, 2005, 36: 299–319; Miguel A. Granada, "Johannes Kepler and David Fabricius: Their Discussion on the Nova of 1604," in *Change and Continuity in Early Modern Cosmology*, ed. Patrick J. Boner (Dordrecht: Springer, 2011), pp. 67–92.

9 [Johann Valentin Andreae], *FAMA FRATERNITATIS R. C. Das ist, Gerucht der Bruderschafft des Hochlöblichen Ordens R. C. An alle Gelehrte und Heupter Europas. Beneben deroselben Lateinischen CONFESSION* (Kassel: Wilhelm Wessel, 1615), p. 55. See the English translation by Eugenius Philalethes (=Thomas Vaughan): "yea, the Lord God hath already sent before certain Messengers, which should testifie his Will, to wit, some new Stars, which do appear and are seen in the Firmament in *Serpentario* and *Cygno*, which signifie and give themselves known to every one, that they are powerful *Signacula* of great weighty matters. So then, the secret hid Writings and Characters are most necessary for all such things which are found out by Men: Although that great Book of Nature stand open to all Men, yet there are but few

However, it was the famous illustration of the *Collegium Fraternitatis* in the so-called 'fourth Rosicrucian manifesto,' published in 1618 by Daniel Mögling (under the pseudonym Theophilus Schweighardt) and bearing the title *Speculum Sophicum Rhodo-Stauroticum*, where both novae in Serpentarius (with the date 1604) and in Cygnus (in this case without indicating the year, clearly as a sign of the uncertain date of its appearance) were shown as heralding and illuminating the divinely inspired reformation that the Rosicrucian Fraternity was proclaiming.[10]

It is not our aim, however, to follow these echoes of Kepler's combined treatment of the novae. Rather, we focus on the successive development of the motif of the Christianization of the constellations, as suggested by Kepler in his report on the nova of 1600. Nevertheless, it would be misleading to consider this idea as the main theme of the report. The above mentioned idea or 'fancy' is more accurately described as a chance observation by Kepler, whose main concern was demonstrating that the highly discussed star in Cygnus was truly a nova. Accordingly, we will first examine Kepler's original notice of its appearance and then his arguments for its novelty, concluding with an overview of subsequent proposals for the Christianization of the constellations among other authors.

7.2 Spreading News of the Star in Cygnus and Kepler's Role

After the position of the nova was published by Willem Janszoon Blaeu in his celestial globe of 1601,[11] Kepler cursorily described the spread of its news on page 308 of *De stella nova*: word of the star arrived at Kassel, where Justus Bürgi

that can read and understand the same", *The Fame and Confession of the Fraternity of R: C: Commonly, of the Rosie Cross* (London: J. M. for Giles Calvert, 1652). The *Fama* had already made a reference to the fiery trigon and the transformations in it: "Trigono igneo [...] dessen Flammen nun mehr je heller leuchtet und gewißlich der Welt den letzten Brand anzünden wird," *ibid.*, p. 11. For the editions and initial responses to the Rosicrucian manifestos, see *Cimelia Rhodostaurotica: Die Rosenkreuzer im Spiegel der zwischen 1610 und 1660 entstandenen Handschriften und Drucke*, ed. Carlos Gilly (Amsterdam: In de Pelikaan, 1995).

10 See Fig. 7.2 and Carlos Gilly, "Las novas de 1572 y 1604 en los manifiestos rosacruces y en la literatura teosófica y escatológica alemana anterior a la Guerra de los Treinta Años," in *Novas y Cometas entre 1572 y 1618: Revolución cosmológica y renovación política y religiosa*, ed. Miguel Á. Granada, (Barcelona: University of Barcelona, 2012), pp. 275–332; José Bouman and Cis van Heertum, eds., *Divine Wisdom, Divine Nature: The Message of the Rosicrucian Manifestoes in the Visual Language of the Seventeenth Century* (Amsterdam: In de Pelikaan, 2014), pp. 99–111.

11 *JKGW*, 1, 307.1–3. The actual date was apparently 1602.

FIGURE 7.2 Daniel Mögling, *Collegium Fraternitatis*, 1618. José Bouman and Cis van Heertum, eds., *Divine Wisdom, Divine Nature* (2014), p. 98.

(1552–1632) worked as the globe- and clockmaker for the Landgrave Maurice of Hesse-Kassel.[12] Bürgi communicated the news to Franz Gansneb Tengnagel (1576–1622), son-in-law to Tycho Brahe, when Tengnagel and Johan Eriksen visited Bürgi "at the beginning of 1602."[13] After leaving Kassel, Tengnagel and Eriksen informed David Fabricius in East Frisia[14] before Fabricius, in turn, informed Melchior Jöstelius (1559–1611), professor of mathematics at Wittenberg.[15] Surprisingly, although Fabricius wrote to Kepler in March urging him to collaborate with Tengnagel in order to "put the common good first, and [together] under Tychonian arms lead the exiled Urania back to her ancestral kingdom,"[16] he mentioned the nova to Kepler neither in this nor in two subsequent letters of 28 April (o.s.).[17] Kepler first received notice of the nova in a letter of 5 May 1602 from Ambrosius Rhodius (1577–1633), Jöstelius' colleague at Wittenberg,[18] who reported for certain that the nova had been observed in Prague and asked Kepler for his published *prodromum* on it or, at the very least, a private communication in a detailed letter.[19]

---

12 According to Kepler, *ibid.*, 307.21–23, the news arrived to Bürgi in Kassel from the Earl of Lippe (in Westfalia), who in turn had been informed by the architect of Prince Maurice of Nassau. In a letter to Fabricius of 18 December 1604 (*JKGW*, 15, no. 308, ll. 34–35), Kepler reported that Bürgi had taken the information from Blaeu's globe. This does not concur with the early date in the year marking Tengnagel's and Eriksen's visit to Kassel.

13 *JKGW*, 1, 309.18. On Eriksen and Tengnagel, see Christianson (ref. 3), pp. 272–273 and 366–372. On Kepler's troubled relations with Tengnagel as the representative of Tycho's heirs and the obstacles he posed for Kepler's *Astronomia nova*, see James R. Voelkel, *The Composition of Kepler's* Astronomia nova (Princeton: Princeton University Press, 2001), Chapter 7 ("The Tychonics").

14 On Fabricius, see Voelkel (ref. 13), especially Chapter 8 ("David Fabricius"), and Granada (ref. 8).

15 On Jöstelius, see Christianson (ref. 3), pp. 297 f.

16 Letter of 13 March 1602 (o.s.), *JKGW*, 14, no. 211, ll. 18–20, as translated by Voelkel (ref. 13), p. 172.

17 Letters of 28 April 1602, *JKGW*, 14, nos. 213 and 214. Kepler seems, however, to acknowledge some debt to Fabricius. See letter no. 239 of 2 December 1602 (*JKGW*, 14, l. 633): "de stella incognita nihil amplius quam tu scio, et per te rescivi" [I learnt of it from you]; letter no. 308 of 18 December 1604 (*JKGW*, 15, ll. 32–34): "De stella anni 1600 nemo plane ad me scripsit [...], omnes a me sunt admoniti, ut ego a te, tu a Byrgio." It is worth noting that in his preface to *De stella in Cygno* (in *Joanni Kepleri astronomi opera omnia*, vol. 2 (Frankfurt-Erlangen: Heyder & Zimmer, 1859), pp. 751 f.), Christian Frisch quotes a letter from Fabricius to Kepler of 16/26 August 1602 (absent from Caspar's edition in *JKGW*), in which the Frisian astronomer elaborates on the nova after first briefly mentioning it in a letter of 1 August. See below, ref. 38.

18 On Rhodius, see Christianson (ref. 3), pp. 339–340.

19 *JKGW*, 14, no. 215. ll. 18–25: "Scripsit hisce diebus noster Fabricius Frisius ad D. Jöstelium praeter caetera etiam de nova stella quam a se aliquoties in pectore cygni visam et

No doubt the news greatly surprised and excited Kepler, who immediately wrote to Herwart von Hohenburg in Bavaria, asking him to request further confirmation from other scholars.[20] In August or September 1602,[21] Kepler finally received from Herwart a letter addressed by Johann Bayer to Marcus Welser, in which the Augsburg astronomer confirmed the presence of the star and spoke of it as newly seen:

> On the new star in the Hen or Swan [...]. When I observed its figure on 7 June and some of the following days, [...] I noticed from the first moment a strange, for me suspicious little star. I remained in doubt about it until the summer solstice. Accordingly, I accurately sketched out with utmost attention the position of the Swan or Bird and the location of

observatam dicit. Procul dubio vobis qui Instrumentis abundatis, aliquid etiam de illius apparitione constabit. [...] Tuum prodromum solicitè expecto. Et si quid aliud, Literae tuae paulò prolixiores luculentissimè significabunt." In *JKGW*, 1, 308.24, Kepler attributed the information to Jöstelius: "Joestelium ad me Pragam."

20 Unfortunately, Kepler's letter is lost. We know of it only from Herwart's response of 28 June, in which the Bavarian Chancellor says that he had received no confirmation of the star. Cf. *JKGW*, 14, no. 217, ll. 16–22: "Was im anderen schreiben [of Kepler] *de nova stella circa pectus Cygni* angeregt worden, hab ich nitt underlassen etlichen *Matheseos studiosis* anzumelden, unnd hette gern gewelt solliche *Observationes* zu bekhomen die ich dem herrn zueschicken mechte. So hab ich aber sovill befunden, das bißheero kein *Mathematicus* diser landsart disen *novam stellam* observiert, biß allererst auf main beschehen anmelden, wie ich dan verhoffe in kurz dem herrn davon *Observationes* zuezuschicken, die aber gleichwoll *satis crassae, et multis nominibus mancae et defectuosae*". Among the astronomers addressed by Herwart was Michael Maestlin. Accordingly, we have found among Maestlin's manuscripts in the Württtembergische Landesbibliothek in Stuttgart (Cod. Math. 4° 14b, p. 83) the letter written by Herwart on 28 May 1602: "Clarissime Doctissimeque Vir, Domine et amice honorande. Cum oblatione offitiorum meorum, ad te scribendum esse duxi, id quod ad me quoque diversis ex locis perlatum, Exortam videlicet novam stellam, quae in constellatione pectoris του κυκνου, Cygni seu Oloris visa est. Opinor, atque adeo plane mihi persuadeo, te quoque locum eius, motum, et parallaxin, an et qualem habeat, observasse, Quod si ita se res habeat, rogo etiam atque etiam, uti me quoque observationis tuae tractu temporis certiorem facere non dedigneris, idemque facturus sum vicissim, si aliorum observationes nactus fuero. Hisce bene vale, et officiis meis utere. Datae Monachii 28 May A° 1602." It is highly probable that Maestlin did not respond.

21 We infer this from Herwart's letter to Kepler (no. 219, dated "den lesten *Julii Anno* 1602"), in which the Chancellor writes that he is sending an invalid report on the nova ("Ferrners habt ir beyverwart zu empfahen, was mir wegen *novae illius stellae* einkhomen, so ich Euch wie ubergering unnd schlecht gleichwoll die *observation* ist, dannoch nit verhalten wellen," ll. 7–9), which cannot be the one by Bayer. Herwart adds: "Ich bin von zweyen orten besserer *Observationes* gewertig, da mir die einkhomen, uberschickh ich Euchs gleichfals." There can be no doubt that one of these was Bayer's report.

> the stars within it, according to the classical doctors of astronomy, in addition to my own observation as well. I found, as usual, 17 stars with two formless ones; besides these, *I seized upon another one*, almost of the third magnitude, *at the bottom of the neck where the breast begins, no longer subject to doubt for me, but truly never seen until now.*[22]

Johann Bayer published the following year his famous *Uranometria*, where he located the star (designated as P Cygni) between the end of the neck and the beginning of the breast.

Bayer added the following commentary that conveyed where the new representation of the Crucifixion appeared:

> Star P of the third magnitude, first seen and observed in 1600 and received as a new phenomenon with the almost tacit agreement of everyone, retains the same place to this day [...]. Stars α, β, γ, δ, ε form the effigy of the cross, also known in antiquity among the peasants. Join P to these stars and you shall have the figure of the Crucified with the head inclined.[23]

The star, either a nova or an old star previously unobserved, endured for a number of years. Beyond the publication of the Rosicrucian manifestoes, it remained present in 1619, as Wilhelm Schickard stated in his manuscript treatise on the comet of 1618:

---

22 "De novo Gallinae seu Cygni astro [...]. Septimo Junii, et aliquot diebus seqq. σχεδιογραφιαν huius [...] contemplatus, ab initio statim, peregrinam quandam, mihique suspectam notavi stellulam, circa quam, ad solstitium aestivum usque, dubius haesitabam. Oloris igitur seu Avis situm, stellarumque illius positum, secundum classicos Astronomiae Doctores, mea etiam superaddita observatione, quanta potui attentione, ad unguem delineavi. Inveni secundum vulgaria 17 stellas, cum duabus informibus, praeter has *aliud in fine colli, principio Pectoris, mihi iam amplius non suspectum, sed revera nunquam hactenus visum astrum deprehendi*, magnitudinis quasi tertiae," *JKGW*, 1 ("De stella ... in Cygno"), pp. 308 f.; my italics. Bayer's report is found in Mss Pulkovo, III, 88r–89r. Unfortunately, the partial transcription by Caspar contains no date. Frisch (ref. 17, 2, pp. 755 f.) quotes from two letters of June 1602, with observations on the nova by Johann Georg Brengger and Georg Henisch (also present in Mss. Pulkowo III), which arrived in the hands of Herwart from Augsburg through Marcus Welser and were forwarded by the Bavarian Chancellor to Kepler.

23 "P. tertii fulgoris stella, Anno MDC. primùm conspecta et observata, omnium fermè tacito consensu, pro novo phaenomeno recepta, eundem adhuc hodie retinet situm [...]. Stellae α, β, γ, δ, ε Crucis effigiem, antiquitus, etiam agricolis cognitam, constituunt; his P. adiungas licet, Crucifixi inclinantis formam habebis," Johann Bayer, *Uranometria: Omnium asterismorum continens schemata, nova methodo delineata, aereis laminis expressa* (Augsburg: Christophorus Mangus, 1603), fol. 17r–18v.

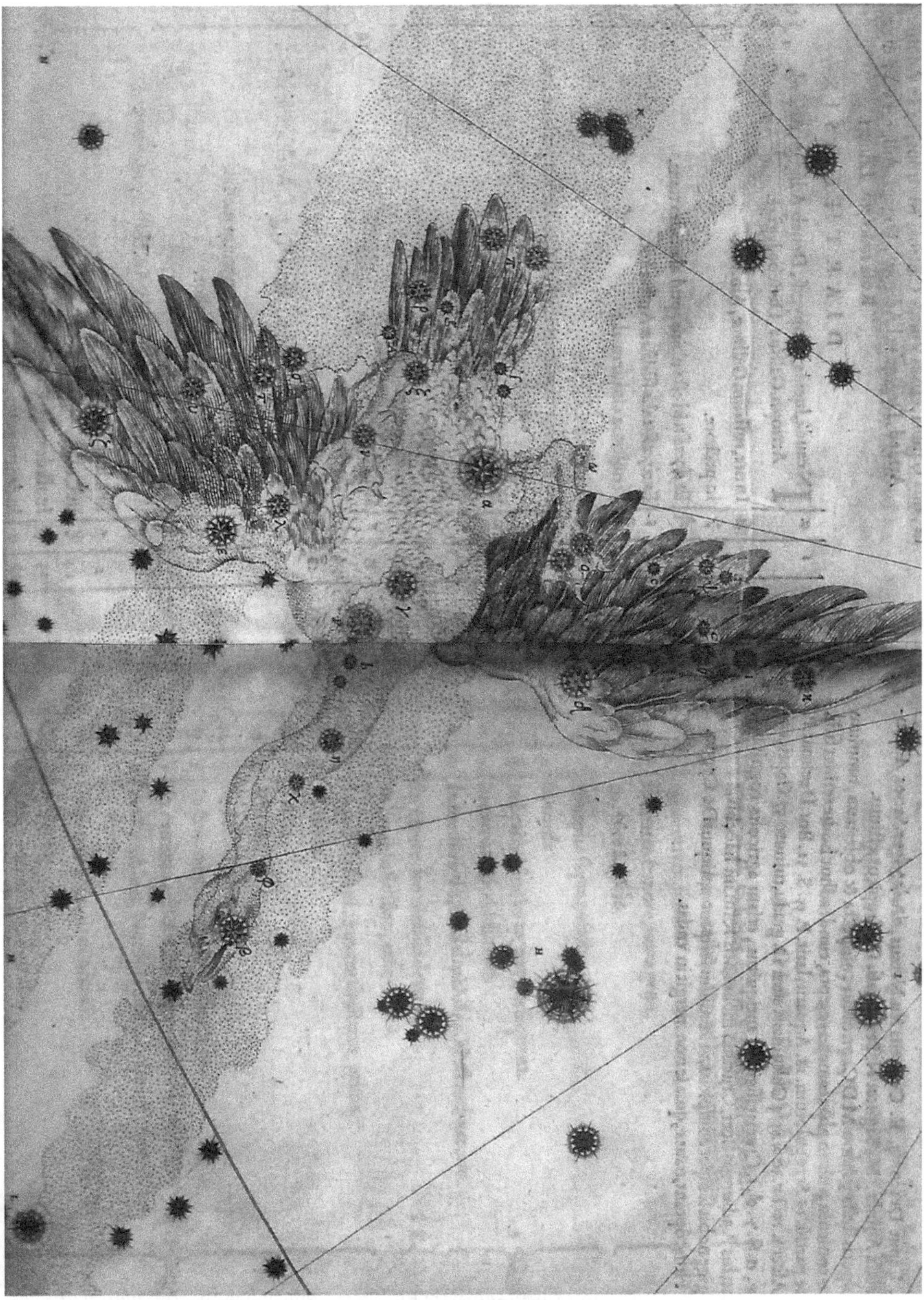

FIGURE 7.3 Constellation of Cygnus. Johann Bayer, *Uranometria* (Augsburg, 1603).
SOURCE: BAYERISCHE STAATSBIBLIOTHEK

> The third [after the novas in Cassiopeia and Cetus] appeared around the year 1600 and endures until the present day, so that one can still see it all night long in the bright sky. It is thus more than eighteen years old now and continues to possess the same mass and position as it did when it first appeared; so there is no way of knowing whether it may still one day disappear again or endure down to the last day with the other [stars].[24]

Schickard later wrote to Kepler (on 2 April 1623, o.s.) that the star had disappeared the previous winter:

> Did you observe the disappearance of the nova in the breast of the Swan? Either my eyes deceive me extraordinarily or it has altogether passed away. For it never appeared during this entire winter, though I frequently sought it with the utmost diligence through the telescope, and summoned some other persons endowed with acute vision to whom the location could be easily shown, since it formed the inclined head of the dying Christ, with the other greater stars representing the cross, according to the imagination of the vulgar.[25]

Nevertheless, in another letter from 11 June (o.s.), Schickard affirmed—contrary to Kepler's conviction, as stated in his lost response to Schickard's previous letter, that the nova was practically extinguished in 1612—that it was still

24 "Der dritte [after the novas in Cassiopeia and Cetus] ist ohngefehrlich *Anno* 1600. entstanden, und wehret noch auff den heuttigen Tag, also daß man ihn alle Nacht am hellen Himmel noch sehen kan. Ist also nun mehr uber die 18. Jahr alt, und bleibt in seiner Grössin und Ort bestendig, gleichsam wann er von Anfang da gestanden wer; also daß man noch der Zeit nit wissen kan, ob er noch künfftig wider möchte vergehn, oder gar biß an den Jüngsten Tag mit den andern wehren." Wilhelm Schickard, *Cometen Beschreibung In zwen underschidliche Partes abgetheilt, deren Erster Von denselbigen ins gemein: der Ander Von allen Insonderheit, sonderlich aber denen drey Jüngsten, In abgeloffenen 1618 Jahr erschienen, aussführlich handelt*, manuscript at the Württembergische Landesbibliothek, Stuttgart, Chapter 23, p. 65.

25 "[…] animadvertisti ne Disparitionem novae illius in pectore cygni? aut insigniter me oculi fallunt, aut prorsus evanuit. Nam totâ hac hyeme nusquam comparuit, quamvis diligentissimè per conspicilia saepius quaesierim ego, et alios visu pollentes adhibuerim, queis locus facilè indicari potuit, quòd cum reliquis grandioribus crucem repraesentantibus, inclinatum quasi caput morientis Christi, vulgari imaginatione efformaret," Wilhelm Schickard, *Briefwechsel*, vol. 1: 1616–1632, ed. Friedrich Seck (Stuttgart-Bad Cannstatt: Frommann Verlag, 2002), no. 38, p. 132, ll. 17–20. Schickard added a draft of the presumed cross with the nova (= the head of Christ) at the intersection of the two arms of the cross. A transcription is also given in *JKGW*, 18, no. 944, ll. 72–80.

visible through a telescope: "post proximas ad te literas (in quibus iam totaliter disparuisse FALSO credebam) adhibito tubulo, iterum exile vestigium eius vidi."[26] And Schickard reaffirmed this in his *Astroscopium*, published later the same year:

> And how many are there among us who observe the death of that prodigious little star in the breast of the Swan, born 23 years ago? When it first appeared in the year 1600, it matched its peers of the third magnitude; now, it seems to me that it scarcely amounts to the sixth honour, and [...] thus hastens toward total extinction. This work of God, clearly miraculous and presaging the end of the world, is most worthy of every consideration.[27]

The star reappeared, however, in 1655 and remained visible until 1662. Another outburst took place in 1665 and was followed by numerous fluctuations.[28] Since 1715, P Cygni has remained a fifth-magnitude star, with only minor fluctuations in brightness. It is therefore not a supernova (like the stars of 1572 and 1604), but a variable star, like the nova first observed by Fabricius in the constellation of Cetus in 1596.

Kepler's report appeared only in 1606. As causes for his delay, Kepler adduced that the expected extinction of the star had not come about and the mathematicians to whom he had written requesting their opinions had not responded.[29]

---

26 Schickard, *Briefwechsel*, 1 (ref. 25), no. 39, p. 133; *JKGW*, 18, no. 953. ll. 12–14.

27 "Et quotusquisque est inter nos, qui prodigiosae illius in pectore Cygni, hodie viginti tres annos natae stellulae, interitum animadvertat? quae cum Anno 1600. primum exoriretur, tertiae magnitudinis socios aequiparabat, nunc mihi vix sexti honoris esse censetur; et [...] adeoque ad totalem extinctionem properare. Quod sane miraculosum et ruentis mundi praesagum DEI opus, consideratione omnino dignissimum est," Wilhelm Schickard, *Astroscopium: Pro facillima Stellarum cognitione excogitatum, & Commentariolo illustratum*, s. l., 1623 (we quote from the 1698 edition, Stuttgart and Leipzig: Johann Herebord Klosius), Preface, pp. 7–8. See also *ibidem*, § 17, pp. 27–29. Though possibly a not very reliable observer, Julius Schiller affirmed in his *Coelum stellatum christianum* (1627) that he had seen the star several times in 1625 and in June and August 1626. See below, ref. 107.

28 See Michael Hoskin, "Novae and Variables from Tycho to Bullialdus," in *Idem*, *Stellar Astronomy: Historical Studies* (Chalfont St Giles: Science History Publications, 1982), pp. 22–28.

29 Kepler explicitly mentions Blaeu in *De stella nova*: "Quid verò literis meis acciderit, ignoro: responsum quidem hactenus ab illo accepi nullum," *JKGW*, 1, 307.11–13. Cf. Kepler's letter to Blaeu in *JKGW*, 15, no. 282 (letter of 7 February 1604), ll. 1–11: "Certiorem te reddo, Caesaream Maiestatem cupere de incognita Cycni stella certi quippiam edoceri. [...] in mandatis accepi uti ad te scriberem, teque per amorem veritatis [...] certiorem ut redderes

Moreover, the nova of 1604 had appeared, demanding for obvious reasons his complete attention, while the permanence of the star in Cygnus still begged the question of its novelty.[30] Finally, the report was published, closing the first part of *De stella nova* and accompanied by a dedicatory letter to Baron Johann Friedrich Hoffmann, who had generously provided Kepler with the quadrant and sextant he had used to observe the star while Brahe's instruments remained unavailable "by some inextricable fate."[31] In this way, Kepler kept his promise to his patron to dedicate to him the report he had in fact completed in 1602.[32]

7.3 Kepler's Defence of the Novelty of the Star in Cygnus

The central concern regarding the nova in Cygnus and the most salient subject of dispute (as Kepler always acknowledged) remained whether it was a nova or an ancient star originally created by God with the rest of the heavens, never identified nor observed until the present time. Secondarily, and only if the star was in fact identified as a nova, did the issue of its significance come under consideration. As a result, Kepler's main aim in the report was to settle the dispute with empirical evidence and reasons arguing sufficiently for the novelty of the star.

If the majority of astronomers were silent or uncertain, Kepler knew two who inclined to interpret it as an original star. In *De stella nova*, Kepler only mentions Michael Maestlin (1550–1631), but through his correspondence he also heard from David Fabricius. As Kepler summarized the arguments his

---

Caes. Maiestatem, an verè tibi sis conscius, aliquando illam abs te non esse visam; an verò solummodo quibusdam argumentis, coniecturis et fide tabularum nitereris: et quo die, qua occasione primum omnium in eam incideris."

30 *JKGW*, 1, 295.16–26: "Nimirum finem exspectare volebam novi Phaenomeni [the nova of 1600]; intereaque ad praecipuos Mathematicos scripsi, collecturus eorum suffragia. Cum autem neque stella extingueretur, nec ea promptitudine responderent Mathematici, quam sperabam: tempus interea lapsum est; donec post biennium et amplius, ingens illa nova stella, qua de hoc libello [*DSN*] egi, effulgeret in pede Serpentarii: cuius miraculi magnitudo, cum luculentam aliquam disputationem poscere videretur: aequum videbatur, ut *parva Cycni stella, novitatis non adeò confessae*, sicuti claritati cedebat nuperae illi in Serpentario, sic locum etiam ei primum in libro, disputationisque capita concederet: ipsa si quid haberet peculiare, cognituque dignum, eius explicationis locum in libri calce nancisceretur"; my italics.

31 In his words, "quodam inextricabili fato," *ibid.*, 294.16. The reason was indeed the problems so frequently posed by Tycho's heirs. *Ibid.*, 294.24–29.

32 *Ibid.*, 294.29–31: "Quae itaque in hac stella hisce tuis instrumentis observaveram, quaeque de ea dicenda essent Astronomo; *illo anno* [1602] *perscripta*, et paucis aliquibus communicata, cogitabam edere, Tuoque Nomini dedicare"; my italics.

teacher at Tübingen had shared with him in a letter of January 1605,[33] Maestlin suspected that he had previously seen the star despite some uncertainty and no observational evidence. As a result, he suggested that it was not a nova, as it was inferior in magnitude to the novae of Cassiopeia and Serpentarius, as well as longer lasting. The fact that the star was absent from the received star catalogues (that of Ptolemy and the more recent tables in Tycho's *Progymnasmata*) was no obstacle, as it was well known that the catalogues omitted many stars, even those with a magnitude similar to that of the star in Cygnus.[34] Among the examples given by Maestlin, Kepler only mentioned the one near the second star in the tail of Ursa Major.[35]

Fabricius, who was initially silent on the star and later conducted an intense exchange with Kepler on the nova in Serpentarius,[36] expressed an increasing tendency to interpret the star in Cygnus as an old star, even if he had been the first observer in 1596 of another nova, the similarly variable star 'mira Ceti.' In his first account (1 August 1602), Fabricius told Kepler that he had originally seen the nova in March and observed neither the slightest hint of change nor measure of parallax over the following months.[37] In subsequent letters,

33 Maestlin remained silent until 1605. In a letter to Fabricius of 4 July 1603 (*JKGW*, 14, no. 262, ll. 215–217), Kepler reported: "Mestlinus mussat, post omnes machinas, quibus ab ipso responsionem flagitavi, tacet iam triennio. Puto nihil illi constare." In this same letter, however, Kepler proudly states (ll. 1046–1047) that "Stella Cygni in globos iam irrepsit novae titulo observatore Jansonio, qui anno 1600 visam ait." See ref. 1 above and Fig. 7.1.

34 *JKGW*, 1, 297.40–298.4: "Maestlinus quidem, confessus ambigere se, inclinat tamen eò, ut, quò diutius suam examinet memoriam; hoc magis putet, sibi etiam olim visam. At statim subiicit argumenta [...] scilicet, quod duret tantos annos; quodque Ptolemaeus multas fixas omiserit." For the more extended argument by Maestlin, see his letter of 28 January 1605 in *JKGW*, 15, no. 322, ll. 56–74. Maestlin concluded, however, that he was prepared to give up in the face of more accurate observations ("certioribus tamen observationibus nihil detractum velim, quibus lubens cedam").

35 *JKGW*, 1, 299.24–27: "Hic Moestlinus producit stellulam, quae super mediam trium in cauda Ursae maioris visitur, Equitatorem seu Aurigam dicunt agricolae, quasi qui super medio equo sedeat: quae etsi non contemnendae magnitudinis, omissa tamen est a Ptolemaeo et Hipparcho." Maestlin means Alcor (80 Ursae Maioris), which forms with its companion star Mizar (ζ Ursae Majoris) a binary system. Kepler responded (see the following lines on the same page) that the vicinity of these stars was the cause for Alcor only being visible by those with very acute eyesight, whereas the nova in Cygnus and that in the breast were of the same magnitude and sufficiently distant from one another to be clearly distinguished.

36 On this interchange in correspondence as well as in *De stella nova*, see Granada (ref. 8).

37 David Fabricius to Johannes Kepler, 1–5 August 1602, *JKGW*, 14, no. 221, ll. 560–563: "Vidi illam [novam stellam] primo in Martio, et nullam varietatem in illa animadverti hactenus, nullam parallaxin habet, quia distantiae sextantis invariati semper eaedem in locis

however, Fabricius expressed a growing preference for identifying it as an ancient star: "I am strongly perplexed by the new star. If it is truly a nova, why does it endure so long with the same magnitude? This was not observed in the nova of Cassiopeia, since it decreased gradually, until it finally vanished;"[38] "the new star (if it truly was a nova) still appears at the same location in which I first saw it;"[39] "about the new star, I am very doubtful whether it is truly a nova, even if it may not be found on any globes or tables [...]. Surely, if it were not a nova our imagination would have deceived us terribly and earlier astronomers would have acted negligently for not having noted this sufficiently clear star, whereas they had observed the unformed stars around this location that are much smaller."[40] Finally, in a letter of 7 May 1603, Fabricius shared his opinion with Kepler that all novas and comets were obscure stars, originally present in the heavens from the first creation and miraculously illuminated by God at precise moments in order to signify something important to mankind.[41]

---

extra refractionem. De tempore primae apparitionis nihil certi hactenus expiscari potui." In the letter of 16/26 August quoted by Frisch (ref. 17) but absent from the edition by Caspar, Fabricius wrote more broadly: "Invenio eam [stellam] nunc 4. magnitudinis stellis vix comparari posse, quae antea in Martii initio tertii honoris sidera nonnihil etiam superare videbatur. [...] Notabile est, quod nova haec stella constituta sit in forma crucis cum praecipuis in Cygno stellis. [...] color rutilus ostendit eam Martialis naturae esse. Forma tamen crucis ostendit magnam spem gratiae divinae et liberationem, propter filium Dei in cruce passum. Sed de huius stellae apparitione, loco, motu, forma, effectibus, publico scripto per Dei gratiam suo tempore agere constitui, ubi et alterius cuiusdam novae stellae (quae anno 96. [...] apparebat) et a me observatae descriptio dabitur, nec non meas de futura magna coniunctione ♄ et ♃ in ♐ cogitationes et observationes addam."

38 "De nova stella valdè perplexus sum. Si verè nova est, cur tam diu perduret eadem magnitudine, quod in nova apud Cassiopeam non fuit observatum. Decrevit enim illa paulatim , donec tandem evanesceret." David Fabricius to Johannes Kepler, 24 September 1602, *JKGW*, 14, no. 227, ll. 25–27.

39 "Nova stella (si modo verè nova fuerit) adhuc apparet in eodem quo primum vidi, loco permanens." David Fabricius to Johannes Kepler, 4 November 1602, *JKGW*, 14, no. 233, ll. 30–31.

40 "De nova stella plurimum dubito, an verè nova, etsi in nullis globis vel tabulis inveniatur [...]. Certè si non esset nova, miserè decepisset nos nostra imaginatio, et negligenter fecissent priores astronomi, quod satis claram stellam non notassent, cum informes multò minores circa illum locum observarint." David Fabricius to Johannes Kepler, 8 December 1602, *JKGW*, 14, no. 240, ll. 125–129.

41 David Fabricius to Johannes Kepler, 7 May 1603, *JKGW*, 14, no. 257, ll. 42–48: "Ego puto novas stellas qualis illa, quae anno 72 fulsit, erat, et hodierna in Cygni pectore esse putatur, item cometarum corpora, non tunc primum fieri incipere, quò ad essentiam, cum se nobis conspiciendas praebent; sed in ipso mundi primordio una cum caeteris stellis conditas esse, et ab illis sola lucis carentia se distinxisse. Ratio est, quod illis extraordinarie Deus ad aliquid significandum, certis temporibus uti constituerit." Cf. *ibidem*, 24 June 1603,

For his part, Kepler wrote Fabricius in February 1604 stating that, "moved by trustworthy arguments, I have no doubt that the star is a nova."[42] Numbering ten, the arguments appeared in his report according to the following synthesis:[43]

1. The nova was absent from the star catalogues of Hipparchus and Ptolemy, nor was it mentioned in the description of Cygnus in Hipparchus' commentary on Aratus' *Phaenomena*.[44] It was also absent from both catalogues by Tycho Brahe,[45] the first comprising 777 stars reckoned to 1589 (closed in 1592, but printed in 1602 in the *Progymnasmata*) and the second, increased to 1,000 stars in 1595–1596 and printed by Kepler in the *Rudolphine Tables* (1627). The 200 additional stars, omitted from the earlier catalogues, were all in the northern hemisphere, many of them of the fifth or sixth magnitude. In Cygnus, the new catalogue added eight stars (five of which were of the sixth magnitude),[46] but not the nova of 1600 near the breast. We can thus understand the following comment by Kepler:

   > Do you actually believe that while he [Brahe] attended to and examined each of the fixed stars, especially the breast of the Swan, he would have let this nova, nearest to it and of the third magnitude, silently go without mention if he would have seen it? [...] It is therefore clear that this modern star was not seen by Tycho.[47]

JKGW, 14, no. 261, ll. 163–176. This is the opinion Fabricius later affirmed when the nova of Serpentarius appeared. For Kepler's criticism of this opinion, see Granada (ref. 8).

42 Johannes Kepler to David Fabricius, February 1604, *JKGW*, 15, no. 281, ll. 212–213: "Novam esse nil dubito, motus argumentis fide dignis."

43 I present the arguments in a rather systematic and very abridged form.

44 *JKGW*, 1, 298–299. Giovanni Battista Riccioli (1598–1671), who relied mainly on Kepler's *Narratio* for his report on the nova in his *Almagestum novum* (1651), wrongly blamed Kepler for confusing the Hipparchus from Bithynia, author of the commentary on Aratus (his only preserved work), with the astronomer Hipparchus of Rhodes. Cf. *Almagestum novum* (Bologna: Vittorio Benacci, 1651), vol. 2, p. 167a: "Hipparchus Bithynus (quem tamen per imperitiam Keplerus hoc loco eundem facit cum Hipparcho Rhodio)."

45 *JKGW*, 1, 301. For Tycho's catalogues, see Victor E. Thoren, *The Lord of Uraniborg: A Biography of Tycho Brahe* (Cambridge: Cambridge University Press, 1990), pp. 294–300, 383, and the still useful J. L. E. Dreyer, *Tycho Brahe: A Picture of Scientific Life and Work in the Sixteenth Century* (Edinburgh: A. & C. Black, 1890), pp. 227, 265, 347 ff.

46 Cf. *Tabulae Rudolphinae*, in *JKGW*, 10, 109–110.

47 *JKGW*, 1, 301.14–26: "An putas illum [Brahe], dum singulas fixas, adeoque et pectus Cygni obit et lustrat; vicinissimam hanc novam, tertiae magnitudinis, incommemoratam et tacitam praetermissurum fuisse, si illam vidisset? [...] Clarum itaque est, non visam esse Tychoni hanc modernam."

2. The nova was also absent from the accurate description of the Milky Way and its partition in two branches, precisely at Cygnus, given by Ptolemy in the *Almagest*, Chapter VIII, 2.[48] Ptolemy describes with utmost precision the path of the Milky Way through Cygnus, the margins of the milk and the positions of the stars inside and outside it. It is impossible that the new star, positioned precisely at the point where the two branches fork (see the image of the Swan in Bayer's *Uranometria* in Fig. 7.3 above), could have escaped the very accurate observation of Ptolemy: "But if this star appeared at that time in the Milky Way, and indeed close to the margin as it does today, do you believe that it could happen that Ptolemy overlooked it?"[49]
3. No mention of the star appears in the records, now in Kepler's hands,[50] of the many star observations made by Brahe and his assistants between 1582 and 1597. *Pectus Cygni* (γ Cygni) is recorded many times, with its distance from other stars measured, as the only star in the breast with no others in the vicinity. Since the distinction between both stars should have been made in order to avoid confusion, its absence allows us to conclude that the nearby nova was not present in the heavens at that time.[51] It is also clear from the records that several small stars outside the Swan, omitted by Ptolemy, were discovered and incorporated into the catalogue. Arguably, then, the nova (similar in magnitude to the star in the breast) had been absent from the heavens.[52]
4. An observation by Brahe of the comet of 1580 furnishes Kepler with a fourth argument. In his description of the observation, Brahe states that the line connecting the comet with the head of Hercules stretched forth "to the middle star in the body of the Swan, as far as can be seen."

48 See Ptolemy, *Almagest*, translated and annotated by G. J. Toomer, with a foreword by Owen Gingerich (Princeton: Princeton University Press, 1998), pp. 400–404.

49 *JKGW*, 1, 301.4–7: "Quod si ut hodie, sic iam tunc, haec stella in via lactea, et quidem proxime Marginem apparuisset: fierine potuisse credes, ut Ptolemaeus illam omitteret?"

50 *Ibid.*, 301.27–29: "Sextò, versatur sub meis manibus Protocollum observationum Tychonis (sic appellavit Chaos observationum, uti illae singulis noctibus conscriptae fuere sine ordine) [...]"

51 *Ibid.*, 302.2–5: "Quod si tunc, uti iam, geminae fuissent in pectore, aequalis propemodum magnitudinis: an non seipsum confudisset, neglecta distinctione: an non ipsa haesitatio, utra sit vera pectoris, ipsum ad Ptolemaei abacum invitasset inspiciendum?"

52 *Ibid.*, 304.38–305.1: "Rursum igitur vides, non omissas circa Cygnum informes, si quae sunt a Ptolemaeo praeteritae. Cur ergo haec omissa, si tunc fulsit? Minimè gentium omissuri erant stellam superfluam tertii honoris, si tunc fulsisset."

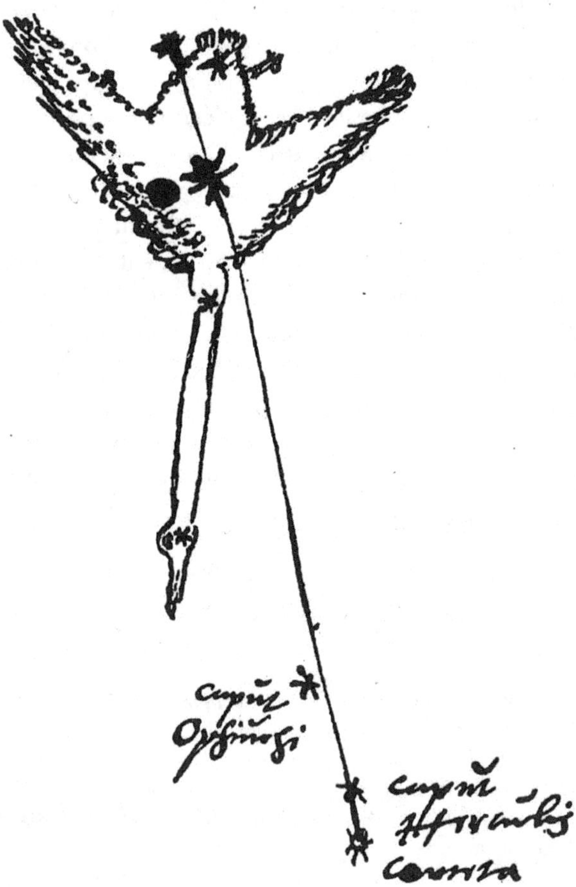

FIGURE 7.4 Tycho's figure of the comet of 1580 with respect to Cygnus. *TBOO*, 13, 324.

Accordingly, Kepler says that only one star was visible in the breast of Cygnus, and this is confirmed by the figure drawn by Brahe.[53]

5. To conclude, Kepler appeals to contemporary witnesses. Besides the above mentioned Blaeu (the first discoverer) and Bayer, he summons Justus Bürgi, the clockmaker of Emperor Rudolph II from 1604 and of

53 *Ibid.*, 306.29–33: "[Tycho] pergit: 'Videbatur autem ea linea, per Cometam in [...] caput Herculis, protendi (ubi producta esset) in eam quae media est in corpore Cygni, quoad visum.' Si duae tunc fuissent: discrevisset, quia addit, 'quoad visum.' Haec ibi. Additur et pictura, ubi solitariè pingitur media in corpore." The passage quoted by Kepler corresponds to the observation of 25 November, in *TBOO*, 13, 324 f. (see Fig. 7.4).

Landgrave Maurice of Hesse at Kassel when news of the nova in Cygnus first arrived from the Netherlands in 1601.[54] In 1592, Bürgi had produced for Landgrave Wilhelm of Hesse-Kassel (1532–1592) a planetary clockwork that incorporated a stellar globe in silver indicating the positions and magnitudes of the stars as established by the court astronomer Christoph Rothmann in earlier years.[55] Although this globe is lost,[56] we know from Kepler's testimony that Bürgi turned to it after first learning of the appearance of the nova and attempted to determine its age. Bürgi confirmed that the new star was not inscribed in the globe (where the stars and their positions were represented in accordance with the Kassel star catalogue elaborated by Rothmann),[57] and marked it with ink. Some years later, already in Prague, the repair of the globe (previously sent as a gift to the Emperor in Prague) gave Bürgi the opportunity to verify that the globe only showed the old Ptolemaic star in the breast. "Thus," Kepler concludes, "Bürgi confidently declares that it is truly a nova."[58]

54 On Bürgi, see Ludolf von Mackensen, *Die erste Sternwarte Europas mit ihren Instrumenten und Uhren: 400 Jahre Jost Bürgi in Kassel* (Munich: Callwey, 1979).

55 On the programme of stellar observation at Kassel conducted by Christoph Rothmann (ca. 1560–1601) in collaboration with Bürgi, see Jürgen Hamel, *Die astronomischen Forschungen in Kassel unter Wilhelm IV*, Acta Historica Astronomiae, 2 (Frankfurt: Harri Deutsch, 1989); Miguel A. Granada, Jürgen Hamel and Ludolf von Mackensen, *Christoph Rothmanns Handbuch der Astronomie von 1589: Kommentierte Edition der Handschrift Christoph Rothmanns "Observationum stellarum fixarum liber primus," Kassel 1589*, Acta Historica Astronomiae 19 (Frankfurt: Harri Deutsch, 2003).

56 On this globe, see John H. Leopold, *Astronomen, Sterne, Geräte: Landgraf Wilhelm IV. und seine sich selbst bewegenden Globen* (Lucerne: Fremersdorf, 1986), pp. 175–176.

57 Though it was unknown to Kepler, it is interesting to note that the nova was also absent from the very accurate star catalogue compiled at Kassel by Christoph Rothmann accounting for 383 stars. See the luxurious presentation copy in 2° Ms. astr. fol. 7 at the Murhardsche Bibliothek Kassel, fol. 12, where 17 stars are registered with the old star "In pectore" in the fourth place. (We are grateful to Jürgen Hamel for providing a copy of this text). No doubt this would have further confirmed for the Imperial Mathematician the novelty of the star near the one "in pectore." See also Jürgen Hamel, "Die Begründung der modernen astronomischen Beobachtungstechnik an der Kasseler Sternwarte im 16. Jahrhundert: Der Kasseler Sternkatalog 1586—eine Fehleranalyse der ihm zugrunde liegenden Positionsmessungen," in *Nouveau ciel, nouvelle terre: La révolution copernicienne dans l'Allemagne de la Réforme, 1530–1630*, ed. Miguel Á. Granada and Édouard Mehl (Paris: Les Belles Lettres, 2009), pp. 67–84, on pp. 71–73.

58 *JKGW*, 1, 307.14–308. 9.

The above arguments allow Kepler to conclude that "the novelty of this star is sufficiently proven; and he for whom these are not worth believing might as well doubt the testimony of all mankind."[59] At this moment in 1606, and before the nova in Cetus was seen for a second time by Fabricius in 1609, astronomers contemplated only the possibilities of an original star present in the heavens from the creation of the world or a total nova appearing for the very first time. In any case, the subsequent discovery that the presumed nova was in fact a variable star, which might have intermittently appeared in the past, changed nothing with respect to the central concern: the repeatedly verified mutability of the heavens and, as a consequence, the crisis of the classical view of the cosmos.

7.4 The Christianization of the Constellations

While Kepler's vision of the crucified Christ, with the nova appearing as his heart, clearly cannot be credited as the sole cause of ensuing claims of the Christianization of the stellar vault,[60] it probably served to stimulate some in Germany. Let us briefly examine four proposals of this kind that were presented in Germany after 1606, the first three by authors directly connected to Kepler and all of them familiar with his *De stella nova*.

The first is Helisaeus Roeslin (1545–1616) and his widely ranging treatise *Mitternächtige Schiffarth* (1611). Roeslin was closely familiar with Kepler's *De stella nova*, as witnessed by his polemic with the work in the *Historischer, Politischer und Astronomischer naturlicher Discurs* (1609). It was in the *Discurs* that Roeslin first presented his views on the nova.[61] The more miscellaneous

59 *Ibid.*, 309.29–31: "Satis opinor contestatam esse Novitatem huius stellae: nam si cui haec non sufficiunt; illum credibile est, ne totis quidem gentibus affirmantibus, fidem habiturum."

60 There had been many attempts in this vein ever since the Church Fathers, as documented by Wolfgang Hübner, *Zodiacus Christianus: Jüdisch-christliche Adaptationen des Tierkreises von der Antike bis zur Gegenwart* (Königstein: Hain, 1983).

61 Helisaeus Roeslin, *Historischer / Politischer und Astronomischer naturlicher Discurs Von heutiger zeit Beschaffenheit / Wesen und Standt der Christenheit und wie es ins künfftig in derselben ergehn werde / Aus anleitung dero von Anno 1600. her am hohen Himmel erschienenen grossen Wunderzeichen / sonderlichen dess Cometens anno 1607. genommen* (Strasbourg: Conrad Scher, 1609). On Roeslin's revealing exchange with Kepler after 1606, see Miguel A. Granada, "After the Nova of 1604: Roeslin and Kepler's Discussion on the Significance of the Celestial Novelties (1607–1613)," *Journal for the History of Astronomy*, 2011, 43: 353–390, on pp. 371 f. on the nova in Cygnus.

Mitternächtige Schiffarth deals with many issues, among them the question of the Northeast Passage to China, sought by Dutch sailors and referenced in the title. A further subject, stimulated by the recent discovery of the four Medicean planets and published by Galileo in his *Sidereus nuncius* (1610), involved the novas. First came the implications of the nova in Cygnus.

According to Roeslin, Cygnus (also known as Gallina and Crux) represents and reveals Christ as well as the state of the Christian religion. This is clear from Matthew 23:37, where Christ says to Jerusalem: "how often would I have gathered thy children together, even as a hen gathereth her chicks under her wings," and also from the fact that, like a dying swan singing for the final time, Christ uttered from the cross his last seven salubrious words before he died (Luke 23:46). In addition, the constellation was customarily called the Cross in Germany, since the Milky Way depicts the vertical trunk of the cross between the stars in the beak and the tail, while the bright stars in both wings and the breast draw the horizontal arm of the cross.[62] Now, the appearance of the nova, representing the heart of the swan, casts the figure of the Crucifixion. Roeslin recalls that this had been anticipated by Kepler in *De stella nova*, with the nova representing the heart of Christ on the cross that was previously deprived of his presence.[63] Roeslin thus transforms Kepler's 'fancy' into an actual presence of Christ in the heavens, affirming that he had *seen* this before reading Kepler's work. Roeslin further adds that the bright stars in the wings (ε, δ in Bayer's *Uranometria*) represent Christ's hands nailed to the Cross, the star in the beak (β) his feet and the one in the breast (γ) his head, while the Milky Way provides a linen cloth covering the body of Christ as far as his feet.[64]

This new, "clear image" ("Solch scheinbar Bild"), formed by the four ancient stars and the novel fifth one, signifies to Roeslin the condition of the various

62 Helisaeus Roeslin, *Mitternächtige Schiffarth Von den Herrn Staden inn Niderlanden vor XV. Jaren vergebenlich fürgenommen, wie dieselbige anzustellen, daß man daselbst herumb in Orient und Chinam kommen möge, zu sonderen der Christenheit, sonderlich Teutschlands Nutzen und Wolfart* (Oppenheim: Hieronymus Gallart, 1611), pp. 98–99.

63 *Ibid.*, p. 99: "der newe Stern denen in Flügeln und Bauch an der grösse fast gleich [...] und sich an das Herz des Schwanens gesetzet/ nicht weit von diesem der im Bauch des Schwanen stehet/ da gibt es ein Gestalt/ als wann der *Salvator* am Cretuz hanget/ dann zuvor war es ein Creutz ohn ein *Salvator*, jetzt aber ein Creutz mit dem *Salvator*, wie dann *Dominus Keplerus* solches in seinem Tractat auch melden thut."

64 *Ibid.* pp. 99 f.: "Da dann der newe Stern das Hertz des *Salvators* gibt/ der im Bauch des Schwanens gibt das Haupt/ so sich etwas zu der Seiten neigt/ die zween zu eusserst in den Flügeln geben beyde Händ mit ausgespannen Armen an das Creutz gehefft/ die weisse Stras des Himmels gibt ein Gestalt eines Leinwats/ damit der Leib biß zu den Füssen umbwickelt ist/ der Stern in *cauda Cygni* stehet zu oberst am Stammen des Creutzes."

Christian Churches and confessions at the present time as well as the near future. In a first approach, the four surrounding stars are taken to represent respectively the Roman Church (*pectus Cygni* / the head), the Christian church of the mythical Prester John in Aethiopia (the star in the left wing / right hand), the trinitarian Protestants in Northern Europe (the star in the right wing / left hand), and the Greek Orthodox Church in the East (the star in the beak / in the feet of Christ). The other stars around the cross represent heathen religions, while the brightest star (α; cauda Cygni / the top of the cross) represents Islam as an antitrinitarian religion that accepts Christ as a prophet.[65]

In a second approach, however, Roeslin focuses on the German empire as a universal monarchy, with the four stars around the nova identified as the Roman Church (the head, *pectus Cygni*), Calvinism (left hand), Lutheranism (right hand) and Anabaptists, and the Antitrinitarians ("die neuen Arrianer") represented by *cauda Cygni*.[66] Like the hen gathering her chicks under her wings, the nova-heart of Christ in the middle represents Christ's calling to universal tolerance and freedom of religion as the sign of the new era opened by the fiery trigon:

> In the middle of the four stars found in the figure of the Swan, the new star that appeared in the heart of the Swan or in the very heart of the Saviour shows that He is God to all and they must find unity in Him; from Him, just as from the heart, the force flows outward to all the members [...] and can be seen from it how outrageous it is when one part afflicts the other with hate and condemnation for reasons of faith;[67]
>
> Just as the figure of the Saviour remains firm with the four stars in the sky above and does not fail, so must the four religions of Christendom endure and complete their course below in the world in the Empire according to this and no other way. The fifth star, newly introduced in the very heart of the Saviour, shows that Christ the Lord lies at the centre of

65 *Ibid.*, pp. 100–102.

66 *Ibid.*, pp. 104–106.

67 *Ibid.*, p. 103: "die vier Sternen im Bild des Schwanens fürgetragen/ zwischen welchen sich mitten eingesetzt hat der newe Stern im Hertzen des Schwanens/ oder auch im Hertzen des *Salvators*, anzuzeigen/ daß er aller gott seye/ und in ihme sie müssen vereiniget werden/ von ihme als vom Hertzen der Außfluß und Krafft gehe in alle Glieder [...] und darauß zu sehen ist/ wie so freffelich Stück es seye/ wann einer den anderm des Glaubens halben hassen und verdammen thue."

> the matter and wishes to act where there is freedom of religion, not bondage or compulsion.[68]

Roeslin claims that this new configuration of the constellation, as well as its meaning, have found confirmation in Galileo's recent discovery of the four planets revolving around Jupiter. In accordance with the astrological role of Jupiter as a symbol of religion, the four moons around it signify the same as the four original stars around the nova in Cygnus, that is, the four parties in Christendom and in Germany.[69] Roeslin concedes the possibility that, just as Galileo and Kepler hold, these new planets are as old as the creation but have only been seen for the first time thanks to the magnifying power of the telescope. If, as Kepler thinks, further secondary planets should be circling Saturn, Mars and Venus, their observation would confirm the 'old' character of the Medicean planets.[70] Otherwise, Roeslin could confirm his own hypothesis that the four Medicean planets were (similar to the nova in Cygnus) recently created by God and produced precisely at the moment when Galileo directed his telescope at Jupiter:

> Yet should no new star be found around the other planets, I wish to consider the ones around Jupiter as entirely new and first placed in the heavens by God in our own time, just as those novelties in Ursa and Cygnus.[71]

In any case, Roeslin finds that the appearance of the four moons of Jupiter at that time reinforces the meaning of the nova as the heart of Christ at the center of the Christian confessions. In response, he calls for nothing less than the closing of the period of religious persecution and the start of a new era of "religious freedom" (*Freiheit der Religion*). In fact, Roeslin concludes, when

68 *Ibid.*, p. 110: "Dann das Bild deß *Salvators* mit den vier Sternen oben im Himmel das stehet feste und fehlet nicht/ So müssen die vier Religionen der Christenheit unten in der Welt im Reich seyn/ und ihren Lauff volbringen/ dieses und kein anders nicht/ und hat sich der fünffte und newe Stern deß Hertzens deß *Salvators* mitten eingesetzet/ anzuzeigen/ daß der HERR Christus mitten im Handel sey/ und wircken wölle wo Freyheit der Religion sey/ nicht wo Zwang und Trang sey."

69 *Ibid.*, p. 118.

70 *Ibid.*, p. 119.

71 *Ibid.*: "So sich aber umb die ubrige Planeten kein newer finden solte/ so will ich diese umb den *Iovem* darfür halten/ daß sie gantz new seyen/ und zu unsern Zeiten erst von Gott an Himmel gestelt seyen/ wie diese newe in Ursa und Cygno."

God aims to inform mankind about novelties in religion and the Church, He does so through Jupiter and the Swan.[72]

If Roeslin limited his Christianization of the stellar vault to the constellation of Cygnus, the proposal of Wilhelm Schickard (1592–1635) was far broader and bolder. Schickard presented his sweeping vision in both a manuscript treatise on the comet of 1618 and in his published *Astroscopium* (1623). He probably knew the works of Roeslin, who had graduated earlier than him from the University of Tübingen. Schickard mentions Roeslin in his manuscript treatise several times,[73] but only refers to Roeslin's cometary theory as expressed in his *Tractatus meteorastrologiphysicus*.[74] Nevertheless, Schickard most probably knew Roeslin's ideas about the Swan, since they were already expressed in his widely diffused *Discurs* of 1609. (Kepler had strongly criticized this work in his *Antwort auf Röslini Discurs* from the same year.) Schickard deals with the recent novas in a section of his cometary treatise (§ 23), where he discusses the novas in Cassiopeia, the Whale, the Swan and the Serpent Bearer, illustrating his views with magnificent aquarelles. Two of the paintings depict the nova in the Swan and its transformation into the Crucifixion.[75] After mentioning that the star is still present in the sky nineteen years after its initial appearance with the same (third) magnitude, Schickard continues with a long passage worth quoting in entirety:

> The star appears in the breast of the Swan, just as the ancient Chaldeans and the inventors of astronomy distributed the stars. This distribution is not kept today through skill or subtlety but from age and antiquity alone,

72 *Ibid.*, p. 120.

73 Wilhelm Schickard, *Cometen Beschreibung In zwen underschidliche Partes abgetheilt, deren Erster Von denselbigen ins gemein: der Ander Von allen Insonderheit, sonderlich aber denen drey Jüngsten, In abgeloffenen 1618 Jahr erschienen, aussführlich handelt*, 1619 (preserved at the Württembergische Landesbibliothek in Stuttgart, Cod. Math. 4° 15b, Nr. 8), pp. 27 f., 41, 84, 107, 111, 132.

74 Helisaeus Roeslin, *Tractatus meteorastrologiphysicus. Das ist, Aus richtigem lauff der Cometen / zusammenleuchtung der Planeten / etlicher Herrn Nativiteten / Natürliche Vermütungen und eine Weissagung. Von Veränderungen der Regimenten in Königreichen unnd Fürstenthumben der Welt und andern wichtigen Wunderlichen / Politischen / Weltlichen unnd Kriegssachen / So sich Kurtz künfftig innerhalb wenig Jaren begeben werden / von jetzt an biß 1604. nach Christi Geburt unnd etwas hernach / Allen Christlichen Potentaten unnd hohen Oberkeiten unnd jedermenniglichen zur Warnung unnd Auffmunterung* (Strasbourg: Bernhard Jobin, 1597).

75 See Schickard (ref. 73), pp. 64–67, with the images of the Swan and the Crucifixion on pp. 65 f.

for it is known that such a distribution of the sky through figures was already in use before the times of Job and Isaiah. Accordingly, I wished to assign it another figure in the form of the Crucified Saviour, since the five brightest stars of the Swan clearly convey a cross, as one evidently sees if both figures are compared. This is especially apparent in the heavens themselves when the Cross appears directly overhead between sunset and the Great Bear, something that not only farmers, shepherds and watchmen realised and recognised, but also several more recent astronomers who, to commemorate the great act of our salvation, make it out to be a crucifix. To this in particular the new star certainly makes its own contribution, for it does not appear precisely in the middle of the wooden beams but a bit to the side, giving the impression of a downcast head. Thus, I do not wish to draw it to the superstitious interpretation shared by our superstitious astrologers, especially since paganism displeases me so profoundly and I consider us Christians a great disgrace for still holding to the pagan figures of Hercules, Erichtonius, Perseus, Andromeda, Cassiopeia and others, as well as the fabled Pegasus and so many more, while a God-loving Christian heart should curse and eradicate the coarse commemoration of everything taken from their fanciful Merlin's Bible and reeking of the long accursed Ovidian *Metamorphoses*. For this reason, I have often wished for a Christian Reformation and even sought it by my own person from the history of creation, the Apocalypse of John, and other things, hopefully not without success. The matter does not depend, however, on the private judgement of a single person but must be made unanimously by popular consensus of all astronomers should it otherwise be approved for use.[76]

---

76 *Ibid.*, pp. 65 f.: "[the star] steht in der Brust des Schwanens, wie die uhralte Chaldeer und erste *inventores* der Astronomi solche *stellas* distribuirt haben. Welch ihr Außtheilung man noch heuttigs tags, nit von der Kunst und Zierlikeit, sonder allein der Antiquitet und Alters halben noch behelt, dieweil gewiß ist, daß vor Jobs und *Esaiae* Zeitten, solche Außtheilung des Himmels, durch Bilder, schon in *usu* gwest. Darnach hab ich ihn auch auff eine andere Gestallt, in form eines *Crucifixi Salvatoris* sezen wöllen, dieweil die fünff helleste Sternen des Schwanens, gar deuttlich ein Creutz geben, wie man *ex collatione utriusque figurae* augenscheinlich sihet. Und sonderlich am Himmel selbsten, wan er *inter occasum et septentriones* nidergehet, solch Creutz gerad ubersich steth, daß es auch nit allein die Baursleutt, Schäffer und die Wächter, merkhen und darfür ansehen; sonder auch ettliche *recentiores Astronomi*, zur Gedächtnuß des grossen Werks unserer Erlösung, ein *crucifix* drauß machen. Darzu sonderlich der newe Stern auch seins Theils wol hilfft, weil er nit gar in Mitte der Hölzer, sonder ein wenig beiseyts stehet, gleichsam ein Anzeigung

For Schickard, then, the images of the stellar constellations were invented by the Chaldeans and owe their authority to antiquity and old age alone. Unlike Kepler, Bayer or Roeslin, who restricted Christianization to the Swan, Schickard aimed to cover the entire stellar vault. The reason was not any presumed emergence of real Christian figures in the heavens, but a decision stemming from sincere Christian shame over the survival of "heathen images" and an attempt to reconstitute the constellations in harmony with true religion. This "Christian reformation" could only be the result of "universal consent among astronomers [*publico omnium Astronomorum consensu*]."

The cometary treatise was written in German and addressed to the Duke of Württemberg, who had requested it from Schickard. The report was not destined for print and thus remained in manuscript. By contrast, Schickard published the *Astroscopium* (1623) in Latin.[77] Here, besides indicating that the star was still present though greatly diminished in magnitude and close to extinction,[78] Schickard again addressed the issue of Christianizing the constellations. Now, after describing the origin of the stellar images as the result of the imagination of Mediterranean peasants, sailors and ancient scholars ("eruditi") who aimed to immortalize heroic figures like Hercules and Perseus,[79]

des genaigten Haupts gibet. Welches ich dann mit nichten zur abergläubischen Außlegung ziehen will, wie die *superstitiosi Astrologi nostri*: sonder weil mir das heidnische Wesen so höchlich mißfellt, und uns Christen für ein sondere Schandt halte, daß wir die heidnische Bilder *Herculis, Erichtonii, Persei*, item *Andromedae, Cassiopeae* etc. so dann auch des erdichten *Pegasi* etc. und anderer vil mehr, noch behalten sollen; daß doch alles auß ihrer fabulosischen Mehrlins Bibel genommen, und nach deren lengst verdamten ovidianischen *Metamorphosi* stinkt, deren greweln Gedächtnus ein christlich Gottliebendes Hertz verfluochen und außrotten solte. Derentwegen ich hierinnen offt ein *Christianam Reformationem* gewünscht, auch eigner Person *ex Historiâ Creationis, vel Apocalypsi Johannis* etc. solche, verhoffentlich nit *infeliciter* versucht: aber das Ding ist nit *unius privati arbitrii*, sonder müste einhelliglich, *publico omnium Astronomorum consensu* geschehen, solt es anderst *usu* approbirt werden." On this cometary treatise, see Horst Mauder, "Die Kometenschriften von Schickard und Kepler," in *Zum 400. Geburtstag von Wilhelm Schickard: Zweites Tübinger Schickard-Symposion 25. bis 27. Juni 1992*, ed. Friedrich Seck (Sigmaringen: Thorbecke, 1995), pp. 151–166. On Schickard's reform of the constellations, see Wolfgang Hübner, "Die Christianisierung der Sternbilder in Schickards *Astroscopium*," *ibid.*, pp. 131–150, on pp. 132 f. for the Swan and 137–148 for his reform in general.

77 Schickard (ref. 27). On the different editions of this work, see Friedrich Seck, ed., *Wilhelm Schickard 1592–1635: Astronom, Geograph, Orientalist, Erfinder der Rechenmaschine* (Tübingen: Mohr, 1978), Appendix 1 ("Werke von W. S."), pp. 396–398.

78 Schickard (ref. 27), pp. 28–29, as indicated above, p. 157.

79 *Ibid.* (§§ 20–23), pp. 35–41.

thus giving birth to an arbitrary system of heathen character and convention,[80] Schickard reaffirmed his wish for Christian reform. He claimed that reconfiguring the constellations was in accordance with Scripture and rather easy to perform: "I would like some Christian to reconfigure this [stellar vault] and transform it into sacred images: this not only conforms to divine law, [...] but is also most easy to accomplish."[81] While his aim applies to the entire stellar vault, Schickard advances only a series of examples: Perseus with the head of Medusa becomes David with the head of Goliath; Hercules with his stick is suddenly Samson with the jaw.[82] In other cases, the change will simply assign Christian references to the images: Gemini shall be understood as Esau and Jacob; Aries as the lamb sacrified by Abraham; the Whale as that of Jonah; the Fishes those multiplied by Christ; the Triangle the Trinity.[83] The images and their use shall thus be preserved while avoiding any trifles or abuses ("Ita manerent imagines, abolerentur saltem nugae; retineremus usum, abusum tolleremus").[84] Just as other peoples (e. g. the Arabs) have modified the constellations and their names, or classical antiquity applied a variety of names and images to a single asterism, Schickard claims that Christians are more justified to exercise the same liberty, doubtless because he considers Christianity the true religion. Here, Schickard claims to change the Swan into the Cross of Christ: "I add only a single difference according to my taste [*ad meum gustum*]: peasants and shepherds transform the rustic Swan into the cross of the Savior; the nova beautifully adorned this configuration by representing, as it were, the inclined head of the dying Christ."[85]

Schickard's *Astroscopium* was received by at least two German scholars who enthusiastically expressed the same aim. The first was Jacob Bartsch (ca. 1600–1633), who in 1630 would become Kepler's son-in-law. On 27 March 1624, Bartsch sent from Strasbourg a letter to Schickard, in which he extolled Schickard's work and announced that, for his own part, he was enclosing

80 *Ibid.* (§ 24), p. 42: "ista, multis pudendis fabulis involuta, nihil nisi Ethnicismus sapiunt."

81 "Optarem, ut ea Christianus aliquis refingeret, et in sacras imagines transformaret: quod non tantum divinae legi conforme esse, [...] sed et factu facillimum," *ibid.*

82 *Ibid.*

83 *Ibid.* (§ 26), pp. 45–46. For precedents among Christian writers from antiquity, see Hübner, *Zodiacus Christianus* (ref. 60).

84 *Ibid.*

85 "Unicam [variationem] addo, quae ad meum gustum est. *Ex Cygno rustici et opiliones efformant Crucem Salvatoris*, quam configurationem pulchrè ornavit Nova, repraesentando inclinatum quasi caput morientis Christi," *ibid.* (§ 27), p. 52; note that Schickard interprets the nova differently from Kepler and Roeslin, namely as the head of Christ rather than his heart. See also Hübner (ref. 76), p. 148.

his newly published *Planispherium stellatum*, where he had adopted (with due acknowledgment of priority) Schickard's proposal of christianising the constellations.[86]

Bartsch included Schickard and his *Astroscopium* among the authors and works on which he had drawn to arrive at his own achievement.[87] In Chapter 4 ("De Stellis Fixis, seu Signis Coelestibus Stellatis," pp. 40–66), Bartsch increased the number of constellations from 48 to 73: 27 in the northern hemisphere and 34 in the southern, in addition to the 12 zodiacal constellations.[88] In his description of all the constellations (except only those recently added in the southern hemisphere; from 19, Grus, to 34, Dorado or Piscis Hispanicus),[89] Bartsch mentions the names ascribed to them by the different traditions and adds his own proposal for a Christian appellation, coinciding with Schickard in the short series of Christian names and meanings suggested by the Tübingen astronomer.[90] The new representations proposed by Bartsch relate to the Old and New Testaments (Lyra is that of David and Cepheus is King Solomon,[91] while Pegasus and Equuleus are the two asses with which Christ entered Jerusalem),[92] normally with a positive or holy reference. In some

86 See Bartsch's letter in Schickard, *Briefwechsel* (ref. 25), vol. 1, no. 55, p. 146: "Mitto autem exemplar Planisphaerii mei Stellati, seu Vice-Globi Coelestis in plano delineati [...] quaedam, praecipue nova biblica et Arabica Stellarum atque Asterismorum nomina, ex ipsius Astroscopio mihi commendatissimo [...] desumta, meis, non tamen quasi Autor essem ipse, sed cum honorifica ipsius mentione inserui."

87 Jacob Bartsch, *Planisphaerium stellatum sive vice-globus coelestis in plano delineatus* (Strasbourg: Marcus ab Hayden, 1624), sig. A4 r. Cf. also pp. 4 and 50 f. for further mentions of Schickard's careful correction of the Arabic star names.

88 *Ibid.*, pp. 51–66.

89 *Ibid.*, p. 66.

90 On several occasions, Bartsch mentions as a predecessor to his project a certain Barthasius (pp. 52, 54, 56, 59, 60, 62, 63; for the first time on p. 51 wrongly written "Batassio"). He is referring to the French poet Guillaume Du Bartas, who in his *Second Semaine* (1584), Book VIII ("Les Colomnes"), verses 460–554, had already pleaded for a Christian reformation of the constellations. Bartsch referred to some of the changes adopted by Du Bartas, which he most probably knew from the German edition that included a German translation by Tobias Hübner: *La seconde Sepmaine de Guillaume de Saluste Seigneur du Bartas /Die andere Woche* (Cöthen: Johann Röhnern, 1622). We are grateful to Isabelle Pantin for having confirmed that Du Bartas and the *Second Semaine* are the author and work referred to by Bartsch. On this, see now Isabelle Pantin, "Les fables des astres: Continuité et mutations de 'l'affichage céleste' à la Renaissance," in *Savoirs et fiction au Moyen Âge et à la Renaissance*, ed. D. Boutet and J. Ducos (Paris: Presses de l'Université Paris-Sorbonne, 2015), pp. 363–378, on pp. 373–377.

91 *Ibid.*, pp. 55 and 53.

92 *Ibid..*, pp. 54 f.

cases, however, Bartsch assigns negative references: Bootes can be identified with Nimrod, Dracus with "draco infernalis" and Hydra with "the serpent of Paradise."[93] Needless to say, Bartsch also sees in Cygnus the cross of Christ.[94]

Schickard's *Astroscopium* was also warmly received by Julius Schiller (1580–1627), a Catholic doctor in law from Augsburg engaged in the same project. In his thank-you letter of 4 October 1623 (n. s.), Schiller informed Schickard that he had begun his own work on the Christianization of the heavens some fifteen years earlier and hoped to publish it shortly.[95] Schiller further added that Johann Bayer, the famous author of the 1603 *Uranometria* and fellow jurist in Augsburg, was supplying him with the stellar cartography from a corrected version of his work, while Schiller introduced new images and names in order to purge the heavens as completely as a second Augean stable. Schiller adopted the following pattern for his reform: persons and symbols of the New Testament for the constellations of the northern hemisphere, the twelve apostles instead of the zodiacal signs, and persons and symbols of the Old Testament for the southern hemisphere. In the same vein, the reform extended to the planets according to the following descending geocentric order: Adam, Moses, Joshua, Christ, the Virgin Mary, John the Baptist and Eve.[96]

93 *Ibid.*, pp. 52, 65.

94 *Ibid.*., p. 55. Of particular interest is the description of Cygnus, with an oscillation between Kepler's and Roeslin's identification of the nova with the heart of Christ and that of Schickard with his head: "Christianus substituat imaginem Salvatoris, qui seipsum Gallinae, ut sidus hoc etiam dicitur, assimilat Matt. 23 v. 37, et crucem suam portat, Ioh. 19 v. 17. [...] Vel sit crux Christi eidem affixi: cauda enim summitatem stipitis, rostrum eius plantam, anco uterque transversarium, et cor [Cygni] Crucifixi caput repraesentet: vel *quia circa huius seculi initium nova quaedam satis conspicua stella, pectore minor, rostro maior, configurationem istam pulchrè ornavit, quae minutula iam ad interitum vergit, vel pectus, vel inclinatum Christi morientis caput notet*"; my italics.

95 Schickard, *Briefwechsel* (ref. 25), vol. 1, no. 43, p. 136, ll. 15–17: "Ei enim operi (horis quidem et temporibus subsecivis) iam annis plus minus quindecim, humili conatu invigilo, illudque DEO bene favente, brevi in lucem proditurum spero."

96 *Ibid.*, ll. 28–39: "Ego autem monstrorum illud et ferarum, fabulosorum hominum chaos, et velut alterum Augiae stabulum (salvis quidem et illaesis asterismis) omnino repurgandum, et, sive Sanctorum Deoque acceptabilium personarum, sive aliorum fidei nostrae mysteriorum celebratione (inter quae, cum gustu dominationis tuae, omniumque piorum, in constellatione Cygni, signum S. Crucis Christi) reparandum putem, sic tripartito caelo, ut zodiacus, Duodecim Christi apostolos, inde meridionale hemisphaerium, Veteris testamenti memoranda, Septentrionale vero, novi flores, referat [...]. Atque ne gentilis Mars, cum Venere etc., per Sanctam illam Apostolorum coronam discurrat, collegium Planetarum, memoriis Sanctorum Adami, Moysis, Josue, Christi, Mariae Virginis, Joannis Baptista, et Evae, offerendum duxi."

FIGURE 7.5 Christianization of the planets. Julius Schiller, *Coelum stellatum christianum* (Augsburg, 1627).
SOURCE: BIBLIOTECA DE RESERVA, UNIVERSITAT DE BARCELONA

Schiller wrote again a year later, now boasting that his undertaking had enjoyed the approval of many, with only one criticism whose content (amounting to merely the inconveniences derived from the identification of Christ and the Virgin Mary with the Sun and the Moon) he enclosed and dismissed as unimportant.[97] In the meantime, Schiller had also taken notice of Bartsch's proposal, which, though worthy of praise for its Christian intention, was to be rejected partly due to its adoption of evil references for the heathen

97 Letter of 20 September 1624 (n. s.), in Schickard, *Briefwechsel* (ref. 25), vol. 1, no. 66, pp. 157–159. The report of the criticism is on pp. 158–159. Meanwhile, Schiller must have changed the allocation of the Virgin from Venus to the Moon, which he finally adopted in his published work.

constellations of Draco and Hydra (the infernal dragon and serpent of Paradise respectively).[98]

Schiller's *Coelum stellatum christianum* finally appeared in 1627. It was edited by Jacob Bartsch after Schiller's sudden and unexpected death in May.[99] The frontispiece of this splendid edition displayed the new personification of the planets: in the middle, Christ as the Sun; to his right, the Virgin Mary for the Moon, Adam for Saturn and Moses for Jupiter; to his left, Elias for Mercury, Joshua for Mars and John the Baptist for Venus.[100]

In his technical preface ("Monita sive Notae ad ipsas operis huius tabulas uranometricas"), Schiller duly acknowledges the contributions of the recently deceased Johann Bayer, who provided charts of all the constellations as an improved second edition of his *Uranometria* from 1603.[101] Schiller performed the metamorphosis[102] of the pagan figures of the constellations (employing the stars without form around the asterisms to complete the Judeo-Christian transformation) following the systematic program indicated above: Christian personifications and motives occupy the boreal constellations, the twelve apostles the zodiacal signs, and figures and motives

98 *Ibid.*, p. 157, ll. 35–39: "Bartschium, post admonitionem D[ominationis] T[uae] vidi, cuius inventiones et substitutiones Christianas pro gentilibus, uti in plerisque merito laudo, ita illas, quibus Draconem et Hydram gentilem, Draconi infernali, et Serpenti Paradysi assignat, omnino aversor: Qui enim Christianus ferat, Diabolum, inter Sanctorum hominum memorias, in caelo celebrari?"

99 See the letter of 4/14 July 1627 from Bartsch to Schickard, in *Briefwechsel* (ref. 25), vol. 1, no. 181, pp. 289 f. Bartsch had begun collaborating with Schickard the previous year. He was asked by Schiller from his deathbed, and later by Schiller's heirs, to bring the work to an end and personally see it through press: "ego post octiduum ab Haeredibus [...] rogatus sum, ut ex voluntate Dn. Schilleri defuncti, qu[ae] excudenda et absolvenda adhuc restant, in me susciperem, sicque opus posthumum brevi lucem publicam videre posset," p. 290. Johann Bayer had died by at least May 1625.

100 See also the explanation by Schiller in his long preface, pp. 6a–8a.

101 Julius Schiller, *Coelum stellatum christianum, Ad maiorem Dei Omnipotentis, Sanctaeque Eius tam Triumphantis, quam Militantis Ecclesiae gloriam, Obductis Gentilium Simulachris, Eidem Domino et Creatori Suo, postliminio quasi restitutum* (Augsburg: Andreas Aperger, 1627), pp. 19a–20a. Schiller denounced (p. 19b) the unauthorized pirate reprint of the *Uranometria*, published in 1624, and claimed for his work the character of the second, improved edition of Bayer's stellar cartography.

102 Schiller closed the preface by applying the title of Ovid's poem to his own performance: "*Sicque* (ut Nasonis verba in calce quoque Metamorphoseos huius Christianae, adhibeam) *opus exegi*, Domino quod dante, *nec ignes, nec poterit ferrum, nec edax abolere vetustas*," p. 18 b. Cf. Ovid, *Metamorphoses*, 15, 871–872: "Iamque opus exegi, quod nec Iovis ira, nec ignis / nec poterit ferrum, nec edax abolere vetustas."

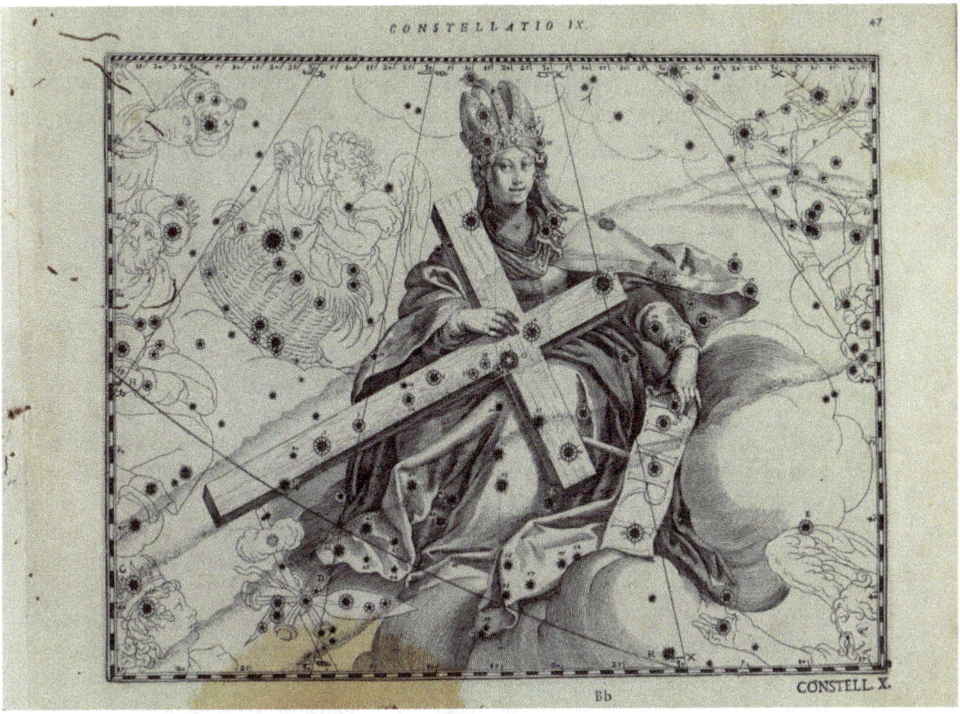

FIGURE 7.6 Transformation of Cygnus into the Cross of Christ. Julius Schiller, *Coelum stellatum christianum* (Augsburg, 1627).
SOURCE: BIBLIOTECA DE RESERVA, UNIVERSITAT DE BARCELONA

of the Old Testament the southern constellations, including the most meridional constellations recently added and now conceded to Aaron, Job, Eve, Abel and the archangel Raphael (the latter at the South Pole, in opposition to the Archangel Michael, who took the place of the Little Bear at the North Pole).[103]

I shall only briefly mention the transformation, as the ninth constellation, of the Swan into the Cross of Christ.[104] While Schiller omits the

103 See the full presentation of the reform in the preface, pp. 8b–17b. Schiller does not forget the Milky Way, transformed (p. 17b) through the suggestion of the Jesuit astronomer John Baptist Cysat (ca. 1587–1657) into *Via omnium sanctorum*.

104 Schiller (ref. 101), p. 47.

Crucifixion,[105] he does not forget to include the nova (indicated by G in Fig. 7.6).[106]

Assembling the unformed stars around the cross, Schiller adds the figure of Saint Helena, mother of Emperor Constantine, traditionally held as having discovered the cross of Christ in 326 AD.[107]

---

105 On the other hand, Schiller finds a place to mention the passage of Matthew 23:37, near and dear to Roeslin: "minime tamen obscurum esse, cur hic pro Cygno [Gallina] ponatur, quia scilicet etiam Graece ορνις saepenumero hanc avem (gallinam) denotet. Velut in Evangelio Matth. 23. per antonomasiam pro αλεκτορις. Pulcherrime ad rem praesentem, Christum scilicet expansis brachiis in Cruce pendentem, queritantem: *Hierusalem Hierusalem etc. quoties volui congregare filios tuos quemadmodum Gallina congregat pullos suos sub alas et noluisti?* (p. 46)."

106 *Ibid.*, p. 47. In his commentary on the constellation (p. 46), Schiller refers to what he has said in his commentary on the previous constellation (Lyra transformed into Christ's manger; see p. 44): "Stella peregrina S. Crucis, Cygni: sive Nova Anni 1600. quae tertii fulgoris stella, dicto anno M.DC. primum conspecta et observata, omnium ferme tacito consensu, pro novo phaenomeno recepta, eundem adhuc hodie (scribit Bayerus in priore *Uranometria* 1603. edita) retinet situm [...]. Eandem idem Joannes Bayerus, antequam lethali morbo corriperetur, mense Ianuario Anni M DC XXV. pristini quidem situs, quantitatis vero non omnino sextae, et talis fere, qualem vicinae tres 27. 28 et 29 [...] prae se ferunt, inter sidera adhuc observavit. Quam et ego in Autumno eiusdem Anni aliquoties, et quidem primo Septembr. meo iudicio infallibiliter deprehendi: nonnihil tamen minorem tribus parvulis istis ei vicinis, modo prorsus disparentem, modo subinde, sicque inconstantem comparentem. Imo si Brahe, Bayerus et reliqui novam recte posuerunt, ut arbitror, arbitror etiam plane, novam me vidisse iterum 28. Iun. Anni 1626. quamvis non admodum serenum fuerit coelum: tandemque nocte serena 18. Augusti eodem anno 26: paulo obscuriorem et magis dubiam iis, quas num. 27. 28. et 29. in Constell. sequ. [Cygnus] notavi. Licet autem haec prae illis minutulis bene apparens est: tamen hoc loco, utpote in via lactea, stellarum maximus numerus, eum qui tubo velit uti, facile confundit: adeo ut eam difficulter iam discernendam, et fugitivi quasi ac per vices tantum percipiendi luminis stellulam, totali disparitioni proximam esse ambigi non possit. Quidnam autem rationi magis et veritati consentaneum sit, dicere, eam penitus extinctam et in nihilum redactam: an vero inusitato et incognito motu in excelsum, unde prius potuit descendisse, reversam, et sic prospectui paulatim deminutam; aliis discutiendum relinquo: iis praesertim, quibus practice investigare libuerit, itane se res habeat, ut supra [...] insinuatum, stellarum scilicet alias aliis in Firmamenti corpore sublimius reconditas videri." If it is rather strange to see Brahe among those considering the star in Cygnus as a nova (Kepler was a better candidate), it is similarly curious to find Schiller considering the diminution and disappearance of the star to be an optical effect of its progressive return to an initial dwelling place in the high firmament. This hypothesis had been advanced to explain the disappearance of the nova of 1572 by John Dee and Cornelius Gemma, and possibly known to Schiller through Brahe's *Progymnasmata*. On these ideas and the critical response they received from Kepler, see the contribution by Dario Tessicini to this volume.

107 *Ibid.*, p. 46: "Iovem Cygnum ex Leda Helenam Troiae excidium genuisse, Poetae fabulantur: et en Cygnum hunc abolet Augustissima Helena, quae S. Crucis Lignum invenit Anno Christi 326." Cf. also Preface, p. 11b: "invenit enim extulitque Sancta Imperatrix Helena

Throughout his work, Schiller proceeds not only as a pious and devout Christian aiming to restore God's glory, but also as an obedient servant of the Roman Catholic Church. Indeed, the Catholic Church is identified with the *ecclesia triumphans et militans*, rising to the heavens *pari passu* as the Christian religion, just as the full title on the frontispiece makes clear and some constellations express explicitly. Schiller thus substitutes for the Great Bear the *Navicula Sancti Petri* (or the small vessel of Saint Peter), identified with the Catholic Church:

> The second, akin to the first, and a brilliant constellation [...] visible to us the whole year even though [it is] in continuous motion, seems to me to square with the small vessel of St Peter as a figure of the Catholic Church: for this [ship] [...], like the untouched [virgin] betrothed to the Almighty, deserved to be located among the more prominent constellations. [...] What happens in this place occurs just as in the fables: established from hatred toward Callisto that it could never descend or bathe in the sea, it gives place, as if in the likeness of prophecy, to the uppermost glory of the Holy Mother Church, in accordance with Christ's word firmly asserting that She shall never hide or sink.[108]

At the same time, the northern constellation of the Triangle is interpreted by Schiller not as the Trinity, as Schickard and Bartsch had suggested, but transformed into the *Mitra Pontificalis*, correctly placed above the head of the first zodiacal sign metamorphosed into Saint Peter, the head of the Christian Church.[109] Schiller also proposes to consecrate Wednesday, the day of the week assigned to Mercury, "to the Catholic Church and her necessities."[110]

This apotheosis of the Roman Catholic Church probably contributed to the scarce fortune of Schiller's proposal among Protestant circles, even if his two tables showing the constellations in both celestial hemispheres were included, with a summary of the complete work, in Andreas Cellarius' (ca. 1596–1665)

Constantini magni genitrix, defossum in terra Crucis Lignum, celeberrimo cultu et miraculis postea exaltatum."

108 "Secunda, ac cognata priori, et praeclara constellatio [...] totoque anno nobis conspicua, licet continuo motu agitata, Naviculae S. Petri, ut typo Ecclesiae Catholicae [...] quadrare mihi visa est: haec enim [...] velut intacta Altissimi Sponsa inter eminentiora sidera collocari meruit. [...] Evenitque hoc loco, ut quod fabulis, in odium Callistus statutum, ne unquam tingui aequore seu occidere possit, id in summam Sanctae Matris Ecclesiae gloriam, et secundum verbum Christi, Eam nullo aevo occultandam supprimendamvè asseverantis, quasi in prophetiae similitudinem cedat," *ibid.*, Preface, p. 9b.

109 *Ibid.*, Preface, p. 13b.

110 *Ibid.*, p. 22 b.

famous *Harmonia macrocosmica* (Amsterdam, 1661).[111] But Catholic circles expressed equally scarce enthusiasm for the reform. As the Jesuit Athanasius Kircher (1602–1680) commented in the 1660 edition of his *Iter exstaticum*,[112] "nomina antiqua tam altas iecere radices, ut ea inviti retinere cogamur" (ancient names have grown roots so deep that we are forced, though unwillingly, to retain them).

111 Andreas Cellarius, *Harmonia Macrocosmica seu atlas universalis et novus totius universi creati Cosmographiam generalem, et novam exhibens* (Amsterdam: Johannes Janssonius, 1661), pp. 161–173.

112 Athanasius Kircher, *Iter exstaticum coeleste* (Würzburg: J. A. and W. Endter, 1660), Itinerarium I, Dialogus I (Praelusio in firmamentum), p. 335.

CHAPTER 8

Epicurean Astronomy? Atomistic and Corpuscular Stars in Kepler's Century

Pietro Daniel Omodeo

8.1 Introduction

At the threshold of modernity, novae and comets increasingly attracted the attention of European cosmologists.* In fact, the parallax determination of their distances and cosmic trajectories undermined well established vistas on worldly order.[1] They challenged the scholastic image of nature, as their superlunary location cast doubt on the double physics of the Aristotelians separating the realm of generation and corruption below the Moon from the ethereal perfection of the superlunary heavens, subject only to local motion and by no means to material change. Debates on new stars and comets paralleled the rise of novel planetary theories, which aimed to answer the crisis of the Ptolemaic and Aristotelian cosmos. In this context, Tycho Brahe issued his *De mundi aetherei recentioribus phaenomenis* (1588), which stands out as a clear instance of the convergence of the two tendencies: 1. to recast post-Copernican planetary theory into a geocentric framework and 2. to revise celestial physics in the light of comets' measurements and the dissolution of the celestial spheres or orbs. Such dissolution went hand in hand with the acceptance of the paradoxical-sounding expression 'celestial novelties,' no longer a contradiction in terms but rather a measurable reality.[2] The transformation

* This essay was written in the framework of the ERC consolidator project *Early Modern Cosmology*, which received funding from the European Union's Horizon 2020 Research and Innovation Programme (GA 725883).

1 The literature on the subject is extensive but, for the purposes of the present discussion on the philosophical impact of celestial novelties, it should be sufficient to mention the following sources: *Novas y cometas entre 1572 y 1618: Revolución cosmológica y renovación política y religiosa*, ed. Miguel Ángel Granada (Barcelona: Universitat de Barcelona, 2012); *Celestial Novelties on the Eve of the Scientific Revolution, 1540–1630*, ed. Dario Tessicini and Patrick Boner (Florence: Olschki, 2013). Also, see Miguel A. Granada, *Sfere solide e cielo fluido: Momenti del dibattito cosmologico nella seconda metà del Cinquecento* (Milan: Guerini, 2002).

2 In this respect, the exchanges between Christoph Rothmann and Tycho Brahe on comets and celestial fluidity are of the utmost relevance. See, among others, Miguel A. Granada, Adam

 | DOI:10.1163/9789004437272_010

of the conception of the heavens made the ideation of an alternative physics urgent. A new natural philosophy should account especially for the motion of planets through a fluid ether in the absence of material transporters.[3] The cosmological shift also opened up a theoretical space for the development of physico-mathematics; this is the task that Johannes Kepler took upon himself in paradigmatic works such as the *Astronomia nova* (1609). Arguably, planetary debates directly inspired the celestial physics of Kepler, Descartes and their likes, though controversies on comets were a secondary but no less important path towards a new mathematical science of nature. Eventually, the planetary and cometary themes converged in the work of Halley and Newton.[4]

Recent scholarship has pointed to the link between Kepler's theory (bringing together heliocentrism with novel laws and geometries of planetary motion) and the discussion of celestial novelties (first, his edition of Brahe's posthumous work on the nova of 1572, and secondly the discussion of the nova of 1604).[5] The theoretical challenge for Kepler—just as for Brahe one generation earlier—was to account for celestial novelties without renouncing the geometrical harmony of the heavens. In the concluding sections of *De stella nova in pede Serpentarii* (1606), Kepler cautioned astronomers against neo-Epicurean interpretations of the nova as testimony that nature is ruled by chance:

> Only those philosophers to whom confusion is pleasing rather than order, who have debased the most beautiful world with intolerable atoms, deserve this, [namely] that neither are they moved by the beauty and congruence of that which they call chance nor do they allow others to take pleasure [in them].[6]

Mosley and Nicholas Jardine, *Christoph Rothmann's Discourse on the Comet of 1585: An Edition and Translation with Accompanying Essays* (Leiden-Boston: Brill, 2014).

3 Cf. Granada, "'A quo moventur planetae? Kepler et la question de l'agent du mouvement planétaire après la disparition des orbes solides," *Galilaeana*, 2010, 7: 111–141. Also, see William H. Donahue, *The Dissolution of the Celestial Spheres, 1595–1650* (New York: Arno Press, 1981) and Mary S. Kelly, *Celestial Motors: 1543–1632* (Ph.D. Dissertation, University of Oklahoma, 1964).

4 This is Édouard Mehl's point in his article, "Théorie physique et optique des comètes de Kepler a Descartes," in *Novas y cometas* (ref. 1), pp. 254–274. For a more recent study of early-modern cometary debates, see Anna Jerratsch, *Der frühneuzeitliche Kometendiskurs im Spiegel deutschsprachiger Flugschriften* (Stuttgart: Franz Steiner Verlag, 2020).

5 In particular, see Patrick J. Boner, "Kepler's Copernican Campaign and the New Star of 1604," in *Change and Continuity in Early Modern Cosmology*, ed. Patrick J. Boner (Dordrecht: Springer, 2011), pp. 93–114.

6 Cf. *De stella nova*, JKGW, 1 285.40–286.3: "Soli hi philosophi, quibus pro ordine placet confusio, qui mundum pulcherrimum inter inamoenas atomos abiecerunt, hoc sunt meriti, ut pulchritudine et congruentia huius, quem dicunt, casus, nec ipsi moveantur, nec alios delectari

As Kepler explained, his words should not be read as an implicit acknowledgment of the beauty of chance (*pulchritudo casus*) but, on the contrary, as an attack on chance itself. For him, *casus* is opposed to natural causality and goes together with impiety:

> For what is chance? To be sure, it is a very detestable idol, and nothing but an offence to the highest and almighty God and to the most complete [*absolutissimi*] world He created. A blind and brash motion is ascribed to it as its soul and an infinite chaos as its body. Hence, the eternity, the omnipotence and the creation of the world, which belong to God, are impiously transferred [to nature].[7]

Kepler's "opposition to unnamed Epicureans," as Patrick Boner has stressed,[8] raises the question of the contemporary dissemination of theses linked to the Epicurean legacy. Certainly, Kepler picked up polemical themes that were widespread among the followers of Philipp Melanchthon at Protestant universities.[9] However, the question remains: Who were the dangerous Epicureans of Kepler's time, what theories did they put forward and what accounts of the celestial novelties did they advance?

My present purpose is to discuss the natural-philosophical dimension of the problematic evidenced by Kepler's concerns, but I would also like to go beyond Kepler and trace some general lines underscoring the presence of Epicurean themes in the comet controversies of the sixteenth and seventeenth centuries. I will specifically consider atomistic and corpuscular motifs connecting the debates on new stars and comets before and after Kepler. Atomism and the corruptibility of the heavens—theses that were reinforced by observations of celestial novelties—were at odds with the harmonic vision of the world that Kepler had embraced from the outset of his astronomical enterprise, as is clearly evidenced by the theological speculations of *Mysterium cosmographicum*

---

patiantur." English translation by Patrick J. Boner from *idem*, "Kepler v. the Epicureans: Causality, Coincidence and the Origins of the New Star of 1604," *Journal for the History of Astronomy*, 2007, 38: 207–221, on p. 211.

7 *De stella nova*, *JKGW*, 1, 284.18–23: "Casus vero quid est? Nimirum idolum est detestabilissimum, et nihil aliud nisi contumelia summi et omnipotentis Dei, et quem ille condidit, Mundi absolutissimi: Cui motus caecus et temerarius est pro anima, chaos infinitum pro corpore. Huic et aeternitas, et omnipotentia et Mundi creatio, Dei propria, transcribuntur nefarie."

8 Boner, "Kepler's Copernican Campaign" (ref. 5), p. 216.

9 The anti-Epicurean polemic is recurring in Philipp Melanchthon and Paul Eber's *Initia doctrinae physicae*, first published in 1549. The criticism begins in the dedicatory letter. Cf. *Initia doctrinae physicae* (Wittenberg: Johannes Lufft, 1549), f. A5r: "Non enim a Democrito aut Epicuro, aut aliis, qui doctrinam corrumpunt, ars petenda est."

(1596).[10] In turn, atomistic theories rehabilitated infinitism and cosmological homogeneity, while undermining faith in divine providence and the harmony of Creation. Among the atomists of Kepler's time, one might mention mathematicians and scholars in mechanics such as Giovanni Battista Benedetti and Galileo Galilei, both of whom were persuaded that the existence of a physical void (and atoms) was necessary for motion.[11] That Galileo's atomism could be accused of heresy is well known and should not be discussed in further detail.[12] Here, I will focus on atomistic and corpuscular *philosophers* who contributed to the astronomical debate and the understanding of comets.[13]

Among the atomistic tendencies that might have worried Kepler, one should refer to the Lucretian revival induced by Giordano Bruno, whose philosophy was held in high esteem by many in Kepler's Prague.[14] I will briefly examine Bruno's cometary theory and its reception by the lesser known English philosopher Nicholas Hill, author of the Epicurean and Brunian booklet *Philosophia Epicuraea Democritana Theophrastica proposta simpliciter non edocta* (Paris, 1601, and Geneva, 1619).

As a second phase of the atomistic discussion of celestial novelties in the seventeenth century, I would like to turn to the cometary theory of Pierre Gassendi in his *Syntagma philosophiae Epicuri* (1649), as well as to René Descartes' corpuscular physics, one generation after Kepler. These philosophers' views bear witness to a cultural tendency directed towards the physicalisation of comets at once continuing and opposing the Keplerian line of inquiry. Most importantly, Descartes connected the causal explanation of the

10 See among other publications Peter Barker and Bernard R. Goldstein, "Theological Foundations of Kepler's Astronomy," *Osiris*, 2001, 16: 88–113; Charlotte Methuen, *Kepler's Tübingen: Stimulus to a Theological Mathematics* (Brookfield: Aldershot, 1998). See also my essay, "La contingente geometria del cosmo nella Dotta ignoranza cusaniana: Cusano e Keplero a confronto," in *Filosofia Arte Scienza in Cusano e Leibniz*, ed. Antonio Dall'Igna and Damiano Roberi (Milano: Mimesis, 2014), pp. 215–226.

11 On Galileo, see Paolo Galluzzi, *Tra atomi e indivisibili: La materia ambigua di Galileo* (Florence: Olschki, 2011). On Benedetti, see Pietro Daniel Omodeo and Jürgen Renn, *Science in Court Society: Giovanni Battista Benedetti's* Diversarum speculationum mathematicarum et physicarum liber (*Turin, 1585*) (Berlin: Edition Open Access, 2019).

12 Pietro Redondi, *Galileo Heretic* (Princeton: Princeton University Press, 1989).

13 Note that Stillman Drake interprets Galileo's atomism as an anti-metaphysical stance, differently than most of the authors I discuss here. Cf. his "Introduction" to *The Controversy on the Comets of 1618*, trans. Stillman Drake and C.D. O'Malley (Philadelphia: University of Pennsylvania Press, 1960): pp. vii–xxv.

14 I discuss this context in the conclusion of my article, "Astronomia, filosofia e teologia nel tardo Rinascimento tedesco: Heinrich Julius di Braunschweig e il soggiorno di Giordano Bruno in Germania," *Giornale Critico della Filosofia Italiana*, 2011, 90: 307–326.

origin and dissolution of novae, cometary theory and planetary theory on the basis of a corpuscular matter theory.

As a further development of the atomistic and corpuscular debates surrounding celestial novelties, I will deal with the interpretation of celestial novelties in the context of Cartesian astrology as developed by Johannes Placentinus, a lesser known professor of mathematics at Frankfurt (Oder) and court mathematician of Brandenburg. Placentinus' theories bear witness to a hybrid reception of Copernican and Keplerian astronomy and astrology as well as Cartesian philosophy in the second half of the seventeenth century.[15]

8.2 Epicurean Comets by Giordano Bruno and Nicholas Hill

Bruno's eclectic natural philosophy picked up key elements from the Epicurean conception, most importantly the atomic constitution of matter. Yet, his atomism, widely known and debated in Kepler's Middle Europe, was also marked by elements that are absent from (if not radically at odds with) the doctrines of the ancient atomists. The most important difference between Bruno's version of atomism and the classical tradition is the rejection of chance. In his vision of the world, chance is substituted by the universal vicissitude of all things. The infinite variety of beings coming to being and passing away is the apt expression of God's omnipotence.[16] Rather than chance, it is based on a 'principle of plenitude' according to which the Scholastic distinction between God's *potentia absoluta* and his *potentia ordinata* (the latter serving as the source of the Creation) should be wiped away in the name of an actual infinity of nature.[17] Furthermore, Bruno embedded his speculations on the atom in a wider treatment of threefold minima: metaphysical, physical and gnosiological. The material atom of the natural philosopher shares one and the same essence with the mathematical point and the metaphysical unity—which is

15 For an introduction to the life and thought of this lesser known middle-European Copernico-Cartesian, see my essay, "Central European Polemics over Descartes: Johannes Placentinus and His Academic Opponents at Frankfurt on Oder (1653–1656)," in *History of Universities*, 2016, 29: 29–64. On Dutch Copernico-Cartesianism, see Rienk Vermij, *The Calvinist Copernicans: The Reception of the New Astronomy in the Dutch Republic, 1575–1750* (Amsterdam: Koninklijke Nederlandse Akadamie van Wetenschappen, 2002), esp. pp. 142–148.

16 Michele Ciliberto, *La ruota del tempo* (Rome: Editori Riuniti, 1986).

17 Arthur Oncken Lovejoy, *The Great Chain of Being: A Study of the History of an Idea* (Cambridge, MA: Harvard University Press, 1936); Miguel A. Granada, "Il rifiuto della distinzione tra *potentia absoluta* e *potentia ordinata* di Dio e l'affermazione dell'universo infinito in Giordano Bruno," *Rivista di Storia della Filosofia*, 1994, 49: 495–532.

minimum and maximum at the same time in accordance with the basic principle, of Cusanian derivation, that the opposites coincide in the absolute.[18] Bruno believed in the infinity and eternity of nature and thereby consciously transgressed the limitations that most Scholastics ascribed to nature. "Hence," as Kepler would say, "the eternity, the omnipotence and the Creation of the world, which belong to God, are impiously transferred [to nature]."[19]Alongside atomism and Cusanian metaphysics, the Copernican element famously plays a central role in Bruno's grand philosophical vision. All stars, plunged into an infinite space, acquire the status of countless suns dispersed across the universe, each one encircled by the trajectories of its own animal-like planets.

Bruno conceives of comets as massive celestial bodies similar to the Earth and the planets circling around any other stellar system. However, they move independently of the trajectories of the other planets. In the fourth dialogue of *De l'infinito universo e mondi* (1584), Bruno refers to comets in order to criticize the Aristotelian theory of natural place.[20] In fact, Aristotle considered comets to be composed of terrestrial matter, which "is dragged upwards as an exhalation by the force of the first mobile to the inflamed realm of fire" (*in forma di exalazione è montata in alto sino all'incentiva* [*incendiata*] *regione del foco come rapita dal vigor del primo mobile*) and is forced to turn around the Earth. Bruno questions that such a body, depicted as "heavy, thick and dense" (*grave, spesso e denso*), could remain suspended in the air and not descend according to the theory of natural place. He specifically refers to the comet of 1577, on which he possibly wrote the lost treatise *De' segni de' tempi*.[21] What explains the motion of comets in the Aristotelian framework? Aristotle would argue that comets are dragged (*rapite*) by the circular motion of the air, but Bruno is not satisfied with this answer. He objects that all planets should be dragged (*rapiti*) in the same manner. Observations show that comets possess their proper motion independently of the daily rotation and motion of other planets. Moreover, the element around the Earth is "such a liquid air and subtle body" (*sì liquido aere*

18 For further details, see my essay, "Minimum und Atom: Eine Begriffserweiterung in Brunos Rezeption des Cusanus," in *Die "Modernitäten" des Nicolaus Cusanus: Debatten und Rezeptionen*, ed. Tom Müller and Matthias Vollet (Bielefeld: Transcript, 2013), pp. 285–304.

19 *De stella nova* (ref. 7).

20 Giordano Bruno, *De l'infinito*, in *Dialoghi filosofici italiani*, ed. Michele Ciliberto (Milan: Mondadori, 2000), pp. 414–416. For a detailed discussion of Bruno's cometary theory, see Dario Tessicini, *I dintorni dell'infinito: Giordano Bruno e l'astronomia del Cinquecento* (Pisa-Rome: Serra, 2007), Chap. 4, "Le comete." Also, see *idem*, "cometa," in *Enciclopedia Bruniana e Campanelliana*, ed. Eugenio Canone and Germana Ernst (Pisa-Rome: Serra, 2010), pp. 34–43.

21 Bruno (ref. 20): "[…] come per quarantacinque giorni continui a tempi nostri n'è vista una."

e sottil corpo), so it is not apt to hold and transport the huge body of a comet. According to Bruno, "the truth concerning the nature of comets" (*la verità della natura delle comete*) is that they are a "kind of celestial body" (*specie di astro*). They are heavenly bodies that draw closer and move away from us. Through their motion, they appear to increase in size and become inflamed when they move toward us and diminish in size and extinguish their fire when they leave our proximity.[22]

Bruno resumed and deepened his cometary theory in the 'Frankfurt poem' *De immenso et innumerabilibus* (1591). In Book 4, Chapter 13, he expounded the thesis that "there are more planets than appear [to us]" (*quod plures sint planetae quot apparent*).[23] Comets are indeed planets. Bruno contends that they sometimes appear without a tail, as was the case with the nova of 1572; and he refers to Helisaeus Roeslin's hypothesis that the comet of 1577 was the same celestial body.[24] According to Bruno, the tail does not mark a difference between planets and comets, as we read in *De immenso* [6, 20]:

> One should not deem them to have a different substance than the other planets simply because they have a tail. In fact, as I have already argued from observations made in our time and in the past, comets sometimes display a larger tail, sometimes a smaller one, sometimes a tiny one, and sometimes none at all. This can occur either along the whole arc of their circuit in which they are visible to us or in some part of it; in fact, they are found to differ from planets for no other reason than their rare appearance. [...] For planets appear tailed from some regions but their tail, as one may understand, can diminish or even disappear completely.[25]

---

22 *Ibid.*: "[...] et essere tale astro che col proprio moto avvicinandosi et allontanandosi verso e da questo astro, per raggione di accesso e recesso [avvicinamento ed allontanamento], prima par che cresca come si accendesse, et poi manca [decresce] come s'estinguesse [...]."

23 Giordano Bruno, *Opera latine conscripta*, ed. Francesco Fiorentino (Naples: Morano, 1884), vol. I 2, p. 65.

24 *Ibid.*, p. 70. Cf. Jürgen Hamel, "Die Vorstellung von den Kometen seit der Antike bis ins 17. Jahrhundert: Tradition und Innovation," in *Georg Samuel Dörffel (1643–1688): Theologe und Astronom*, ed. Johannes Richter (Plauen: Vogtland, 1994), pp. 97–122, on pp. 113–116 ("Die Kometentheorie von Helisäus Röslin").

25 Bruno (ref. 23), p. 226: "Propter eam ergo, quam habent caudam, non sunt credendi alius substantiae, quam planetae reliqui: quandoquidem, ut in superioribus notavimus ex observatis nostro et alio tempore, et cometae interdum apparent maiore, interdum minore, interdum minima, interdum nulla cauda. Et hoc vel toto circuitus sui arcu, quo nobis sunt conspicui, vel parte; et quaedam non alia ratione differre comperiuntur a planetis, praeterquam quia non sunt ordinariae apparentiae. [...] Planetae enim ex aliqua regione caudati videbuntur, quorum cauda sicut intendi potest, et remitti, ita et omnino tolli."

Comets' motions are as perfect as those of the planets; for instance, the trajectories of Venus and Mercury (4, 13):

> Hence, it is proved that the comet is a heavenly body [...], in the seventh place, from the perfection of its motion, superior to the perfection of the lunar motion and comparable with that of Venus and Mercury, with whose motion it had a strong similarity.[26]

However, the fact that the 'poles' of the comet's motions are different than those of the planets running along the ecliptic make the former invisible to us for long periods of time. Comets, in conclusion, are Earth-like bodies (6, 20):

> The appearance of a comet is the solar light reflected by the celestial body's watery surface, which is opposite our eyes and with which our line of vision [*radius noster visualis*] forms an angle with the solar ray [*radius*]. As an entity [*subjectum*], a comet is a planet that circles around the Sun just like the Earth. Its matter [*substantia*] is composite. The only difference with the famous planets, in fact, is relational: they [the comets] are rarely visible for the sole reason that their circular path does not bring them to such an opposition between our sight and the Sun that it reflects its light in a mirror-like [frontal] manner.[27]

Among Bruno's readers and followers, the English philosopher Nicholas Hill deserves particular mention. This "philosopher-physician of the Elizabethan age"[28] belonged together with Thomas Harriot to the so-called 'Northumberland circle' of scholars who coupled an empirical and mathematical approach to nature with anti-Aristotelian views and an adherence to the Copernican system.[29]

26 *Ibid.*, pp. 74–75: "Colligitur ergo cometam illum, astrum esse. [...] Septimo a motus perfectione, super perfectionem motus lunae et cum perfectione motus Veneris et Mercurii conparabili, cum quorum etiam motibus analogiam habebat."

27 *Ibid.*, p. 225: "Cometae apparentia est lux Solis reflexa in aqueam, ad oculosque nostros oppositam, astri superficiem, in qua scilicet radius noster visualis cum solari radio angulum efficient. Subiectum cometae est planeta, Solem circumscurans non minus, atque aliter, quam Tellus, substantia quaedam composita est; ab istis vero famosis planetis sola relatione differens: quoniam ea de caussa raro apparent, quia eorum circulus non venit ad eam oculorum nostrorum et Solis oppositionem, ut specularem reddat lucem."

28 According to Sandra Plastina's expression. See her "Introduzione" to Nicholas Hill, *Philosophia Epicuraea Democritana Theophrastica proposta simpliciter non edocta* (Pisa-Rome: Serra, 2007), pp. 11–75. I will quote from this edition.

29 Cf. Robert H. Kargon, *Atomism in England from Harriot to Newton* (Oxford: Clarendon, 1966).

Harriot derived from Bruno his infinitist atomism that brought together speculations about natural *maxima* and *minima*.[30] Hill was even more keen on Bruno's philosophy than Harriot, as is evidenced by his ethical-philosophical tract, *Philosophia Epicuraea Democritana Theophrastica*, first printed in Paris in 1601.[31] The tract aimed at revitalizing the Epicurean and Brunian legacy. Hill defended a Christianized atomism connecting Democritean physics and divine providence. He presented it as a philosophy that was neither new nor old, "philosophia nec nova nec vetus."[32] Specifically, he dealt with comets in theses 379 and 380, in two paragraphs that closely follow in the footsteps of Bruno's text:

> 379. The appearance of a comet is the solar light reflected by the celestial body's watery surface, which is opposite our eyes and with which our line of vision [*radius noster visualis*] forms an angle with the solar ray [*radius*]. As an entity [*subjectum*], a comet is a planet that circles around the Sun just like the Earth. Its matter [*substantia*] is composite. The only difference with the famous planets is its ὀλιγοφανεία, that is, the rarity or faintness of its appearance, as its circular path seldom comes to such an opposition of our sight to the Sun as to reflect its light in a mirror-like [frontal] manner. [...]
>
> 380. The comet's tail is a vaporous substance, which does not belong to the solid part of the celestial body; it reflects the light of the Sun in a hair-like form; its quantity, density and rarity elongates, reduces, extends and diminishes the tail.[33]

---

30 See *Thomas Harriot: An Elizabethan Man of Science*, ed. Robert Fox (Aldershot: Ashgate, 2000). Also, see John Henry, "Thomas Harriot and Atomism: A Reappraisal," *History of Science*, 1982, 20: 267–303.

31 Among the studies on Hill, see especially Sandra Plastina, "Nicholas Hill: 'The English Campanella?'," *Bruniana & Campanelliana*, 1998, 4: 207–212, and "'Philosophia lucis proles verissima est': Nicholas Hill lettore di Francesco Patrizi," *Bruniana & Campanelliana*, 2004, 10: 175–182. I deal with Hill's Brunian cosmology in *Copernicus in the Cultural Debates of the Renaissance: Reception, Legacy, Transformation* (Leiden: Brill, 2014), pp. 372–377.

32 Hill (ref. 28), p. 80.

33 *Ibid.*, pp. 152–153: "379. Cometae apparentia est lux solis reflexa in aqueam, et oculis nostris oppositam astri superficiem, in qua radius visualis cum solari radio angulum efficit. Subiectum cometae est, planeta solem circumcursans, a famosis planetis differens ὀλιγοφανείᾳ, i. exiguitate seu tenuitate apparitionis, illius circulo raro ad eam solis et oculorum oppositionem deveniente, ut specularem reddat lucem. [...] 380. Cauda cometae est substantia vaporosa ad astri partem solidam non spectans, reflexam solis lucem refrangens, et in crinitam formam ducens, cuius copia, densitas, et raritas caudam intendit, remittit, extendit, et minorat."

This is, in fact, a summary of Bruno's position. After Paris, the booklet was reprinted in Geneva in 1619 and can be seen as a witness to the European circulation of Bruno's ideas as well as a certain revival of eclectic Epicureanism. As for the Keplerian polemics against the philosophers of chance, it should be remarked that, although Bruno's ideas spread atomism and an image of the world in which plenitude takes the place of order, neither Bruno nor his followers embraced a vision of the world ruled purely by chance. Rather, they believed that divine providence is better secured by the assumption of infinitism than by geometrical proportion. As a matter of fact, Hill explicitly insisted on the importance of providence.[34] These views are certainly far from Kepler's but cannot be reduced to the Epicurean legacy tout court.

8.3 Post-Keplerian Atomistic and Corpuscular Celestial Novelties: Pierre Gassendi and René Descartes

The champion of Epicureanism one generation after Bruno and Hill, Pierre Gassendi founded his natural philosophy on ground far different from that of his predecessors. His views on comets were correspondingly distant from theirs.

However, before I deal with Gassendi's distinct Epicurean views on comets, I would like to stress the complexity and ambivalence of his link to Kepler's astronomy.[35] One could mention that, following Kepler's instructions in the *Admonitio ad astronomos* found in his final ephemerides (1630), Gassendi observed Mercury's solar transit in 1631 and thus established his renown among European astronomers. Additionally, Gassendi embedded the proof of Mercury's heliocentric path together with his disquisitions on comets and new stars in a neo-Epicurean vision of the world, as expressed in the *Syntagma philosophiae Epicuri* (1649), in particular the chapter *De cometis et novis sideribus* (*pars secunda, sectio secunda, liber V*).

Gassendi's lengthy discussion of comets, rather than pronouncing a clear-cut position on the debates, is a doxographic overview of ancient and modern opinions about their origin, nature, motion and effects. Gassendi remarks that the nature of comets seems to hover between that of heavenly bodies and that

34 Cf. Pietro D. Omodeo, "Perfection of the World and Mathematics in Late Sixteenth-Century Copernican Cosmologies," in *The Invention of Discovery, 1500–1700*, ed. James Douglas Fleming (Farnham-Burlington: Ashgate, 2011), 93–108.

35 For an insightful discussion, see Kuni Sakamoto, "The German Hercules's Heir: Pierre Gassendi's Reception of Keplerian Ideas," in *Journal of the History of Ideas*, 2009, 70: 69–91.

of sublunary ones.[36] Depending on the different theories brought forward by natural philosophers and astronomers, they might possess a fiery nature and originate in the uppermost cloudy sphere, originate and move in the ethereal region in orbs such as those of Mercury and Venus, or travel freely through the immense heavens. The variety of opinions is so vast that Gassendi declares himself incapable of identifying the right one. Rather, he deems it expedient to follow Epicurus' decision not to decide:

> Do you think that you can easily solve this issue if Epicurus did not want to determine which one of the two opinions was true? Indeed, the matter is very obscure. Whatever opinion you choose, no matter how you argue, so many difficulties will arise that it is expedient to suspend one's own judgment on this issue more than on anything else.[37]

Evidently, Gassendi's path to Epicureanism implies a sort of agnosticism on scientific matters that remain mysteries. Epicurus' teaching induces him to moderation and to suspend his own judgment on the nature of comets. Gassendi prefers to limit himself to observational reports rather than speculating on whether comets are fiery bodies arising from the terrestrial realm or produced in the ethereal region, whether their tails are an inflammation or an optical effect, as Cardano believed, and so on and so forth.

> Thus, since we see difficulties arise everywhere, it will be better not to embrace any definition and be satisfied to recount the facts [*historia*] about the comet that recently came to light when it first appeared to us. By solely dealing with that which can be drawn together by means of comparison with other [comets], we shall determine the most likely opinion.[38]

---

36 Pierre Gassendi, *Opera omnia* [Lyon, 1658] (Stuttgart-Bad Cannstatt: Frommann, 1964), vol. 1, p. 700a.

37 *Ibid.*, 701b: "An putas vero, nisi Epicurus determinare voluit utra sententiarum vera esset prae alia, res determinari facilius iam potest? Res sane perobscura est, et utramcumque delegeris, quomodocumque eam exposueris, tot occurrunt difficultates, ut non videatur usquam magis, quam heic continendus assensus."

38 *Ibid.*, p. 704b: "Cum videamus ergo ubique difficultates occurrere; planius est, ut nihil definiamus contentique simus heic animi gratia narrare historiam Cometae illius, qui ut nuperus exstitit, sic visus nobis primus fuit, et, coniiciendo solum, attexere, quid ex collatione ipsius cum caeteris, dicere quis quadam cum specie verisimilitudinis possit. Is Cometa sub finem anni MDCXVIII. et sequentis initium universae Terrae apparuit."

The lack of clarity in Gassendi's position on the nature of comets stands in stark contrast with the atomistic theories of those eclectic revivers of Epicurus preceding him. This is all the more striking, since Gassendi proves so knowledgeable about the scholarly literature on the subject, including neo-atomistic views à la Bruno and Hill. Although neither of the two predecessors is explicitly mentioned, their theses are reported in the doxography. Gassendi traces their opinions back to ancient schools different than the Epicureans. For instance, the planetary nature of comets is ascribed to the Pythagoreans, a notion derived from Aristotle. Gassendi additionally reports that some thinkers refused to consider comets to be *planetae* similar to those of our system; instead, they treated them as wandering stars travelling across the ether like *stellae*. Such a view, Gassendi remarks, is based on an analogy between comets and fish that are visible only when they rise to the surface but disappear as soon as they plunge to the depths again. In one passage where Gassendi explains the cometary theory of Kepler, Gassendi mentions him among the moderns who rely on this well known analogy:

> There are others, among whom Kepler is the most prominent, who believe that the ether [*aura aetheria*] condenses in some places producing orbicular masses. Comets originate from these conglomerates, once they are set in motion and imbued with light. They move through the immensity of the ether just like fish in the vastness of the sea. While we are on shore, we only see the fish that do not pass far from us. In the same manner, we can only see from the Earth those comets whose paths are not distant. Those visible to us receive their light from the Sun. Their tail and its direction, opposite the Sun, can be explained considering that it [the tail] is nothing other than an effluvium from the head, produced by the force of the light that traverses it, until the comet is extinguished, due to the continuous outflow.[39]

39 *Ibid.*, p. 704a: "Sunt alii, quos inter Keplerus praecipuus, qui velint auram aetheriam variis in locis concrescere in moleis quasdam orbiculareis, et hasce moleis seu concretiones motione accepta, luceque imbibita, Cometas evadere; neque ferri minus frequenteis per immensitatem aetheris, quam pisceis per vastitatem maris; tametsi, ut eos solum pisceis qui dum stamus in littore, non procul praetereunt, videmus; ita ex Cometis ii soli apparent, qui a nobis in Terra degentibus non procul semitam instituunt: Istos qui a nobis videntur, accipere a Sole lucem, et caudam idcirco gestare, ipsamque a Sole aversam, quod ipsa nihil aliud sit, quam effluvium ex capite vi lucis traiicientis factum; adeo ut Cometa effluxu continuo demum exsolvatur." Cf. Patrick J. Boner, "Kepler on the Origins of Comets: Applying Earthly Knowledge to Celestial Events," *Nuncius*, 2006, 21: 31–47.

All natural explanations of comets possess some shortcoming in Gassendi's eyes. The thesis of their ethereal origin is dubious in his view because it is not evident that something can be generated from the very subtle matter of the heavens and, even if this were admissible, it would be hard to explain why the cometary conglomerate should be set in motion. Gassendi mentions Kepler's theory of comets' linear motion as well as the circular theory held by others. The Galileo-Grassi controversy over whether the trajectory of comets is linear or circular looms large over this issue.[40] Gassendi limits himself to noting that the circular theory is not supported by empirical evidence, since "no comet is recalled to have come to us twice."[41]

In the section "De effectibus Cometarum in hisce inferioribus" ("On the effects of comets in this lower realm"), Gassendi rejects astrological interpretations of comets as harbingers of war, plague and famine. These are pernicious *fabulae* for Gassendi and should be rejected as superstitions just as Epicurus banned the myths of ancient religions.[42] The meteorological phenomena that accompany the appearance of comets, especially winds, should not be explained by astrological means but simply through the action of light, as is the case with other celestial bodies.[43] The lasting influence of Pico's criticism of astrology is clearly at hand.[44]

Concerning the interdependency of winds and comets, Gassendi discusses this 'observational evidence' in connection with explanations of the origin of comets from the inflammation of sublunary matter in the superior spheres of the air.[45] He also specifies that the concomitance of strong winds (*magna ventorum vis*) and comets is frequent but not always given. I shall soon return to a variation of the same meteorological topos in connection with Placentinus' heterodox reception of Cartesian cometary theory.

One may find more affinity with Renaissance atomists such as Bruno and Hill in Descartes' corpuscular cometary theory than in Gassendi's Epicurean

40 Drake and O'Malley (ref. 13).

41 Gassendi (ref. 36), p. 704b. For Kepler's theory of comets' linear motion, see *ibid.*, p. 706b: "nullus Cometa ad nos bis accessisse memoratur." Cf. James A. Ruffner, "The Curved and the Straight: Cometary Theory from Kepler to Hevelius," *Journal for the History of Astronomy*, 1971, 2: 178–194, on pp. 178–186.

42 Gassendi (ref. 36), p. 712.

43 *Ibid.*, p. 711b: "Certe seu Venti, seu aliae impressiones in aere ad Cometarum seu exortum, seu duratione consequantur, id tribuendum videtur seu luci, seu qualitati alii, ut in Astris caeteris fit."

44 On the developments of the astrological polemics involving comets during the sixteenth century see, among others, Cesare Vasoli, "Andreas Dudith-Sbardellati e la disputa sulle comete," in *idem*, *I miti e gli astri* (Naples: Guida, 1977), pp. 351–387.

45 Gassendi (ref. 36), p. 701b.

Syntagma. Descartes already drafted a natural explanation of comets and their motions in Chapter 9 of *Le Monde*, "De l'origine, et du cours de planetes et des cometes en general, et en particulier des cometes" ("On the origin and course of planets and comets in general, and of comets in particular").[46] Descartes saw comets as celestial bodies crossing different planetary systems in a multicentric in-*de*-finite universe that bore a resemblance to Bruno's infinite universe. Such a legacy is controversial, as Descartes never wished to be connected with the doctrines of the 'impious' Bruno. He explicitly refused cornerstones of Bruno's cosmology such as the existence of a physical void and spatial infinity, which he substituted, however, with ambiguous doctrines such as his theory of the three elements, including a most subtle ethereal one, and that of the indefinite dimensions of the world. In spite of these cautionary measures, Descartes' opponents did not fail to stress the intellectual liaison with compromising forerunners, and they compared his cosmology with that of Bruno.[47] Many among them, in particular his Jesuit opponents, perceived the ambiguity of his positions (including that of the "immobility" of the Earth "transported" by celestial streams around the Sun) as a dissimulation strategy aiming to disguise the fundamental theses at odds with orthodox views that should be upheld for philosophical and theological reasons.[48]

Descartes gave his most detailed treatment of novae and comets in the third book of the *Principia philosophiae* (1644). There, he developed a fully fledged theory that assumes a continuity between various celestial bodies and the reciprocal convertibility of stars, comets and planets. In particular, Descartes offers a corpuscular explanation of the sudden appearance of novae such as that of 1572. The analogy between the Sun and the other stars (reminiscent of Bruno) leads him to assume that the spots observed on the surface of the Sun are common to all stars. In his view, sunspots are impurities emerging on the surface of the starry body through the motion of its subtle celestial matter. In some cases, the entire disc of a star can be obscured by a huge dark spot: "And consequently, it can happen that one and the same spot extends over the entire

46 René Descartes, *The World and Other Writings*, ed. Stephen Gaukroger (Cambridge: Cambridge University Press, 2004), pp. 37–41.

47 Cf. Saverio Ricci, *La fortuna del pensiero di Giordano Bruno* (Florence: Le Lettere, 1990), esp. Chap. 3.

48 Cf. Roger Ariew, "Censorship, Condemnations, and the Spread of Cartesianism," in *Cartesian Empiricism*, ed. Mihnea Dobre and Tammy Nyden (Dordrecht: Springer, 2013), pp. 25–46.

surface of a star and remains there for a long time before being destroyed."[49] As a consequence, stars can appear and disappear:

> Indeed, it can even happen that the spots which cover some star become [with the passing of time] so dense as to entirely conceal it from our view: thus seven Pleiades could formerly be counted, though now we see only six. On the other hand, it can also happen that a star which we have not seen before unexpectedly shines forth with a great light in an extremely short time.[50]

The entire universe is subject to change and transformation according to a principle of cosmological homogeneity. The *tourbillions* of subtle matter deputed to substitute the Scholastic spheres and their function of transporting the planets around their centers, in this case suns, can be altered and dissolve into anything else. If a cosmic whirl dissolves, the star at its center enters another whirl and becomes either a comet or a planet.[51] If it traverses the whirl and moves beyond its boundaries to another system, it becomes a comet; if it begins circulating around the center of the whirl, it becomes a planet:

> [...] if this globe is so solid that, before descending to the point at which the parts of the vortex move the most slowly, it acquires a degree of agitation equal to that of those parts among which it is located; it descends no further, and will proceed into other vortices, and become a comet. On the other hand, if it is not sufficiently solid to acquire so much agitation, and therefore descends below that point [at which the parts of the vortex move the most slowly], it will remain a certain distance from the

49 Descartes, *Principia philosophiae*, 3, 102, trans. Valentine Rodger Miller and Reese P. Miller, *Principles of Philosophy* (Dordrecht-Boston-London: Kluwer Academic Publishers, 1982), p. 139. Cf. *Oeuvres de Descartes*, ed. Charles Adam and Paul Tannery, vol. 7 (Paris: Cerf, 1905), pp. 151–152: "Hincque potest contingere, ut aliquando una et eadem macula sopra totam superficiem alicuius sideris se extendat, ibique diu permaneat priusquam dissolvi possit."

50 Descartes (ref. 49), 3, 104; *Principles of Philosophy* (ref. 49), p. 140. Cf. *Oeuvres de Descartes* (ref. 49), p. 152: "Quinetiam fieri potest, ut aliquod sidus tot et tam densis maculis involvantur, ut visum nostrum prorsus effugiat: sicque olim Pleiades numeratae sunt septem, quae iam sex tantum conspiciuntur. Itemque fieri potest, ut aliquod sidus, nobis antea non visum, brevissimo tempore atque ex improviso, magna luce affulgeat."

51 Descartes (ref. 49), 3, 115.

> star which occupies the center of this vortex; and will become a planet revolving around it.[52]

Stars, novae, comets and planets are not essentially different according to Descartes. They are similar celestial bodies made out of the same matter and subject to the same physical laws. What is more, they are interchangeable and can be transformed into one another. Thus, instead of Gassendi's Epicurean skepticism, Descartes embraces a 'solid' corpuscular vision of comets as celestial bodies traversing the heavens and in all respects akin to the other celestial bodies, *in primis* the planets of the solar system and the Earth itself. As a matter of fact, his cometary theories fall much closer to those of atomists of Kepler's time such as Bruno and his supporters. God's Providence is not banned from the Cartesian *monde* in the name of chance. Rather, it guarantees the existence of nature after its beginning and the laws underlying its dynamics. Still, such a mechanistic view is distant from that of Kepler, since it renounces harmonic proportion together with final causation. According to Descartes, cosmological order is the by-product of matter and its interactions according to the laws of nature rather than the expression of a divine design inspired by geometrical archetypes. Hence, it should be no wonder that his natural philosophy could raise suspicion among others to surreptitiously revive materialist visions that were deemed irreconcilable with piety. The link between Brunian and Cartesian cosmology would become a commonplace of intellectual historiography. The German historian of philosophy Jakob Brucker, for instance, stressed this relationship by arguing, in the fifth volume of his *Historia critica philosophiae* (Leipzig, 1742–1744), that Bruno's "Epicuro-Pythagoreanism" (*systema physicum Epicuri Pythagoreismo incrustatum*) was the most important source of Descartes' cosmology.[53]

---

52 *Ibid.*, 3, 119; *Principles of Philosophy* (ref. 49), pp. 150–151. Cf. *Oeuvres de Descartes* (ref. 49), p. 168: "[…] si globus in illo vortice descendens adeo sit solidus, ut priusquam pervenerit ad terminum in quo partes vorticis omnium tardissime moventur, acquirat agitationem aequalem agitationi earum partium, inter quas versatur, non ulterius descendit, sed ex illo vortice in alios transit, et est Cometa. Si vero minus habeat soliditatis, atque idcirco infra terminum illum descendat, ibi postea ad certam distantiam a sidere, quod illius vorticis centrum occupat, semper manens, circa ipsum rotatur, et est Planeta."

53 Jakob Brucker, *Historia critica philosophiae* (Leipzig: Christoph Breitkopf, 1766), vol. 5, chap. 2, pp. 12–62, "De Iordano Bruno Nolano," on pp. 31–33 and p. 37.

8.4 Novae and Comets from the Viewpoint of Post-Cartesian Astrology: The Case of Johannes Placentinus

After dealing with atomistic, Epicurean and corpuscular approaches to comets in the time of Kepler by two generations of scholars, I would like to consider later developments. As has been argued, Descartes' corpuscular explanation of novae and comets impacted the astrological and meteorological discourses of the sixteenth century. In the Netherlands, his mechanical philosophy contributed to the marginalization of astrology, as it offered alternative and more powerful, integrated explanations of celestial and terrestrial phenomena and their interconnection solely based on material and causal interactions.[54] In Germany, the reception of Descartes was complicated by the fact that its lively astrological tradition led some scholars to undertake a Cartesian reform of astrological doctrines, including those related to comets. The most striking case is the work of Johannes Placentinus, Cartesian professor of mathematics at Frankfurt (Oder). Here, I would like to examine Placentinus' "physical and astrological" report on the "frightening, unusual and damaging winds" that blew in December 1660 and January 1661, *Physicalischer und Astrologischer Bericht Von Denen erschrecklichen, ungewöhnlichen ... und schädlichen Winden* (1661).[55] The title page made clear that the physical-astrological perspective consists, first, in the reliance on "philosophical truths" (*auß Philosophischer Warheit auffgesetzet*) and second, in the explanatory connection of meteorological and heavenly phenomena. The philosophical *Wahrheit* mentioned in the title refers to the author's commitment to Cartesianism. The astrological interdependency of the heavens and earth is evinced by the thesis that the windstorms of 1660–1661 were caused, or at least strengthened, by a new comet and a nova (*neuer Fixstern*) recently observed in the constellation of the Whale.[56]

Placentinus dedicated his publication to the Electoral Prince Friedrich Wilhelm (*Meinem Gnädigsten Churfürsten und Herren*). The preface begins with classical references to Strabo and Pliny, who wrote about winds and the relative pagan god, Aeolus. Placentinus points out that, for Christians, it is the biblical God and eternal king of the world that should be held responsible

54 See Rienk Vermij, "The Marginalization of Astrology among Dutch Astronomers in the First Half of the 17th Century," in *History of Science*, 2014, 52: 153–177.

55 I discuss Placentinus' Cartesian astrology in more detail in "The Mechanization of Astrology in the Context of German Cartesianism," in *Astrology and Anti-Astrology in the Renaissance*, ed. Ovanes Akopyan (in press).

56 On the variable star Mira Ceti (especially around 1660), see Robert A. Hatch, "Discovering Mira Ceti: Celestial Change and Cosmic Continuity," in *Change and Continuity* (ref. 5), pp. 153–176.

for nature's providential unfolding. It is the challenge of the "lovers of natural and astrological arts" (*die Liebhaber der Natürlichen und* Astrolog*ischen Künste*), he argues, to discover its regularities. Evidently, Placentinus embraces Melanchthon's conception of astrology as natural theology, although his scientific 'paradigm' has shifted from Aristotelian hylomorphism to Cartesian mechanism.[57]

Placentinus stresses the main scientific concern at the outset of his report. "Where did these unusual, dreadful, and damaging winds originate from?"[58] The answer to this question requires a preliminary clarification of a series of general propositions, or *Sätze*. The first proposition pertains to hydraulics or, more precisely, to the behavior of fluids (*ein fliessendes Ding*), which according to Placentinus comprise not only the watery element but air and fire as well. This proposition asserts that fluids flow more quickly through narrower passages. Just as sea tides can be explained as a consequence of the pressure exerted on the ether by the Moon forcing terrestrial waters to flow more quickly as it revolves around the Earth, winds similarly blow more strongly in narrow alleys. The second proposition concerns the solar warming of the air surrounding the Earth. Placentinus claims the warming is caused by aerial motions through rarefaction and the production of vapors. The third and fourth propositions concern the fluids' natural tendency to expand and flow, and the production of fluids' flow by means of mechanical motions.

The fifth proposition is astronomical. It asserts that celestial bodies and stars can arise, move and experience annihilation. Among more recent instances, Placentinus mentions Brahe's nova of 1572 as well as that of 1604. Furthermore, Kepler is mentioned as the observer of a nova in the constellation of the Swan that disappeared in 1618, the same year in which a comet appeared in the sky:

> In 1604, a new star was seen in the foot of the Serpent Bearer [*in pede Serpentarii*] until it disappeared in February 1606, just as Johannes Kepler

57 Cf. Sachiko Kusukawa, *The Transformation of Natural Philosophy: The Case of Philip Melanchthon* (Cambridge: Cambridge University Press, 1995) and Dino Bellucci, *Science de la Nature et Réformation: La physique au service de la Réforme dans l'enseignement de Philippe Mélanchthon* (Rome: Ed. Vivere, 1998). On the shifts in approaches to astronomy in late-humanistic Protestant universities, see my essay, "Institutionalized Metaphysics of Astronomy at Early-Modern Melanchthonian Universities," in *Wissen in Bewegung. Institution—Iteration—Transfer*, ed. Eva Cancik-Kirschbaum and Anita Traninger (Wiesbaden: Harrassowitz, 2016), pp. 51–78.

58 Johannes Placentinus, *Physicalischer und Astrologischer Bericht Von Denen erschrecklichen, ungewöhnlichen ... und schädlichen Winden* (Frankfurt an der Oder: Salomon Eichorn, 1661), f. A3*v*: "Woher diese ungewöhnlich, erschreckliche, durchgehende, und schädliche Winde entstanden sind?"

> wrote in *De stella nova in pede Serpentarii*. Also, in 1606 a new star was observed in the breast of the Swan. It remained there for four years according to Kepler, but in 1618, according to Krüger, it disappeared with the appearance of a comet.[59]

Placentinus remarks that the dimensions of these disappearing bodies were huge, as they had been calculated by scholars such as the Copernican astronomer Philip Lansbergen. This observation is at the heart of Placentinus' theory that the annihilation of novae provokes great movements of celestial matter and this, in turn, affects the streaming of air around the Earth. In fact, he argues that the disappearance of a nova does not imply its destruction but rather its transformation into a comet; Placentinus takes this thesis from Descartes' *Principia* 3, 119.

Following his general propositions on the physics of fluids, Placentinus explains the natural formation of winds. This section is entitled "Von den natürlichen und gewöhnlichen Winden in gemein" ("On the Natural and Usual Winds in General"). They are thought to be an effect of solar heat or mechanical displacements and collisions of masses of vapors also known as clouds. Placentinus presents here his *Aeolipyla*, a machine devised to produce "artificial winds." It is a metallic sphere of brass or copper, which functions like a pressure cooker: some water is poured into the sphere through a tiny hole; then the instrument is heated with fire; the heat transforms the water into vapor, which comes shooting out of the hole with great force due to the narrowness of the passage. The generation of winds by solar heat is similar.

The next section of the booklet deals with the strong windstorms of 1660–1661: "Von unsern erschrecklichen Winden insonderheit" ("On Our Frightening Winds in Particular"). Placentinus argues for the anomaly of these winds. As he claims, they traversed Europe, reaching all the way to Turkey. Hence, their causes were not local but 'universal.' They must have been astronomical or, rather, physico-astrological, according to the following account:

> Hence, I should add another cause to those given to account for these great winds. I would like to briefly mention that a star disappeared from the sky in the East and was transferred to another place in the heavens

59 *Ibid.*, f. B2r: "*Anno* 1604 ist [...] ein neuer Stern in pede Serpentarii gesehen worden, biß anno 1606 in Februario derselbe verschwunden ist, als davon schreibet Johannes Kepplerus, *De nova Stella in pede Serpentarii. Item anno* 1606 [*sic*] ist in der Brust des Schwans ein neuer Stern, welcher vier Jahre gestanden, observiret worden, teste Keplero, welchen man aber anno 1618, da ein Comet erschienen nicht mehr gemercket hat, Crügerus."

> (according to my fifth proposition), shortly before the winds enraged and rampaged [the land]. Assuming that this star is [...] a *stella obscura* whose magnitude is only to be compared with that of the Earth, that star left behind a large space reckoned as massive as the Earth. This [large space] could not remain empty on account of the *fuga vacui* [i.e., nature's flight from void]; the air and mist around the Earth flowed from the West [...] with great violence for several hours until the space was filled. They must have produced those universal and continuous winds (according to my third proposition).[60]

In order to corroborate his views, Placentinus printed a letter from the world-renowned Gdańsk astronomer (*Weltberühmter Astronomus*), Johannes Hevelius, describing a comet and a nova that had been briefly visible in the constellation of the Whale.[61] Placentinus lamented the poor meteorological conditions in Brandenburg at the time of the comet. The clouds had prevented him from sighting the celestial novelties described by Hevelius. In addition to providing a witness account from an expert observer, Hevelius' report reinforced Placentinus' theses. The nova's disappearance coincided with the blowing of the great windstorms. Consequently, Placentinus had no doubt that a physico-astrological connection between the two phenomena could and should be established.

The strictly astrological treatment occupies the last part of the treatise on windstorms. For Placentinus, there is nothing mysterious about the action of

60 *Ibid.*, ff. B4*r*–*v*: "Und darumb nebenst diesen angezogenen Ursachen, noch eine andere seyn muß, welche diese grosse Winde verursacht habe; Und solche wil ich alhier kürzlich setzen, und sprechen, daß vor derselben Zeit, ehe die Winde gewütet und getobet, ein Stern am Himmel im Orient [...] sey verschwunden, und in einem andern Orth des Himmels transferiret worden, nach dem fünfften Satz, Und gesetzt, dieser Stern sey [...] stella obscura welche der Erd-Kugel nach der grösse nur zu vergleichen sey: so wird dennoch derselbe Stern einen grossen Raum, welcher der Corpulentz der Erden gleich zu schätzen ist, gelassen haben, und derselbe hat nicht können leer seyn ob fugam vacui; sondern die Lufft und Dünste umb die Erden, haben von Westen [...] mit grosser Gewalt, und etliche Stunden lang, biß der Raum ist erfüllet worden, Fliessen, und solche Universal und durchgehende Winde verursachen müssen, vermöge des dritten Satzes."

61 *Ibid.*, f. C2*v*, "Vom Cometen. Welcher den 3. Febr. Styl. N. erschienen, und vom Neuen Fix-Stern im *Balena* oder Wallfisch, der von dem Weltberühmten *Astronomo*, (tit.) Hr. Johan Hewelcken, etc. in Dantzig *observiret* worden." I discuss the correspondence between Placentinus and Hevelius elsewhere, in "Asymmetries of Symbolic Capital in 17th-Century Scientific Transactions: Placentinus's Cometary Correspondence with Hevelius and Lubieniecki," in the proceedings of the conference "Institutionalization of Science in Early Modern Europe," ed. Giulia Giannini (in press).

celestial bodies on terrestrial phenomena. Cartesian physics offers a mechanistic explanation of astrological influence. In the concluding pages of the *Physicalischer und Astrologischer Bericht*, Placentinus turns to astrology to cast wind horoscopes and assess the consequences of celestial conditions not only on elementary phenomena on earth but also on human affairs. The final section, entitled "Unser *Astrologi*scher Bericht," first investigates the "astrological causes" (*Astrologishe Ursachen*) and then interprets their "meaning" (*Bedeutung*) by means of special horoscopes.

At 6:00 AM on 16 November 1660, the Sun, Mars and Mercury, the "common signifiers of winds" (*communes ventorum significatores*), were in conjunction. Moreover, at the time of the strongest windstorms, at 6:00 AM and 12:00 PM, all of the planets were below the horizon. Just as these conditions suggested "something unusual and strange" (*etwas ungewöhnliches und seltzomes*) in nativities cast for the microcosm of man, they were said to anticipate some exceptional events in the macrocosm as well. These and further astrological considerations led Placentinus to assert the extraordinary nature of the recent windstorms.

Placentinus' interpretation (*Bedeutung*) follows in three steps. First, he explains that winds alter the air and purify it, just as purges cleanse the human body. Secondly, since windstorms are heralds (*Vorboten*) of comets according to his theory, they also announce the same negative effects, that is to say, war, plague and misery. Thirdly, exceptional winds also reveal that the *Jüngster Tag*, the end of the world and the Final Judgment, are coming. A physical explanation is not lacking, although it is intricate: sunspots are signs of the last day; in fact, at the end of the world the entire surface of the Sun will be obscured; according to Cartesian theory, when a star is completely obscured by dark spots, it can be transformed and transferred elsewhere in the form of a comet; the obscuration of the Sun will thus prelude its transformation into a comet and its consequent displacement; as a result of such metamorphosis, terrific winds will be generated, which are an apocalyptic sign according to the Scriptures. Ultimately, Placentinus' Cartesian astrology culminates in a bizarre Cartesian eschatology:

> Hence, these unheard-of winds are also harbingers of the Last Judgment. As I explained before, this [phenomenon] occurs when a star disappears from the heavens, if sunspots [*dergleichen maculis*] entirely cover and hide a heavenly light and the latter is therefore transported to another place, as René Descartes explains in the third part of the *Principles*. Thus, when the great light of the Sun shall be completely obscured by those

spots—together with other earthly and heavenly signs—frightening winds will also originate, and the forces of the heavens will set themselves in motion, as announced by the Holy Bible.[62]

8.5 Conclusion

The nova of 1604 rekindled the question of the corruptibility of the heavens already fueled by debates surrounding the celestial novelties of the 1570s and the invention of planetary systems, such as the Tychonian, which presupposed the fluidity of space. The appearance of unpredictable phenomena such as the novae of 1572 and 1604 also acted to resuscitate materialistic views. In the eyes of Kepler, new and old followers of Epicurus revived ideas of cosmic disorder at odds with the belief that divine wisdom underlies the Creation. Renaissance debates on novae and comets had paved the way for the acceptance of the principle of cosmological homogeneity and for novel conceptions of nature and physics. New theories would account for the possibility of change in the heavens, the very possibility of which had been banned from Scholastic philosophy in the wake of the Aristotelian doctrine of the two physics—in particular, the distinction between the sublunary realm of change and the superlunary one of incorruptible perfection.

In *De stella nova*, Kepler voiced his concern over the resurgence of Epicurean chance. He viewed such a philosophy as radically opposed to his own trust in the geometrical proportion and inner harmony of the heavens—a manifestation of God's providential design. In spite of his concern, I could not trace any relevant Renaissance follower of Epicurean atomism ready to renounce divine providence. Rather, they grasped the guidance of God elsewhere. Instead of mathematical perfection, Bruno and his followers—among whom Hill and possibly Prague courtiers with whom Kepler was familiar—assumed that spatial infinity and the existence of countless beings in the universe better suited

62 *Ibid.*, f. D4*v*: "Also seynd auch Vorboten des grossen Tages, solche unerhörte Winde, welche entstehen, wie oben dargethan, wann ein Stern im Himmel verschwindet, solches aber geschiet, wann ein himmlisches Liecht mit dergleichen *maculis* gäntzlich umbgeben, bedecket, und deßwegen in einem andern Ort versetzet wird: als dieses demonstrieret Renatus des Cartes, *in tertia parte Principiorum*. Wenn also das grosse Liecht der Sonnen, mit denen *maculis* gäntzlich wird tegieret werden, als werden nebenst andern irrdischen und himmlischen Zeichen, auch erschreckliche Wind entstahen, und werden der Himmel Kräffte sich bewegen, wie die heilige Schrift davon meldet."

the Almighty than order and finitude.[63] In this sense, even though Kepler was wrong in assuming that the modern admirers of Epicurus might revive the rule of chance, he correctly stressed that celestial novelties might reinforce “those philosophers to whom confusion is pleasing rather than order,” and “who have debased the most beautiful world with intolerable atoms.”[64] Seen from a Keplerian outlook, Brunian atomism and later corpuscular visions found common ground in their criticism of harmonic finitude and geometrical *concinnitas*.

In this essay, I have considered significant Renaissance and early seventeenth-century instances of cometary theories that relied on atomist and corpuscular philosophies. Bruno and his followers treated comets as celestial bodies that are similar to the planets and the ‘planetary’ Earth. Descartes went as far as assuming the interchangeability of stars, planets, comets and novae, seen as the products of material transformation processes in the heavens. It is remarkable that his cosmology, as well as his planetary conception of comets, resembles the Brunian conception more closely than that of Gassendi. The latter, a renowned French Epicurean, rather expressed an agnostic attitude, based on a skeptical interpretation of Epicureanism. Finally, I considered the Cartesian astrologer Placentinus’ attempt to bring together the legacies of Descartes and Kepler, that is, to connect mechanical explanations of nature with the Keplerian theory of astrological influences. Placentinus interpreted comets as eschatological signs, in spite of his mechanical explanation of these heavenly phenomena. From the viewpoint of the crisis of heavenly harmony, his syncretistic reading of comets as celestial harbingers of calamity is traditionally not disjointed from the move away from the archetypal understanding of cosmic beauty.

In the end, Kepler rightly sensed that phenomena such as the nova of 1604 were leading contemporary cosmologists towards a philosophy of contingency and disorder, abandoning the principle of harmony of classical astronomy.

---

63 On the Imperial Chancelor Johannes Matthaeus Wacker von Wackenfels as one source of Epicurean sympathy in Prague, see the contribution by Dario Tessicini to this volume.

64 Cf. *JKGW*, 1, 285.40–286.3: “Soli hi philosophi, quibus pro ordine placet confusio, qui mundum pulcherrimum inter inamoenas atomos abjecerunt, hoc sunt meriti, ut pulchritudine et congruentia hujus, quem dicunt, casus, nec ipsi moveantur, nec alios delectari patiantur.” English translation by Patrick J. Boner from *idem* (ref. 6), p. 211.

CHAPTER 9

The Correspondence of Clavius, Dal Monte, Magini and Other Italian Astronomers on the Nova of 1604

Matteo Cosci

9.1 Introduction

After the appearance of the 'new star' in October 1604, the most prominent astronomers operating in Italy immediately turned to their epistolary contacts to exchange information and opinions about the unexpected phenomenon. Through their letters, they actively petitioned and proposed news about the nature, position and significance of the *stella nova*. An extremely rare astronomical event of such great impact as the emergence of a bright new body in the sky soon ignited widespread curiosity for trading updates and judgments from both mathematicians and natural philosophers who witnessed it, if not to win renown for the best or most timely explanation, then in some instances to reassure the public of its presumed absence of dangerous implications for everyday life. Only the appearance of the new star of 1572 could have provided a relatively analogous case study, though even if it represented an important reference for comparison, it had not been understood in a single or definitive way by the time of the *nova* of 1604.

The present contribution aims to retrace and review the interplay of these epistolary exchanges and the extant information they contain. In particular, I will focus on the epistolary contacts of high-profile figures such as Christoph Clavius (1538–1612), Giovanni Antonio Magini (1555–1617) and Guidobaldo Dal Monte (1545–1607). In a subsequent contribution, the relevant correspondence of Galileo will also be taken into account, although it must be said from the start that Galileo played a leading role in the exchange of assessments on the nova and was even one of the first scholars to present a scientific account of it through public claims about its controversial distance in a series of academic lectures at Padua that took place no later than two months after the stellar outburst. Aside from Galileo, relevant sections of those other exchanges are considered here, especially during the early months after the appearance. Each of the three main authors—Clavius, Dal Monte and Magini—will thus be analysed specifically with regard to this stellar phenomenon. In turn, each

 | DOI:10.1163/9789004437272_011

section will be divided into two parts and feature alternate voices in order to recreate in a better way their exchanges over questions they addressed and the answers they received from one another. Some digressions will be necessary during the narrative to identify nameless correspondents, discuss the widespread astrological background that accompanied the astronomical study of the phenomenon, account for some separate documents that were enclosed in the correspondence, and finally to supply some missing links in the chain of communication that, as it turns out, also took place by way of various intermediaries.

With few exceptions, all of the various forms of exchange, mutual connection and paths of postal contact among the following astronomers were largely inspired by their common concern for explaining the great novelty in ways that more or less agreed with the traditional Aristotelian-Ptolemaic cosmology. Our purpose here is to show how the exchanges of leading Italian astronomers were directly or indirectly interconnected and how they shared a series of common assumptions. At the completion of my overview, it will thus be clear how an actual network of scientific correspondents was operating at that time, both among clergymen and academic scholars, both within the peninsular boundaries and beyond the Alps. This network was efficient enough to exchange information and data in a fast and effective way, particularly given the diversity of views and methodological assumptions. In fact, the circulation of ideas travelled transnationally on the edge of tension felt between more conservative intellectuals and those *novatores* more disposed to renewing traditional knowledge, particularly in the light of the newly observed phenomena that, for many, ackwardly fit the old epistemological framework that still held sway among the majority of natural philosophers.

Many words have already been expended on the importance of epistolary exchanges in the Renaissance as a literary genre. In fact, letters have been correctly considered a central means of spanning the long distances and levels of social hierarchy across European society, as vehicles of the cultural diffusion of ideas, meta-biographical documents that were coded with rules of their own, etc.[1] But the analysis of epistolary literature as a primary source of study proves

1 See, for instance, *Letter Writing in Late Modern Europe*, ed. Marina Dossena and Gabriella Del Lungo Camiciotti (Amsterdam-Philadelphia: John Benjamins, 2012); Judith R. Henderson, "Humanist Letter Writing: Private Conversation or Public Forum?," in *Self-Presentation and Social Identification: The Rhetoric and Pragmatics of Letter Writing in Early Modern Times*, ed. Toon Van Houdt, Jan Papy, Gilbert Tournoy, and Constant Matheeussen (Leuven: Leuven University Press, 2002), pp. 17–38.

to be remarkably significant in this context for a series of further reasons that will be briefly outlined below.

First, mapping the ongoing network 'in action' allows us to trace the circulation of astronomical knowledge of that time and appreciate its extensiveness and pervasiveness. Second, this allows us to enter a forum of lively debate and participate in the most controversial discourses, appreciating the theses that had once been advanced in a form that was set halfway between official exposition and unofficial working hypotheses for the interpretation of the given phenomenon. In general, we may regard the letters among scientific correspondents as a sort of semi-public pre-print, that is, as accounts that were provisional and revocable and yet in some cases very close to their final formulation; they were, of course, addressed to their explicit recipient, but they were often conveyed to a circle or team of associates with whom they were soon shared after their reception. To offer the most telling example, letters with scientific content that were addressed to a specific member of the *Collegio Romano*, especially if they contained or requested observational data, were immediately conveyed to other colleagues from the same institution. And, as we shall see, Guidobaldo Dal Monte often made reference to the participation in epistolary discussion of voices that could not be reduced to the simple sender/receiver couple, thereby highlighting a procedure that was custom and routine. As a result, letters had a *relatively* private dimension; scientific letters in particular could be commented upon and quoted, they could be preserved for reference for a long time and later made public after some more or less selective editing.[2] (An example is the epistolary treatise *De nova stella disputatio* (1605) by Jan van Heeck, which was heaviliy edited by Federico Cesi for a publication[3] that

2 See Adam Mosley, Nicholas Jardine, and Karin Tybjerg, "Epistolary Culture, Editorial Practices, and the Propriety of Tycho's *Astronomical Letters*," *Journal for the History of Astronomy*, 2003, 34: 421–451, on pp. 422–425: "Early-modern letter writers enjoyed a larger audience than we might expect. [...] Letters were copied and circulated widely in manuscript, and frequently they found a resting place in a collection of letters: that is, in an archive or a letter-book. [...] [L]etters could also be valued because of what they revealed about current or past events (as, for instance, astronomical precedents), the individuals who wrote or received them, or the scholarly enterprises in which they were engaged. Indeed, in the era before the genre of the essay had fully emerged, and before scholarly journals were established, the letter became a favoured medium of academic discourse. Early-modern scholars were well aware that manuscript letters could be circulated, collected, and eventually printed [...] [S]ince a letter could be just as much of a forum for displaying both literary ability and scholarly expertise as a treatise, writing letters was a core component of intellectual life, much like the writing of articles today."

3 Jan van Heeck (Johannes Heckius), *De nova stella disputatio* [Prague], ed. Federico Cesi (Rome: Aloysius Zanetti, 1605). For a modern transcription of the original text, see Barbara Caredda, *Aspetti e momenti del dibattito astronomico nella prima Accademia dei Lincei*

went far beyond the author's will.)[4] The form of the letter as such was suitable for the creation and consolidation of spheres of influence and fields of contacts, even over great distances in space and time, besides being a means of giving and receiving new or recent information.[5]

The date of composition, in particular, is a fundamental aspect when there are measurements at stake, just as is the specification of the place from which information comes, that is, the location from which the measurements have been taken; in all of this, letters performed an essential informative function. Hence, the analysis of indications such as the date and place of dispatch will be considered as an integral part of the following survey.

One final clarification must be anticipated here in order to highlight the recurring scheme underlying the questions accompanying the study of the great novelty of 1604. For many, concerns about the star present a sort of 'pattern,' that is to say, a logical consequentiality that can pass by the modern reader unnoticed but is rather crucial. As will be seen, in fact, the discussion among experts tended to focus mainly on the problem of establishing in the most precise way the *parallax* or angular distance from the earth[6] of the new object as it appeared to observers. Why, then, did the quarrel almost always start or end with the discussion of this aspect in particular? The measure of parallax might seem like just one observable feature among many others, including colour, twinkling or magnitude; but reaching a consensus, if any, on this point was considered to be of fundamental importance for many scholars. The reason is that parallax was an indispensable measurement for establishing the precise altitude, or distance from the earth, of any new object. As we shall see, not everyone was inclined to accept parallactic calculation as a reasonable or reliable method, but in a universe imagined to be a globe made up of concentric

---

(*1603–1616*) (PhD Dissertation, University of Cagliari, 2008), Appendix 1, pp. 215–251, *veprints.unica.it/482/1/b_caredda_td.pdf.*

4 See Saverio Ricci, "Federico Cesi e la *nova* del 1604: La teoria della fluidità del cielo e un opuscolo dimenticato di Joannes van Heeck," *Atti dell'Accademia Nazionale dei Lincei: Rendiconti*, Classe di scienze morali, 1988, 43: 111–133, then as: "'... *et iam Aristotelis denegant coelestia*': Federico Cesi e la *nova* del 1604," in *"Una filosofica milizia": Tre studî sull'Accademia dei Lincei*, ed. *idem* (Udine, Campanotto: 1994), pp. 7–31; *idem*, "Il caso Heckius," in *I primi lincei e il Sant'Uffizio: Questioni di scienza e di fede* (Rome: Bardi, 2005), pp. 207–234.

5 Gábor Almási, "Humanistic Letter-Writing," in *European History Online* (Mainz: Institute of European History, 2010), section 1: 16; http://www.ieg-ego.eu/almasig-2010-en.

6 See Norriss S. Hetherington, "Parallax," in *The Oxford Companion to the History of Modern Science,* ed. John L. Heilbron (Oxford: Oxford University Press, 2003), p. 616; Tofigh Heidarzadeh, *A History of Physical Theories of Comets, from Aristotle to Whipple* (New York: Springer, 2009), pp. 39–41; Steven Weinberg, *To Explain the World. The Discovery of Modern Science* (New York: Harper Collins, 2015), esp. no. 20 on diurnal parallax, pp. 321–323.

spheres, such a method would have allowed for an understanding of the altitude or spherical level to which the new star of 1604 should have been assigned. According to Aristotle, not all levels were qualitatively homogeneous, and the sharp distinction marked by the sphere of the Moon still held sway. Below the lunar sphere, there was the region of change and elementary alteration, beyond which extended the ethereal region of eternal inalterability of the planets and stars. Measuring parallax would have helped astronomers establish with confidence whether the new appearance should have been considered below or beyond the sphere of the Moon. Reaching an agreement on this point would have allowed scholars to establish subsequently whether the nova was to be intended as an actual star if it turned out to be beyond the Moon, or a tailless comet or even some sort of meteorological phenomenon if it turned out to be below it. The first option would have been contrary to Aristotelian knowledge, albeit well established for centuries, while the second would have confirmed that tradition. The problem, however, was that those who obtained the measurement of a supralunar altitude were not always inclined to accept, on the basis of their own evidence, that it should have been a star (the example of Dal Monte shall be emblematic), thus underscoring the endurance of the Aristotelian scholastic tradition, arguably to the detriment of the empirical phenomenology of observations and their mathematical value. Establishing whether it was a comet or a star would have allowed observers to offer a better assessment of the potential repercussions, both theoretical and practical, that such a new appearance might have made on their worldview. For if it had been agreed that it was a fixed star, it represented a second, third or fourth such anomaly following the nova of 1572 and those lesser known novelties of 1596 and 1600. Moreover, the constellation that turned out to 'host' the new star would have to be reconfigured according to a novel system of influences (for horoscopes and *genitures*) that would now have to be recalculated in the light of the new astrological balance—a change that would have been very hard to accept for many. Instead, if the object was a comet then it would have been a more reassuring option at the level of 'maximal systems,' but it would have certainly been just as troubling for its direct meteorological consequences, since comets were believed to be announcements of droughts, storms and other calamities, as well as for its indirect effects on people, since comets were also interpreted as signs of imminent epidemics, wars and revolutions. Therefore, the problem of establishing the parallax often stood before opting for one of the two alternatives. But even after a conclusion seemed to be reached on the matter, many questioned the measurement of parallax as such, a datum that at that time was widely deliberated and doubted in the absence of more precise instrumentation.[7] To avoid this problem and to make the measurement as precise and undebatable as possible, some collected and compared observations from geographical points far removed from one another to obtain a higher angle of perspective verification. Such a strategy allowed observers to understand whether the celestial object under study was actually static or moving along some sort of trajectory. In addition, the result would have allowed scholars to determine whether it could be attributable to a short cometary contrail in slow motion or to the relative immobility of a star. Covering in person the great distances from one observation point to another was highly difficult at this time due to the simultaneous coordination required between different places. Such a contingency again explains the utility and necessity of a *written letter*, a type of document that by its nature was 'geo/chrono-localized' by virtue of the date and place of posting with which it was always marked. For this simple reason, signed letters were written and employed as the proper formal format for sharing updated *Berichte* or scientific observation reports. In this way, one's own results could be compared retrospectively with those of others, including the measure of parallax, the ascertained moment of appearance and every other observable feature, regardless of removal(s) in space and time among the correspondents.[8]

9.2 Clavius' Correspondence on the New Star (1)

An early node in the correspondence network concerning the new star can be found between the illustrious Jesuit mathematician Christoph Clavius, then the head of the Collegio Romano,[9] and Galileo, at that time a relatively young mathematician and astronomer at Padua. Why was such a great authority in

7 For one example of a skeptical voice and his vigorous struggle with Kepler, see the contribution by Patrick J. Boner to this volume.

8 See Adam Mosley, *Bearing the Heavens, Tycho Brahe and the Astronomical Community of the Late Sixteenth Century* (Cambridge: Cambridge University Press, 2007), pp. 165 and 295: "By means of correspondence, early modern astronomers were able to form a community of practitioners with common interests, a powerful strategy for overcoming the difficulties associated with the construction of a universally valid astronomy from any one centre [...] The bridging of the geographical interval between individuals, and of the intellectual gaps between different institutional environments, was an important consequence of the production and circulation of the letters, books and instruments [...]."

9 See *Cristoforo Clavio e la cultura scientifica del suo tempo*, ed. Paola Vasconi (Rome: Gangemi, 2015); *Christoph Clavius e l'attività scientifica dei Gesuiti nell'età di Galileo*, ed. Ugo Baldini (Rome: Bulzoni, 1995).

touch with Galileo on this issue? Although they knew each other, Galileo was not yet widely known among astronomers, while Clavius was an established scholar renowned internationally in the field.

To offer some context, it should be recalled that at that time astronomical subjects were not Galileo's main concern and that he had not built his famous telescope yet. Moreover, Galileo had not yet spotted the star himself, as he was busy with the preparation of his ordinary lectures and some problems primarily on mechanics. The first stargazer who gave notice to Galileo was Ilario Altobelli (1560–1637), an enthusiastic astronomer and colleague who sent him a timely series of detailed sightings from Verona at the end of November 1604. Altobelli continued to update him during the following months. Galileo probably used first-hand information provided by Altobelli as reference material for his commissioned keynote lectures at public meetings that took place at the University of Padua (Studio di Padova) between the end of November and the beginning of December of the same year. Those lectures proved very succesful and provoked a lot of discussion.[10] Their echoes and polemical repercussions would soon reach Rome.

The link with Clavius is constituted by the following letter addressed to Galileo some weeks after his public lectures in Padua. Galileo had previously met Clavius during his visit to Rome in 1597, and since then he had managed to maintain epistolary contact with him.[11] The main aim or pretext of Clavius' letter on this occasion was to offer his correspondent some of his recent publications, such as his *Practical Geometry*,[12] his *Clock*,[13] together with a short compendium,[14] as well as a copy of his less recent essay on the *Astrolabe*,[15]

---

10 This background has been considered together with Altobelli's letters to Galileo in Matteo Cosci, "Le fonti di Galileo per le lezioni e studi sulla nuova stella del 1604," *Archives Internationales d'Histoire des Sciences*, 2018, 68: 6–71. On Altobelli, see also *Ilario Altobelli: scienziato, teologo, corrispondente di Galileo Galilei*, ed. Alessandro Giostra, Francesco Merletti and William R. Shea (Camerano: Empatiabooks, 2011); Alessandro Giostra, "Ilario Altobelli e il cosmo medievale," in *Atti del V Convegno Internazionale su Francesco d'Appignano*, ed. Domenico Priori (Jesi: Terra dei Fioretti, 2011), pp. 263–277.

11 See Christoph Clavius, *Corrispondenza: (1570–1592) Lettere e testi*, vol. II.1, ed. Ugo Baldini and Pier Daniele Napolitani (Pisa: Univ. di Pisa, Dipartimento di matematica, 1992), pp. 115–121, letter nos. 42–45. For a selection of Clavius' (digitalized) letters, see now http://claviusontheweb.it.

12 Christoph Clavius, *Geometria practica* (Rome: Aloysius Zanetti, 1604).

13 Christoph Clavius, *Horologiorum nova descriptio* (Rome: Aloysius Zanetti, 1599).

14 Christoph Clavius, *Compendium brevissimum describendorum horologiorum horizontalium ac declinantium* (Rome: Aloysius Zanetti, 1603).

15 Christoph Clavius, *Astrolabium* (Rome: Bartholomaeus Grassus, 1593).

in case Galileo had not already received it, in exchange for one of those compasses "for geometrical-military use" that Galileo was profitably hand-crafting in his Paduan workshop.[16] Then, as if *en passant*, Clavius wrote near the end:

> Here [in Rome] there have been many whispered rumours about the new star, which we [at the Collegio Romano] have found to be at 17° Sagittarius, with a northern latitude of about 1½°. I would be happy to learn of your observation(s), in case you have made any. Magini wrote to me that he had also observed it to be at the same degree; and the same is confirmed [to me] from Germany and Calabria as well.[17]

There is nothing in Clavius' correspondence that gives us any clues about the information arriving "from Germany." It is probable, however, that those sources were not addressed directly to him but to his Roman assistants or collaborators, such as his brethren Odo Van Maelcote (1572–1615) and Christoph Grienberger (1561–1636), and that Clavius later learned of the novelty through them.[18]

On the other hand, the information that came to Clavius "from Calabria" refers to an extant letter that had arrived in Rome two months earlier from an unidentified physician and mathematician from the village of Paterno.[19] The author wrote to announce the appearance of the great novelty in the constellation of Sagittarius right where Mars and Jupiter were meeting. Moreover, he confirmed that from his point of observation the appearance took place on 9 October and not before, and that he was certain of the date because he had kept his eyes on the area of the sky where the expected planetary conjunction would occur. The sight of three points of light so close to one another, and in particular the sudden appearance of the surprise source, provoked "great wonder" in him, he wrote.[20] The author added that he carried on his observations for the following few nights, noticing from the start an incremental

16 On this instrument, Galileo would soon publish his brief treatise, *Le operazioni del compasso geometrico et militare* (Padua: Pietro Marinelli, 1606). See *OGG*, 2, 365–424.

17 Christoph Clavius to Galileo Galilei, 18 December 1604, in Clavius (ref. 11), no. 240, pp. 136–137: "Qui è stato un gran bisbiglio della stella nova, la quale abbiamo trovata nel 17 grado di ♐, con latitudine borea di gradi 1 ½ incirca. Se V. S. ha fatto qualche osservatione, mi farà piacere d'avisarmi. Il Magino mi scrive d'haverla anco lui osservata nel medesimo grado; et così anco scrivono di Germania e di Calabria." (*OGG*, 11, 121).

18 See Clavius (ref. 11), no. 10, p. 75.

19 Anonymous to Christoph Clavius, 18 October 1604, in *ibid.*, no. 233, pp. 125–126.

20 *Ibid.*, p. 125.

intensification of the luminosity of the novel body. As time went by, it reached the splendour of Jupiter, and he assumed that the star would soon surpass it if it continued the same incremental progression. The observer then claimed that the phenomenon was undoubtedly "of great significance."[21] Whatever it meant, he observed complete silence about what could have ever been foretold from it. Astrological predictability was a rather onerous task and perhaps one that was not apt for discussion with a member of the Collegio Romano. The letter ended with a request for consultation regarding what could be observed and claimed from Rome.[22]

Just one day after the anonymous letter was written, Giovanni Giacomo Staserio (1565–1635) expressed the same worried tone in his own report to Clavius. A native of Bari, Staserio taught mathematics at the Jesuit college in Naples.[23] His apparent intent was to pronounce the novelty of the phenomenon, as well as ask for elucidation on the appearance of a celestial object that, he said, had certainly never been spotted by his predecessors. Staserio focused his attention on the twinkling, superior to that of the other stars, and colour, inflamed like the reddish hue of Mars. With regard to location, Staserio remarked on the proximity to Saturn. The moment of first appearance is backdated by two weeks, namely to 5 October, but it sounds like a vague approximation. Finally, Staserio compared the size of the star to the well known nova of 1572, of which Clavius was one of the most qualified and authoritative eyewitnesses.[24]

On 6 November, Clavius received another letter from Giovanni Antonio Magini, professor of astronomy at Bologna,[25] whose consultation, as reported, Clavius had already mentioned in his letter to Galileo. Clavius and Magini clearly held each other in the highest esteem. Six years earlier, Clavius had succesfully exercised his influence to support the candidature of Magini for the vacant teaching post at Bologna. Their early correspondence on the nova thus came naturally in response to such a major astronomical event. Magini quickly wrote back to Clavius asking for more details about his observations

21 *Ibid.*

22 We shall return to this letter in the following section to say something further about this otherwise anonymous author.

23 Giovanni Giacomo Staserio to Christoph Clavius, 19 October 1604, in Clavius (ref. 11), no. 234, pp. 126–127.

24 See below.

25 Giovanni Antonio Magini to Christoph Clavius, 6 November 1604, in Clavius (ref. 11), no. 235, pp. 128–129.

on the great novelty. He credited Altobelli with the initial discovery,[26] in particular for having brought to his attention the peculiar intermediate position between Mars and Jupiter. Although Magini had only been able to conduct non-instrumental observations, he still concluded that the bright light was "high above the planets" due to its twinkling and great duration. All things considered, he probably suspected that it was a star among the others and not some kind of meteorological disturbance.

A dozen days later, Clavius replied to Magini.[27] In his response, Clavius conveyed the report of the anonymous physician-mathematician. The date of 9 October seemed to him more plausible than the one of four days earlier claimed by Staserio. In any case, Clavius replied, more precise observations had also been conducted from Rome with the help of positioning instruments, confirming the preservation of the same distance from the surrounding fixed stars, including stars from the constellations of Arturus, Lyra and Cygnus. The conclusion seemed to be that such a new luminous object was to be located "in the heavens,"[28] beyond the elementary region of the sky where, in theory, change was not possible.[29] Once this letter had been received by Magini, he then forwarded it to Altobelli, who in turn informed Galileo of its contents in his own letter of 30 December;[30] Magini probably also forwarded the same information to Franz Tengnagel (1576–1622),[31] advisor to the emperor in Prague and son-in-law of Tycho Brahe, as Kepler later reported.[32] Some days later, Jan van Heeck, a member of the Accademia dei Lincei, wrote from Prague to

26 Antonio Favaro, *Carteggio inedito di Ticone Brahe, Giovanni Keplero e di altri celebri astronomi e matematici dei secoli XVI e XVII con Giovanni Antonio Magini tratto dall'Archivio Malvezzi De' Medici in Bologna* (Bologna: Zanichelli, 1886), p. 284, no. 2.

27 Christoph Clavius to Giovanni Antonio Magini, 18 November 1604, in Clavius (ref. 11), no. 236, pp. 129–130. This letter can also be found as part of Magini's correspondence in Favaro (ref. 26), no. 30, pp. 283–284.

28 *Ibid.*

29 See Edward Grant, "Celestial Incorruptibility in Medieval Cosmology 1200–1687," in *Physics, Cosmology and Astronomy, 1300–1700: Tension and Accomodation*, ed. Sabetai Unguru (Dordrecht: Springer, 1991), pp. 101–127.

30 Ilario Altobelli to Galileo Galilei, 30 December 1604, in OGG, 10, 283–284.

31 See Domenico Bertoloni Meli, "Guidobaldo Dal Monte and the Archimedean Revival," *Nuncius*, 1992, 7: 3–34 (esp. § 4 *The nova of 1604*), on p. 27: "Magini acted as an important link between Tycho's son in law Franz Tengnagel at Prague and Christophorus Clavius in Rome."

32 *De stella nova*, JKGW, 1, 158.15–18: "An aemulatione ductus uno die praevenire voluit allum quendam Italum (puto Altobellum Veronensem vicinuum suum, si bene eorum memini, quae mihi Franciscus Gansneb Tengnaglius S. C. Majest. Consiliarius ex literis J. Antonii Magini retulit) [...]" In his earlier letter of 15 January 1605 to Johann

FIGURE 9.1 Andrea Argoli in his old age. Lorenzo Crasso, *Elogii degli huomini letterati*, vol. 2 (Venice: Combi & La Noù, 1666), p. 269.

Federico Cesi in Rome about the nova and his recent exchanges with Kepler and Tengnagel regarding the novelty.[33] By then, the sprawling network of scientific exchange was already widespread, rapid and relatively efficient only a few weeks after the first outburst of the luminary. We shall return to Magini after saying something more about the anonymous physician-mathematician whose alert signal was among the first to be given.

Georg Brengger, Kepler also reported: "Novam stellam vidit quidam Veronae Fr. Altobellus 9 Octobris." *JKGW*, 15, no. 317, l. 401.

33 Giuseppe Gabrieli, ed., *Il Carteggio Linceo della vecchia accademia di Federico Cesi (1603–1630)* (Rome: Memorie dell'Accademia Nazionale dei Lincei, 1938), p. 51.

9.3 Biographical Identification of One of the First Observers of the New Star

The short letter that opens Clavius' correspondence is important because it gives voice to an author who was able to appreciate the importance of the appearance from the very beginning and signal it to the most expert eyes in Italy. Among Italian stargazers, he was probably the first, along with Altobelli, to announce the phenomenon, identify its exceptional character and offer an earnest account of what he saw (if we accept his date of 9 October, N. S., and not before).[34] Although his identity is not immediately clear, an attempt to name the mystery figure may now be made. For in all probability he was Andrea Argoli (1570–1657), a native of Tagliacozzo in Abruzzo, as will become clear from the following analysis of the available evidence.

Regarding the identity of the anonymous correspondent, we only know what Clavius tells us, namely that he was a physician-mathematician who wrote to him from "Cossenza [*sc.* Cosenza] in Calabria."[35] In fact, the sender specified that he was a physician signing his letter as a "medicus,"[36] and the precise observational details that he provided do not leave much room for doubt that he was highly skilled in mathematics as well. However, in his letter there is *no* indication regarding the provenance of the writing from "Cossenza in Calabria," as Clavius reported by quoting the content of the letter to Magini. That must have thus been a supposition by Clavius, who evidently was not acquainted with the sender before hearing from him. What we see in the text, however, is that the letter was written from "Paterno." Admittedly, there exists a place called "Paterno," namely Paterno *Calabro*, in the province of Cosenza (Calabria); Clavius probably arrived at that village in locating the source of the signallation,[37] even if apparently no record suggests it as a significant cultural or political centre where any high-profile figures lived around the beginning of the seventeenth century. On the other hand, there is another small village in Italy named Paterno *di Avezzano*, which is in the modern province of L'Aquila

34 So concluded Wilhelm H. Walter Baade, "Nova Ophiuchi of 1604 as a Supernova," *The Astrophysical Journal*, 1943, 97: 119–127, on p. 119. Cf. Jacco Vink, "Supernova 1604, Kepler's Supernova, and Its Remnant," in *Handbook of Supernovae*, ed. Athem W. Alsabti and Paul Murdin (Cham: Springer, 2017), pp. 1–22, on p. 2.

35 Christoph Clavius to Giovanni Antonio Magini, 18 November 1604, in Clavius (ref. 11), no. 236, p. 130.

36 *Ibid.*

37 The editors of Clavius' correspondence, who initially thought of Argoli as the potential letter sender, proceeded no further in their attempt to identify him since any historical-biographical match between the scholar from Abruzzo and the southern village could not be found. See Clavius (ref. 11), p. 70, no. 6.

(Abruzzo), and found at that time within the fief of Tagliacozzo. No matter how small, the village of Paterno di Avezzano was of major strategic importance. After serving as a stronghold under the duchy of the Colonna dynasty, it became the extreme outpost for the Kingdom of Naples facing the Papal States in central Italy.[38] Now, without disrespecting the delightful village of Paterno Calabro, it is more probable to look at the latter option as a setting of science where, in fact, Andrea Argoli was a young and active intellectual.[39] In 1604, Argoli was probably still pursuing his research in Paterno di Avezzano.[40]

Argoli was trained in astronomy, mathematics and medicine, which he studied at Naples, perhaps even as a self-learner.[41] His acquaintance with Clavius, his personal introduction as a "physician from Tagliacozzo" (later employed at Alatri, in the province of Frosinone) and his active interest in astronomical matters are all confirmed by a subsequent letter to Clavius—this time signed by the name "Andrea Argoli" – in which, some years later, a consultation on the comet of 1607 was requested, similar to what happened here with the star of 1604.[42] Further positive indications are present as well.

It is worth noting that in the same year Argoli published a text on trigonometry for astronomy, *Problemata astronomica* (1604),[43] and he was working on another, tellingly entitled *De novis stellis nostro aevo genitis*.[44] The latter re-

38 See Giovanni Brancaccio, "Tagliacozzo e il confine tra il Regno di Napoli e lo Stato della Chiesa," in *Tagliacozzo e la Marsica tra Antico regime e Risorgimento*, ed. Franco Salvatori (Roma: Abilgraf, 2005), pp. 101–111.

39 See Pietro Riccardi, *Biblioteca matematica italiana dalla origine della stampa ai primi anni del secolo XIX*, vol. 1 (Modena: Società tipografica modenese, 1870), coll. 48–51; Raffaele Aurini, "Argoli, Andrea," in *Dizionario bibliografico della gente d'Abruzzo*, vol. 1 (Teramo: Ars et Labora, 1952), pp. 422–429; Mario Gliozzi, "Argoli, Andrea," in *Dizionario biografico degli italiani*, vol. 4 (Roma: Enciclopedia Treccani, 1962), pp. 132–134; Loris Premuda, "Medici Abruzzesi nello Studio di Padova," *Abruzzo: Rivista dell'Istituto di Studi Abruzzesi*, 1968, 6: 505–512; Lynn Thorndike, *A History of Magic and Experimental Science*, vol. 7 (New York-London: Columbia University Press, 1964), pp. 122–124; Giuseppe Monaco, *Andrea Argoli: Astronomo e medico del Seicento* (Tagliacozzo: Scacco Matto, 2007).

40 See Giammaria Mazzuchelli, *Gli Scrittori d'Italia, cioè notizie storiche e critiche intorno alle vite e agli scritti dei letterati italiani*, vol. 1, pt. 2 (Brescia: G. Bossini, 1753), "Argoli (Andrea)," pp. 1045–1050.

41 *Ibid.*

42 Andrea Argoli to Christoph Clavius, 9 October 1607, in Christoph Clavius, *Corrispondenza (1606–1612): Lettere e testi*, vol. VI.1, ed. Ugo Baldini and Pier Daniele Napolitani (Pisa: Univ. di Pisa, Dipartimento di matematica, 1992), no. 274, p. 64.

43 Andrea Argoli, *Problemata astronomica triangulorum ope demonstrata per sinus, tangentes et secantes ex sola multiplicatione absque divisione* (Rome: s.e. 1604).

44 A reference to this text can already be found in a letter sent in 1634 from Parma by Guilielmus Wilhamer, S.J., where he refers to Argoli as "*alter Maginus*," as found in "Lettere d'uomini illustri del Secolo XVII a Giannantonio Rocca filosofo e matematico reggiano,

mained in manuscript and was never published, as was the case with his commentaries on Ptolemy's *Tetrabiblos*, which should have contained information on the celestial novelties, as well as some astrological speculations (and for this reason did not receive their licence for imprint). Part of this material would be merged many years later with some parts of Argoli's *Pandosion Sphericum* of 1644, including the observations concerning the new star (e.g. *§* LXII: *De novis stellis*), or appear in the designated section of his *Ephemerides* of 1659 (*§* XVII: *De novis phœnomenis*). These titles alone are sufficient to evince the interest of their author in stellar appearances—something that, even taken in isolation, would suggest as entirely plausible a young scholar requesting consultation on the subject from an accomplished expert such as Clavius.

Moreover, it seems that Argoli had advanced his career with the help of the Collegio Romano, possibly through the direct intervention of Clavius, as was the case with Magini. Through the support of the Collegio, Argoli won tenure to teach mathematics at the University of Naples. In his promotion, a role may have been played by their common interest in astronomical research, or perhaps a personal familiarity between Clavius and Cardinal Bixia, a friend and patron of Argoli, since during his years in Rome Argoli had served as Bixia's librarian.

Finally, Argoli keenly cultivated his studies on stellar and judiciary astrology with some public renown. While he received recognition for this work, Argoli soon encountered problems, eventually losing his position and provoking his dismissal from Naples. Some years later, Argoli was again forced to find the freedom necessary for his heterodox views in Padua, under the protection of the Venetian Republic.[45] There, he could carry on his studies, not only in

con alcune [lettere] del Rocca a' medesimi," *Nuovo Giornale de' Letterati d'Italia*, 1785, 31: 181–277, on p. 191.

45 Lorenzo Crasso, *Elogii d'huomini letterati*, vol. 2 (Venice: Combi e La Noù, 1666), s.v. "Andrea Argoli," p. 270: "[…] maggiormente crebbegli l'amore verso l'Astrologia, dal vedersi da molti curiosi stimato. Ma [Argoli] non potè goder molto la serenità de' suoi studi, poichè ò per cagion della detta Astrologia e della sua libera favella, ò per altra cagion che si fusse, provò non senza pericolo non ordinaria contrarietà … onde per sottrarsi a maggior pericoli, e liberarsi dalle sue agitazioni, si condusse a Venezia da lui stimata sicurissimo porto." The author of the reported witness was a contemporary and personal acquaintance of Argoli (they were both members of the *Accademia degli Incogniti* of Venice), therefore the reference is not just a mere conjecture, but an observation perhaps proposed to Crasso by Argoli himself. See Raffaele Mascella and Danilo Pelusi, "Uno scienziato abruzzese: Andrea Argoli," *Ratio Mathematica*, 2006, 17: 49–52, on p. 49.

astrology,[46] but also exploring bold alternatives to the Copernican system,[47] while holding the very teaching post once occupied by Galileo,[48] with relatively few constraints and a fairly positive reception.[49] It comes as no surprise that both in the anonymous letter of 1604 and in Argoli's letter on the comet of 1607 that the author expresses an interest in the astrological importance of heavenly events, although in a reticent and almost self-censored way. During an intense period of the Inquisition, prudence assured patrons and the public, while caution secured continuity of practice.[50]

For all of these reasons, we may conclude that the anonymous physician-mathematician of Paterno considered among the first observers of the new luminary was Andrea Argoli di Tagliacozzo, and not an otherwise nameless (pseudo-)Calabrian correspondent.[51]

---

46 See Pierluigi Pizzamiglio, *L'astrologia in Italia all'epoca di Galileo Galilei (1550–1650): Rassegna storico-critica dei documenti librari custoditi nella Biblioteca 'Carlo Viganò'* (Milan: Vita e Pensiero, 2004), pp. 158–167.

47 Apparently, Argoli was formulating a system based on a single motion of the Earth, as Micanzio hinted to Galileo: "Il Sig. Argoli è dietro al suo sistema con un moto solo della Terra, ma teme d'incontrar mala ventura, perchè havendoli questi Dei terreni fisso il chiodo, se si vuole muovere un tantino, mettono mano a' fulmini." Fulgenzio Micanzio to Galileo Galilei, 22 July 1635, in *OGG*, 16, no. 3160, p. 296. Cfr.: "In questo campo, l'Argoli si è dimostrato essere un intraprendente pensatore, tanto da provare a contraddire Kopernik attraverso un sistema astronomico di proprio concepimento 'con un moto solo della Terra,' quello di rotazione sul proprio asse. Un lavoro non pubblicato—ufficialmente—perché temette 'd'incontrar mala ventura,' ma più verosimilmente inedito perché inefficace e non in grado di salvare le apparenze riguardo il moto retrogrado di Marte e Venere." Cf. Michel Giovannini, *Accademie e astrologia: Ambiente culturale e relazioni erudite attorno a Pompeo Caimo (1568–1631), tra Udine e Roma* (Michel Giovannini: ebook, 2014).

48 See Antonio Favaro, "I successori di Galileo nello Studio di Padova fino alla caduta della Repubblica," *Nuovo Archivio Veneto*, 1917, 33: 96–182, on pp. 112–113; Antonio Gamba, "Andrea Argoli, il terzo successore di Galileo Galilei," *Galileo*, 1996, 36: 4–7.

49 "Andrea Argoli—who kept astrology on the mathematics syllabus at Padua University until the mid-seventeenth century—adopted the same attitude [as Magini], arguing that astral influences were real, although God could annul them or modify them and although their effects were only 'tendential.'" Ugo Baldini, "The Roman Inquisition's Condemnation of Astrology: Antecedents, Reasons and Consequences," in *Church, Censorship and Culture in Early Modern Italy*, ed. Gigliola Fragnito (Cambridge: Cambridge University Press, 2001), pp. 79–110, on p. 101.

50 See *ibid.*; Elide Casali, "Il diavolo dal mantello stellato e la condanna dell'astrologia," in *Il linguaggio dei cieli: Astri e simboli nel Rinascimento*, ed. Germana Ernst and Guido Giglioni (Rome: Carocci, 2012), pp. 153–167.

51 The only other gentleman from Calabria that was involved in the 'discovery' of the new star was the amateur astronomer Camillo Sasso, who was present in Padua (not in "Paterno") at the time when the professor Simon Mayr spotted the new nocturnal light together with his attendant Baldassar Capra. But Sasso neither had the competence to judge as an expert on astronomical matters, nor does it seem that he possessed any personal or

9.4 The Intermediary Informer and Clavius' Record on the Nova of 1572

The following letter we find in Clavius' correspondence is a request for information from Giovanni (Gioanni or Ioannes) Mellei, a correspondent about whom nothing is known except what can be drawn from the letter itself, namely: (1) that he was writing on behalf of someone else and (2) that the person(s) for whom Mellei was asking Clavius' consultation well knew the Jesuit's earlier contributions to stellar astronomy and alleged *novae* in particular. The indication of Mellei is textual and precise on this latter point:

> I pray that Your Fatherhood humbly bestows on me in just a couple of lines His opinion on this new star, which according to the observations made by many from several different places seems to be in the firmament for all the reasons that Your Fatherhood advances in His *Comento sopra la* Sfera *de Sacro Bosco* to prove that the one of 1572 was in the eighth sphere.[52]

The main reason why so many referred to Clavius for a comparative consultation, relying on his observations and considerations, can now be made clear. From the close of the sixteenth century, Clavius imposed his authority as an international expert on the subject, particularly on celestial novelties and other relevant issues, stemming from his scholarship and frequent updates of his famous publications. At that time, Clavius was one of few living scholars who had closely and competently observed and analysed the nova of 1572. He recounted his advanced results in widely read writings.[53] As Mellei recalled, Clavius' observations were exhibited in the *Commentario alla* Sfera *di Sacrobosco*, a work that was not only praised for didactic purposes but also enjoyed a wide reception among scholars, as proved by its many subsequent editions and reprints.[54] In particular, in the third edition of the *Commentario* (1585) Clavius added to

scientific reason for reporting the novelty to Clavius. (See Alessandro Giostra, "Il carteggio tra Ilario Altobelli e Galileo Galilei," in Merletti, Shea, Giostra (ref. 10), pp. 46–115, on p. 73.

52 "Prego Vostra Paternità che si degni d'accennarmi con dua righe il suo parere di questa nuova stella, quale secondo l'osservatione fatta da molti in diversi luoghi pare essere nel firmamento per tutte quelle ragioni che V. P. aporta nel suo *Comento sopra la* Sfera *de Sacro Bosco* per provare che quella del '72 sia stata nel 8.ª sfera." Giovanni Mellei to Christoph Clavius, 9 December 1604, in Clavius (ref. 11), no. 238, p. 134.

53 Peter Dygebauer, "Clavius und die Nova von 1572," *Sterne und Weltraum*, 1968: 7–22.

54 See Sheila J. Rabin, "Early Modern Jesuit Science. A Historiographical Essay," *Journal of Jesuit Studies*, 2014, 1: 88–104, on p. 95; Ugo Baldini (in collaboration with Juan Casanovas), "La sfera celeste di Cristoforo Clavio," in *Osservatorio astronomico di Capodimonte* (Naples: Almanacco, 1996), pp. 175–182.

his chapter *De quantitate stellarum* a short but important digression on the first great and unexpected stellar appearance that had occurred in 1572.[55] That passage was applauded by Tycho Brahe, who many years later wrote to Clavius to confirm that he agreed with his results and complimented the freedom of thought that Clavius had shown by breaking with the broad consensus of Aristotelian natural philosophy so stongly defended by his own religious order.[56] Clavius' analysis has been highlighted in particular by the studies of James Lattis[57] and Dario Tessicini,[58] and we should briefly make reference to it here as well. In short, it turned out to be the most advanced, and most controversial, interpretation of the great novelty of 1572. As Mellei's words testify for their own part, Clavius' digression served as a starting reference and speculative basis for the majority of all surveys on the nova of 1604.

The text of the additional section was subdivided into two parts: a longer and more detailed *pars destruens*, and a shorter and more punctual *pars construens*, in which Clavius expressed his own opinion after criticizing other concurrent theses. Accordingly, in the first part he opposed empirical arguments against the three most accredited theses among contemporary intellectuals who strove to make sense of the starry insurgence.

Defenders of the first thesis maintained that the nova of 1572 was, in fact, one of thirteen stars already known in the constellation of Cassiopeia (in which it appeared). They argued that one of the above stars apparently increased its relative magnitude for a limited amount of time only. Therefore the change as such, they maintained, should have been relegated to the sublunar world, for the change in its appearance took place as a consequence of interposed vapours or earthly exhalations that apparently condensed just below the sphere of the Moon in the upper region of the air. Thus, the result would not have been an actual increase in the size or brightness of the star: from the subatmospheric point of view, it should have been regarded rather as a sort of optical magnification of one of the pre-existing stars.

---

55 From this edition onwards, Clavius integrated this section by adding his observations of the novae of 1600 and 1604.

56 Christoph Clavius, *Corrispondenza (1597–1601): Lettere e testi*, vol. IV.1, ed. Ugo Baldini and Pier Daniele Napolitani (Pisa: Univ. di Pisa, Dipartimento di matematica, 1992), p. 101. See also the corresponding note in the commentary (vol. IV.2), on pp. 54–55, n. 19.

57 James M. Lattis, *Between Copernicus and Galileo: Christoph Clavius and the Collapse of Ptolemaic Cosmology* (Chicago-London: University of Chicago Press, 1994), pp. 145–156 (notes on pp. 250–252).

58 Dario Tessicini, "Il dibattito italiano sulla nuova stella del 1572 (Con il testo del "Discorso intorno a la stella" di Giuseppe Valdagno, ms Ambrosiana R 95 sup.)," in *Novas y cometas entre 1572 y 1618: Revolución cosmológica y renovación política y religiosa*, ed. Miguel Á. Granada (Barcelona: Universitat de Barcelona, 2012) pp. 43–93, on pp. 56–62.

However, all of the thirteen stars traditionally known to make up Cassiopeia were actually visible if carefully counted. The nova was thus recognized as apart or separate from the number of stars historically documented in stellar charts and catalogues; it could not be one of them.

The second thesis regarded the phenomenon analogously as a fixed star, if not as one of the thirteen in Cassiopeia, then as a pre-existing star that had previously passed unnoticed due to its extremely feeble glimmer, only to emerge suddenly in plain sight. As hypothesized by the first thesis, this pre-existent star had possibly augmented its relative brightness because of the interposition of gaseous exhalations that produced a magnifying effect.

But this hypothesis did not work out either. In fact, if the appearance of the new body was due to an optical effect of magnification induced by an exhalation within the atmosphere, then such an effect should have also been noticed proportionally for the other stars, at least for the closest star to the pre-existing one (namely for the one today called *κ Cassiopeiae*). But this did not happen: the other stars maintained their usual magnitude, and it was impossible to imagine a *localized* effect of gaseous atmospheric magnification that focused specifically on a single star.

The third thesis suggested that it was not a star, but a comet. According to the prevailing framework in force, the position of the novelty had to be located in the air at the upper end of the elemental region and, insofar as it was a comet, not beyond. This option was dictated mainly by the appeal to agree with Aristotelian thought, which did not allow changes or substantial alterations beyond the sphere of the Moon.

To those who defended the doctrine of Aristotle, Clavius replied that it was not possible to conceive a comet, since the luminous body did not show any observable angular divergence during the time it might have taken to cover any apparent distance—something that would be expected for a comet. This conclusion was reached by Clavius after the star had been observed from different places at different times. Clavius personally took care to collect and compare the observational surveys coming from Germany, Spain, France and Sicily, along with his own observations conducted from Rome. In his eyes, there was no doubt that the only detectable movement was just the (apparent) one of the celestial vault as a whole, and the distance was superior to the lunar sphere and almost equal to that of the fixed stars; therefore, it was clear to Clavius that the nova of 1572 could not be a comet; *unless*, he added with a sort of last-minute hesitation, the star was some sort of prodigy or divine miracle, or unless the ethereal substance beyond the Moon could allow for some measure of modification of its apparent structure.[59]

After all, Clavius was perfectly aware that admitting that stars could be generated from nothing, and even admitting that such a generation could occur in a place believed to be constituted by a form of matter that was constitutively immune to any change ("*Quintam quandam essentiam*"), was something that at that time was untenable. No doubt it would have caused serious distress to the Aristotelian cosmology.[60] Nevertheless, Clavius recovered to reinforce the credibility of such an unlikely account by citing the ancient opinions of Ambrose, Basil and Gregory of Nyssa, Christian authors who had maintained that the heavens were changeable and corruptible as much as the sensible earthly reality, so that a modification such as the sudden appearance of a new star would not have compromised either an allegedly eternal fixity or a presumed structural immobility of the ethereal element. (Those references may have been made available to him by Cardinal Bellarmine, a friend and colleague of Clavius at the Collegio Romano who had proposed similar ideas in his *Louvain Lectures* of 1572; Bellarmine made reference to the same sources that Clavius later quoted in his *Commentario*.)[61] Furthermore, Clavius clearly benefited from additional consultation with observations conducted by the mathematician Francesco Maurolico (1494–1575) from Messina,[62] who effectively allowed Clavius to improve the precision of his calculations.

In the end, Clavius confirmed the *stellar* nature of the new body in Cassiopeia and, in light of the distance he determined, finally established its location in the most remote region of the heavens *in the eighth sphere* of the fixed stars:

> I think that the nova, whatever it was, existed in the firmament, where the fixed stars are. For it appeared in the etherial region and not in the elemental ... [as we know] because it showed no parallax ... I believe that the new star stood in the firmament and not in any other celestial orb ...

59 See William Graham Lister Randles, *The Unmaking of the Medieval Christian Cosmos, 1500–1760: From Solid Heavens to Boundless Æther* (Aldershot: Ashgate, 1999), pp. 90–92.

60 Christoph Clavius, *Corrispondenza (1602–1605), Note alle lettere e ai testi*, vol. V.2, ed. Ugo Baldini and Pier Daniele Napolitani (Pisa: Univ. di Pisa, Dipartimento di matematica, 1992), p. 67, n. 3. These editors maintain that a statement like the above one is enough to imply that by this time Clavius had implicitly abandoned the Aristotelian theory of the heavens.

61 See *The Louvain Lectures (Lectiones Lovanienses) of Bellarmine and the Autograph Copy of His 1616 Declaration to Galileo*, ed. Ugo Baldini and George V. Coyne (Vatican City: Vatican Observatory, 1984), p. 9.

62 See C. Doris Hellman, "Maurolyco's 'Lost' Essay on the New Star of 1572," *Isis*, 1960, 51: 322–336.

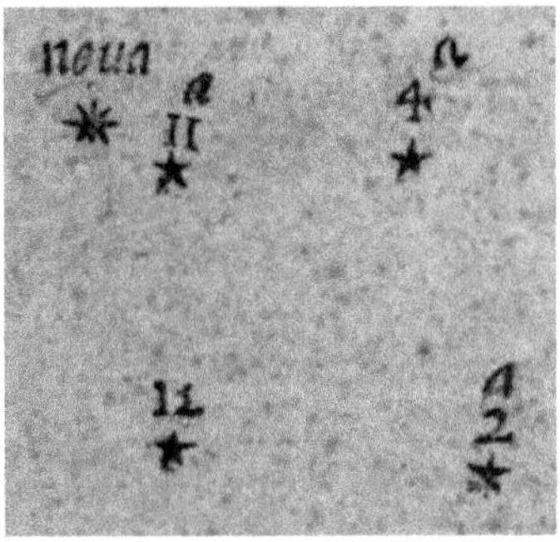

FIGURE 9.2
Relative position of the nova of 1572 in Cassiopeia according to Christoph Clavius, *In Sphaeram Ionannis De Sacrobosco commentarius* (Lyon: Jean de Gabiano, 1607 [5th ed.]), p. 218.

> because neither did I nor did any other astronomer to my knowledge observe any other motion in [the nova] aside from that which we shall observe in the fixed stars.[63]

Thus, for Clavius the nova of 1572 had to be a star that could not exist anywhere but in the ethereal region. Tycho Brahe and Michael Maestlin would also come to the same conclusion.[64] It is very probable, however, that Clavius arrived at the idea independently. In addition, his results enjoyed a reception among the Italian scientific community far superior to that of Maestlin or Tycho because of his magisterial position and the various editions of his widely read works that appeared in swift succession.

In the following editions, then, Clavius included the nova of 1604 within the framework of the same explanatory model that he advanced for the nova of 1572.

Returning to Mellei's original letter of request, the second point worth noting is that, as anticipated, Mellei did not ask for the above information for his own purposes, but on behalf of someone else. As Mellei explained in his letter to Clavius,

63 As translated by Lattis (ref. 57), p. 150: "[...] *stellam illam, quaecunque illa fuerit, in firmamento, ubi stellae fixae sunt existisse. Nam eam in regione aetherea et non in elementari apparuisse ... quod in ea non sit deprehensa aspectus diversitas ... credam stellam illam novam in firmamento, non in alio quovis orbe coelesti, extitisse ... quod neque ego, neque ullus omnino astronomus, quod quidem sciam, alium motum in ea animadverterit, praeter eum, quem in fixis sideribus observabimus.*"

64 *Ibid.* See also Michel-Pierre Lerner, "L'entrée de Tycho Brahe chez les jésuites ou le channt du cygne de Clavius," in *Les jésuites à la Renaissance: Système éducatif et production du savoir*, ed. Luce Giard (Paris: Presses Universitarires de France, 1995), pp. 145–185.

> I have been asked by many gentlemen to write to Your Fatherhood. Therefore, so as not to displease them, I decided to implore You accordingly, knowing that such a grace could be granted by Your Fatherhood, as was the case with other generosities that I received while I was staying in Rome.[65]

Mellei was simply an intermediary correspondent or, to adopt a term employed by Paula Findlen, a "broker of knowledge,"[66] who perhaps hoped to gain some personal profit from the information exchange, or, judging from the servile tone of the letter, to reinforce his relationship with both parties. But who was the original source of the request? Or, more precisely, who were those "many gentlemen" that had expressed interest in Clavius' opinion?

I submit as one highly probable source of such a request Guidobaldo Dal Monte (1545–1607) and his circle of scholars. During the period, Dal Monte and his circle mobilized their contacts to collect as many observations as possible across a broad network of correspondents. In the following section, I explore this hypothesis in further detail.

9.5 Dal Monte's Correspondence on the New Star (1)

Guidobaldo, Marquis Dal Monte and Count of Mombaroccio, was a key figure in early modern Italian culture. From a historiographical point of view, the fact that he aided the career of Galileo has often shifted his own profile and prowess in mathematics and astronomy to the background.[67] It is now recognized, however, that Dal Monte was known in his own time as "a prominent, if not the leading, scientist in Italy."[68]

---

65 Giovanni Mellei to Christoph Clavius, in Clavius (ref. 11), no. 238, p. 134: "Sono stato pregato da molti signori che [io] scrivesse a V[ostra] P[aternità] et io per non mancarli massime, sapendo che V[ostra] P[aternità] era per farmi la gratia sicome stando io a Roma mi haveva fatto d'altre, l'ho voluto supplicarla."

66 Paula Findlen, ed., *Empires of Knowledge: Scientific Networks in the Early Modern World* (London: Routledge, 2019). The quotation refers to the title of the first section of her book (p. 23).

67 See Barbara Aterini, "Guidobaldo Dal Monte (1545–1607)," in *Distinguished Figures in Descriptive Geometry and Its Applications for Mechanism Science: From the Middle Ages to the 17th Century*, ed. Michela Cigola (Cham-Heidelberg: Springer, 2016), pp. 153–180.

68 William R. Shea, "Guidobaldo Dal Monte: Galileo's Patron, Mentor and Friend," in *Guidobaldo Dal Monte (1545–1607): Theory and Practice of the Mathematical Disciplines from Urbino to Europe*, ed. Antonio Becchi, Domenico Bertoloni Meli and Enrico Gamba (Berlin: Edition Open Access, 2013), pp. 95–104, on p. 95.

The first clue in favour of his involvement in the above correspondence with Clavius comes from the place of provenance of Mellei's letter: Pesaro. This was the hometown of Guidobaldo, no more than twenty kilometres from his palace in Mombaroccio (in the province of Pesaro and Urbino), where he lived. Moreover, the letters that Dal Monte wrote in those days were addressed to his pupil and first interlocutor for scientific matters, Pier Matteo Giordani, who at that time resided in Pesaro. One clue of the exchange of astronomical information between Clavius and Dal Monte will be discussed briefly below; for now, we consider what the Marquis wrote to Giordani about the novelty. Their scientific exchange spanned a series of five to six letters that together represent a classic example of the concerns and controversies that the appearance of the new star raised for some of the most learned minds at work at the beginning of the sixteenth century.

This short, albeit important, "*carteggio*" was recovered and published by Arrighi more than fifty years ago,[69] but only recently has it been studied through the contributions of Giostra.[70] Some additional aspects can be further appreciated in the light of new documents, as we shall see, once they are contextualized in the broader framework of dispatches and replies in which they found their original meaning.

Following some formalities and brief discussions,[71] Dal Monte began on 23 November by claiming that, under circumstances of general excitement mixed with great anxiety, was destined to become a self-fulfilling prophecy: "this star or comet will generate much discussion."[72] The disjunction "star *or* comet" should be immediately noted, as it sounds as if one of the two options will rule out the other. For, as we said, according to the Aristotelian world view embraced by Dal Monte, the two kinds of phenomena were believed to belong respectively to two different and mutually exclusive regions of space and fields of study: fixed stars were found in the final sphere of the heavens and as such they were studied by astronomy, while comets were expected to pass only through the sublunar (elemental) region and were studied by meteorology. As

69 Gino Arrighi, "Un grande scienziato italiano: Guidobaldo Dal Monte in alcune carte inedite della Biblioteca Oliveriana di Pesaro," *Atti dell'Accademia Lucchese di scienze, lettere ed arte*, 1965: 12: 183–199. The letters that specifically dealt with the new star (or comet) can be read in the Appendix, on pp. 193–196.

70 Alessandro Giostra, "La *'stella o cometa'* nelle lettere di Guidobaldo Dal Monte a Pier Matteo Giordani," *Giornale di astronomia*, 2003, 3: 46–53; *idem*, "Guidobaldo Dal Monte e i nuovi corpi celesti," in Becchi, Bertoloni Meli, and Gamba (ref. 68), pp. 317–338.

71 Guidobaldo Dal Monte to Pier Matteo Giordani, 12 August 1587, in Arrighi (ref. 69), no. 2, p. 192.

72 Guidobaldo Dal Monte to Pier Matteo Giordani, 23 November 1604, in Arrighi (ref. 69), no. 3, p. 193: "[…] questa stella ò cometa darà da dire assai […]"

a case in point, no one ever thought the novelty might be a star in the elemental region, since it would have sounded completely nonsensical: even approximate distance measures could preclude this option. The trouble was rather that, in the eyes of traditionally trained astronomers, the new body appeared *as a tailless comet, but in the region above the Moon*. And although some amateurs did not hesitate to identify the appearance of a star *tout court*, Dal Monte was extremely reluctant to accept this option, because it ran completely contrary to the structure of the universe as he had known it; thus, he continued to remain at odds with the available evidence, insisting that it *had* to be a comet and could not be otherwise. Every scientific effort made by Dal Monte to account for the appearance was, in fact, an attempt to establish and classify the phenomenon as a tailless comet and not as a star. In his opinion, such an effort would have indeed been an enterprise that truly deserved the praise of those who actually wanted to 'save' celestial phenomena and all of natural philosophy which, up to that time, allowed for the capacity to explain convincingly what appeared before the eyes of observers.

On the other hand, completely replacing traditional cosmology with an entirely unprecedented one *ex novo* seemed utterly inappropriate to Dal Monte. Although many scholars advanced alternative cosmologies based on the great anomaly, its unexpected appearance was an exception that Dal Monte could not accept as sufficient for supplanting the whole system. The world structure, as he knew it, had (almost) always worked effectively up to that moment:

> Then those [astronomers], who want it to be a star, endorse this view just to resolve matters easily and without effort from the difficulties, since it is easy to say that the heavens are corruptible; but if that were actually the case, then it would be something that all of philosophy would find repugnant and one would have to find other starting points.[73]

Initially, Dal Monte seemed to find some support for his hypothesis, or at least so he hoped, in a work written by an author he referred to as "Mazzone," whose text he enclosed in his letter to Giordani. That author, Dal Monte said, "succeded well and … managed to resolve many difficulties … keeping up with Aristotle,"[74] though some days later Dal Monte reversed his informal review after reading the text more carefully, reporting that "he does not actually solve

73 *Ibid.*: "Questi [astronomi] poi, che vogliono che ella sia stella, lo dicano per salvar facilmente, et senza difficultà, le difficultà, perché è facil cosa a dire che 'l Cielo sia corruttibile, ma che 'l sia, questa ripugna à tutta la filosofia et bisognaria trovar' altri principij."

74 *Ibid.*

the difficulties, but rather just gets by as best he can."[75] Dal Monte's reference was probably to Jacopo Mazzoni da Cesena (1548–1598),[76] an astronomer and man of letters who throughout his life endeavoured to reconclile the philosophy of Plato with that of Aristotle. Mazzoni was, in fact, an esteemed intellectual in Central Italy at the time. Despite his adverse opinion on the utility of his work, Dal Monte mourned Mazzoni's absence as many as six years after his death. If Mazzoni were still alive to observe the celestial novelty, Dal Monte declared, he could certainly have proposed something more detailed, and perhaps even have written one of his erudite treatises on the topic. The work by Mazzoni that Dal Monte probably consulted was the *De triplici hominum vita* (1576).[77] There, Mazzoni included a section on the "*mixta imperfecta seu meteorologica*," a subject that squared directly with the interests of Dal Monte as he endeavoured to decide how and where to classify the phenomenon. Throughout his account, Mazzoni closely followed the order of exposition first advanced by Aristotle in the *Meteorologica*. However, he dissociated his view from that of the philosopher at certain points, which he did not hesitate to call "mistakes."

> It follows that fiery appearances of this kind are generated below the Moon, for the exhalation from which they are made is found beneath the Moon and never transcends it. And so it seems that Aristotle was wrong when he located the Milky Way below the Moon, since he referred to it the same causes that he had also assigned to comets.[78]

In comparison to meteorological occurrences, the Milky Way was always identical in appearance and location; phenomena such as comets and shooting stars, on the other hand, were more plausibly attributable to that sector of the natural order in which events such as winds, precipitation and atmospheric

75 *Ibid.*

76 This is the opinion of Arrighi (ref. 69), on p. 187.

77 Jacopo Mazzoni, *De triplici hominum vita, activa nempè, contemplativa, et religiosa methodi tres, quaestionibus quinque millibus, centum et nonagintaseptem distinctae. In quibus omnes Platonis, et Aristotelis, multae vero aliorum Graecorum, Arabum et Latinorum in universo scientiarum orbe discordiae componuntur* (Cesena: Bartolomeo Raverio, 1576), esp. on cc. 289f.–290f.

78 *Ibid.*, c. 289v., q. 3842–3843: "Ex istis omnibus sequitur huiuscemodi ignes infra Lunam generari, quippe cum exhalatio ex qua constant infram Lunam posita sit, nec Lunae orbem transcendat. Hinc etiam apparet malem Aristotelem lacteam viam infra Lunam collocasse, et ad easdem caussas [*sic*] quibus etiam cometas adscripsit retulisse."

changes in general normally took place. In the work of Mazzoni,[79] Dal Monte should have found some of that support he was searching for to present the new appearance as *something that came to be* and, therefore, as something that should have undergone some change to become what it was, whatever it was. It was enough for him to say that the new light in the sky could not be a fixed star, but had to be a comet. In the same vein, Dal Monte invoked the opinion of another author by the name of "Barrocci," perhaps the humanist and painter Federico Barocci (1535–1612) from Urbino.[80] Dal Monte believed that Barocci, if asked, would certainly have agreed with him.

Thereafter, Dal Monte singles out the criterion that could settle the matter. To establish whether the object was a comet or star, Dal Monte determined that it was necessary to understand as precisely as possible the apparent altitude, or depth, of the novelty in the sky; in other words, if it was below or beyond the lunar orb. To that end, however, the measurement taken from just one observational point was not enough, be it Rome, Bologna or Mombaroccio. Surveys had to be carried out from at least two distant points and then gathered together to be considered for proper triangulation. In short, 'parallax' or, as they called it, "*diversitas aspectus*," could be calculated this way. The more numerous and distant the observational points would have been from one another, the more precise and reliable the results would often be. It is not difficult to imagine, then, that the activation (or reactivation) of a correspondence network stemmed in part from the need for those data deriving from other locations. In fact, the epistolary query of Dal Monte to his collaborator, similar to that sent by Clavius to Magini or Altobelli to Galileo, reflects *the mutual need for the latest parallatic measurements* that were recorded in other regions. Moreover, at that time a wide variety of contingencies often affected or altered the surveys, incuding the gathering of clouds or other weather conditions during the night. Sometimes, as with the case of Mombaroccio, the sight of the low horizon was blinded by the profile of the surrounding mountains. These obstacles prevented Dal Monte from performing optimal observations and urged him to contact someone who could watch in his place from more favorable locations.

However, despite the fact that Dal Monte realized the importance of calculating parallax on this matter, he essentially maintained a traditional attitude

79 It has also been proposed that the work under reference here was Mazzoni's manuscript *Discorso a madama serenissima Granduchessa di Toscana* [*Cristina de' Medici di Lorena*] *intorno alla cometa veduta di luglio 1596* (Ms. Urb. Lat. 1363, cc. 1–24). See Bertoloni Meli, "Guidobaldo Dal Monte" (ref. 31), on pp. 29–30: "[There] Mazzoni defended without great convinction the view that comets are generated in the sublunar world and tried to explain the lack of parallax by means of optical refraction."

80 Arrighi (ref. 69), p. 187.

toward mathematics, always giving epistemological priority to natural philosophy or physics and the relevant principles of Aristote.[81] It was a classic case of what William Donahue has called the "invulnerability of traditional opinions to the parallax argument," and "the inherent stability of traditional physics and astronomy."[82] This kind of resistence frequently occurred not only because physics, insofar as it was a branch of natural philosophy, and astronomy, insofar as it was a branch of mathematics, were considered two different and relatively distinct forms of knowledge, but also because the value of numbers was often seen as less significant, almost non-influential and inferior to the value of Aristotelian natural philosophy. For Dal Monte and many of his contemporaries, numbers were still handled and harnessed with scepticism, as if they did not pertain to astronomical matters. Moreover, measurements of parallax were often imprecise, frequently leaving room for judgement, and the logic behind parallatic observation was not always well understood by everyone.[83] In any case, Dal Monte still asked Giordani to forward the following observations in a letter from 6 December 1604:

> ... if I could ever receive these observations made in different places, then I would clarify an opinion that occupies my imagination, [namely] by resolving the phenomenon as a comet and not as a star, for I cannot allow that learned persons should wish to hold the heavens as corruptible at a glance just because they can then say it is a star.[84]

Consequently, Dal Monte asked to contact Homero Tortora, a man of letters and mutual friend in Rome, who should in turn contact Clavius to request his highly coveted observational data. In addition, Dal Monte asked for any observations that might come "from Germany ... from he who [in place of Tycho, i.e.,

81 As was correctly noted by Giostra (ref. 70, n. 2), pp. 319–320.

82 William H. Donahue, *The Dissolution of the Celestial Spheres, 1595–1650* (New York: Arno Press, 1981), p. 35.

83 *Ibid.*, pp. 36–37: "[...] one could quite reasonably deny that parallax measurements could be consistently applied to the upper region of the elements and to the heavens. [...] to insist upon the universal applicability of parallax measurements without proposing a suitable alternative to the peripatetic philosophy which would thereby be destroyed, was seen as wholly unsatisfactory."

84 Guidobaldo Dal Monte to Pier Matteo Giordani, 6 December 1604, in Arrighi (ref. 69), no. 4, p. 194: "[...] se io potessi haver queste osservationi fatte in diversi luoghi mi chiarirei di una opinione, che mi va così per la fantasia, per salvar che ella sia cometa, et non stella, che io non posso acconsentire che persone dotte alla prima vogliono tener' il cielo corruttibile per poter dire che ella sia una stella."

Franz Tengnagel] is currently overseeing everything."[85] Giordani soon fulfilled the request, as we can now confirm from his unpublished correspondence with Tortora.

9.6 The Letters of Tortora to Giordani on the New Star

Fortunately, some of the letters that Tortora wrote to Giordani during this period have survived. Although they are barely legible, they remain extant in their original form,[86] and a couple of passages even cast light on the configuration and course of the epistolary exchange.[87] Significantly, these documents allow us to link the correspondence of Clavius to that of Dal Monte by way of their various intermediaries and letter couriers.

In fact, Matteo Giordani should have written from Urbino to Homero Tortora, who had been living in Rome to supply Dal Monte with the observations collected and culled by Clavius at the Collegio Romano, as Mellei did from Pesaro. In turn, Clavius probably wanted to receive any tentative observations taken by Dal Monte, and he could easily receive them from Mellei or Tortora, since both were in frequent contact with Dal Monte's attendant, Giordani. Giordani and Tortora certainly wrote to one another, as witnessed by their above mentioned letters. At the close of 1604, the two communicated consistently along a line of contact that presumably extended between Rome and Urbino (for not even in this case are the letters from Giordani extant, as far as we know). Thus, Tortora wrote to Giordani in a letter of 24 November 1604:

> I would like to know the opinion of Lord Guidobaldo [Dal Monte] on this nova, whether a star, a comet or whatever, since Father Clavius maintains that it is a new star, as was maintained with the one in Cassiopeia; and he also speaks of having conferred with a scholar from Cosenza and having [collected the evaluations] of many who agree with this opinion, since it was measured and found among the fixed stars, and therefore they hold that the heavens are corruptible; others mock this opinion while

85 *Ibid.*, p. 193. On the correct identification of Tengnagel, which Arrighi did not manage to make in his own analysis, see now Giostra (ref. 70, n. 1), p. 52, and *idem* (ref. 70, n. 2), pp. 323–324.

86 Homero Tortora and Pier Matteo Giordani, *Lettere di Homero Tortora a P. M. Giordani*, Biblioteca Oliveriana di Pesaro, Ms. 415, cc. 35–41.

87 As anticipated by Enrico Gamba and Vico Montebelli, *Le scienze a Urbino nel tardo Rinascimento* (Urbino: Quattroventi, 1988), esp. Chapter 2, "I matematici del Ducato di Urbino nel secondo Cinquecento," on p. 51, n. 30.

> they wait to hear those of the German, English, Spanish, and Oriental observers.[88]

In addition, Tortora was well informed not only of Clavius' scientific judgement, but also about the existence of a growing controversy in Rome about the same dilemma as that of Dal Monte: comet or star? The cutting-edge account of Clavius, though authoritative, should have first struck defenders of the prevailing scientific paradigm as paradoxical and even ridiculous. At the Vatican, any suggestion that the heavens were 'corruptible' was viewed as a sort of bold heterodoxy. Tortora's reference to the informant from Cosenza is evidently Andrea Argoli, who, as we have shown, was actually *not* from Cosenza: as one of the first competent observers, he managed to contact Clavius and trigger the mobilization of experts in astronomy that at that time were practicing in Italy. The early timing of this letter confirms that Clavius was convinced shortly after the appearance that it was a star, almost in concurrence with his first observations and certainly at the time of his correspondence with Magini. His results on the nova of 1572 should have greatly aided him in arriving quickly at this conclusion. Moreover, just one month after the appearance, the mobilization extended to international scholars and stargazers from England, Germany, Spain and the 'Orient' (perhaps including Venetian colonies such as Crete), whose observations were expected to arrive at Rome. When he received this notice, Giordani could perfectly understand the importance of the matters at stake.

Tortora reported again to Giordani in a letter of 22 December. Reaffirming the stellar nature of the bright light defended by Clavius, Tortora claimed in addition that this hypothesis had been corroborated by "demonstrations" and by a similar judgement from some of the German astronomers, probably including Kepler and Tengnagel, or perhaps Johannes Brunowsky, who in turn notified Kepler. Tortora wrote:

> Regarding comets or stars, whatever they are believed to be, I cannot say anything [new] except that today, before a large audience, the Jesuits have performed many demonstrations, following [the hypothesis of] Father Clavius, who believes it to be a star in the eighth sphere in Sagittarius. I do

88 Tortora and Giordani (ref. 86), c. 35v: "[...] desiderarei di sapere l'opinione del Sig. Guidubaldo [Dal Monte] sopra di questa nova, o stella o cometa che sia, perché il P.re Clavio tiene, che sia nuova stella, come fu tenuta da alcuni quella di Cassiopea, et dice di essersi seco confrontato uno di Cosenza [*sic*], et haver di molti che concorrono in q.ta opinione havendo misurato ch'ella si trovi con le stelle fisse, et così tenere che il cielo sia corruttibile, altri si burlano di q.ta opinione, et s'aspettano di sentire quelle de gli Alemanni, Inglesi, Spagnoli, et levantini."

> not know if such an opinion shall be endorsed; however, it is said that in Germany it is held to be a star, and there are letters and demonstrations [from there]. If the presentation held today by the Jesuits at the Collegio Romano is made public, I will send it to His Lordship immediately.[89]

The account in favour of the stellar hypothesis no longer seemed as laughable. Now that many observations had begun to converge and indicate the immobility and distant removal of the light point, it began to gain consent. The witness of a gathering of many ("before a large audience," he wrote) that took place at the Collegio Romano should be taken as a reference to the conference convened by the Jesuit Christoph Grienberger (1561–1636), probably under the personal supervision of Clavius. The text of the formal lesson, entitled *Problema de stellis novis, in quo impressiones aereae ab aethereis discernuntur*, has been recovered and studied by Baldini and is now available for consultation.[90] Grienberger concluded his lesson with as much methodological caution as geometrical certainty that "such a phenomenon, in consideration of its scintillation, motion, and location, conforms to the fixed stars, and *is one of them*. And if we wish to say more, *it is a new star*."[91] Tortora did not go so far in

89 *Ibid.*, c. 40r: "Circa le comete ò stelle che si credonono, non sò dirle altro se non che i Gesuiti pur hoggi, hanno con concorso di molti fatte molte dimostrat.ni suguitando il P.re Clavio che tiene [opinione] che sia stella, et nell'ottava sfera in Sagitario, non so se sarà seguitata; dicono esser tenuta in Germania per stella, et che ne ne siano lettere et dimostrat. ni. Se si darà fuori quello che hoggi hanno fatto sentire al Collegio Romano i Gesuiti, lo mandarò subito a V.S."

90 Ugo Baldini, "La nova del 1604 e i matematici e filosofi del Collegio Romano," *Annali dell'Istituto e Museo di Storia della Scienza di Firenze*, 1981, 6: 63–98; republished in his *Legem impone subactis: Studi su filosofia e scienza dei Gesuiti in Italia, 1540–1632* (Rome: Bulzoni, 1992), Chapter 4, "La nova del 1604," pp. 155–187. Grienberger anticipated a copy of the conference script in the form of a letter to his Jesuit brother, former student and fellow mathematician Francesco Antonio Rubino, who was at that time on mission in Goa, India. The document, dated 22 December 1604, never reached its recipient and is now preserved at the Bibliothèque Nationale de France (Collection Dupuy, 5, *Mémoires et discours sur diverses matières*, VI, cc. 49–53). Cfr. Paula Findlen, "How Information Travels: Jesuit Networks, Scientific Knowledge, and the Early Modern Republic of Letters, 1540–1640 ," in *eadem*, *Empires of Knowledge* (ref. 66), pp. 57–105.

91 As cited in Baldini (ref. 90), p. 77: "Phaenomenon istud, scintillatione, motu ac situ imitari fixas, atque *unam fixarum esse*. Et si cum pluribus loqui velimus, *novam esse*"; my italics. Note also that shortly before the introduction of the telescope some members of the Collegio Romano began to build instruments for observational purposes to meet the new standards of precision established by the work of Brahe. See Ugo Baldini, "The Academy of Mathematics of the Collegio Romano from 1553 to 1612," in *Jesuit Science and the Republic of Letters*, ed. Mordechai Feingold (Cambridge, MA: MIT Press, 2003), pp. 47–98, on p. 58.

his letter as to reach a similar conclusion. Rather, he betrayed a certain degree of scepticism regarding the star hypothesis.

This handful of surviving letters from Tortora to Giordani, including the few lines we have extracted here, deserve a more detailed study in order to recover their most illegible tracts. For now, it suffices to conclude that they prove a clear link, albeit indirect, between Clavius and Dal Monte concerning the nova of 1604. In the end, both sides maintained their initial ideas regardless of their alleged exchange: while Clavius continued to claim that it was a new star, Dal Monte, as we will see in the following section, continued to deny that it could be a star, not to mention a new one, and that it was instead a comet.

9.7 Dal Monte's Correspondence on the New Star (2)

In the first days of its appearance, according to Dal Monte the new luminary looked like a burning, pulsating fire,[92] at least for as long as it could be seen clearly. In subsequent letters, Dal Monte blamed cloudy weather for preventing him from observing what he was determined to prove was a comet. Despite adverse conditions, Dal Monte managed to record the coordinates of the celestial position that he identified when the weather turned more favorable (18½° Sagittarius, with a latitude of 12°15'), and to communicate them to Giordani. Dal Monte asked Giordani to consider them confidentially, at least until he could later confirm the position, hopefully under better conditions. As mentioned above, Giordani should have also been able to forward to his master observations arriving from Germany; noted on the back of the same letter was an "observation from Prague," which at that time was not only home to the court of the Holy Roman Empire, but also a renowned centre of astronomy.[93] On the one hand, Dal Monte welcomed the observation from Prague because it matched his own positional coordinates, while on the other hand he expressed dismay because it ran contrary to his interpretative expectations. In fact, Dal Monte hoped that some degree of parallax could be discerned in order to lower the estimated location within the range of atmospheric phenomena, but such a 'downgrade' was not even allowed by the survey conducted in Prague by

92 Guidobaldo Dal Monte to Pier Matteo Giordani, 6 December 1604, in Arrighi (ref. 69), no. 4, p. 194: "Dirò ben questo, che avendola io guardata ben bene, et durato un pezzo à vederla, io vedevo che ella scintillava tanto forte che non hò mai veduto stella scintillar tanto, et quasi in quel modo che veramente pareva che fusse fuoco et non stella."

93 See Robert J. W. Evans, *Rudolf II and His World: A Study in Intellectual History, 1576–1612* (Oxford: Clarendon Press, 1953 (first ed.), 1984 (second ed.)).

Tengnagel (probably under the supervision of Kepler), who relied on some of the same legendary instruments that had originally belonged to Brahe.[94]

On 20 January 1605, Dal Monte wrote again to Giordani "about this comet, which everyone clamours to be a *star*."[95] For the sake of intellectual honesty, Dal Monte was now willing to concede two points to his many contestants: (1) that the luminous body did not show any detectable "diversity of aspect" (parallax), and (2) that it did not exhibit any particular motion, other than the circular movement of the celestial vault in which it was enclosed. No matter how undeniable, the above two reasons were not enough to convince Dal Monte, who chose to maintain the benefit of the doubt in the face of conflicting evidence. Accordingly, he expressed his double resolution to refer back to the studies of Brahe, which he possessed in his personal library, and to assign the least credit to the astrologers, who seemed to him to rely exclusively on their imagination without ever forcasting anything precise on the matter.

Dal Monte's resistance may be explained by his pressing need to save centuries of astronomical knowledge and expertise from a subversive twist that he might not welcome (and even reasonably fear). But a second explanation for his apparent struggle might be found in the manuscript he had completed just a few months before the appearance of the nova, entitled *De stella magorum* (recently discovered in the Vatican Library).[96] In his unpublished manscript, Dal Monte analysed the famous star of Bethlehem, which according to the Gospel of Matthew had guided the three magi through the desert to reach the place of the holy nativity of Jesus Christ. In his text, Dal Monte examined the issue of identifying, from a more scientific point of view, what kind of astronomical object, if any, the so-called star of Bethlehem was.[97] Of course, this was a standard topic of erudite discussion from antiquity to the Renaissance, engaging the attention of Marsilio Ficino, Nicholas of Cusa, Girolamo Cardano and several other scholars, including Kepler.[98] Curiosity over the matter en-

94 *JKGW*, 1, 209.29–31. On 11 October, Tengnagel conducted an observation "in the imperial garden" in the company of Kepler and some other scholars. On Tycho's instruments, see Allan Chapman, "Tycho Brahe—instrument designer, observer and mechanician," *Journal of the British Astronomical Association*, 1989, 99: 70–77.

95 Guidobaldo Dal Monte to Pier Matteo Giordani, 20 January 1604, in Arrighi (ref. 69), no. 7, p. 195: "[…] sopra questa cometa la qual ogn' un grida stella."

96 Guidobaldo Dal Monte, *De Stella Magorum*, B. U. V. cod. Urb. lat. 1743, cc. 65r–69r.

97 Giostra (ref. 70, n. 2), pp. 326–335.

98 See Michael Walter Burke-Gaffney, "Kepler and the Star of Bethlehem," *Journal of the Royal Astronomical Society of Canada*, 1937, 31: 417–425; Abram J. Sachs and Christopher B. F. Walker, "Kepler's View of the Star of Bethlehem and the Babylonian Almanac for 7/6 BC," *Iraq*, 1984, 46: 43–51; Wolfgang von Chmielewski, "Johannes Kepler and the Star of Bethlehem: A Work in Progress," www.grailgate.com/library/KEPLER-3.

dures today.[99] Referring also to the corresponding interpretations of the Church Fathers, Dal Monte concluded in his essay that the great luminary had been called a 'star' in a general sense only, that it was a divine sign that had not contradicted any meteorological laws, and that it was correctly considered a "miracle" that remained consistent with physical reality as it could be known. For Dal Monte, this was so because it actually was a comet—precisely as he sought to demonstrate in the case of the nova of 1604. Ultimately, he argued that, if God had not wished to subvert the principle of the ingenerability of stars even on the occasion of the birth of his Holy Son, then the same principle would have been subverted with even less justification at a time when no monumental event seemed to be occurring.[100]

Dal Monte was not alone in his view. For instance, Pompeo Caimo (1568–1631), a respected physician from Udine who was working in Rome and was in epistolary contact with Argoli, expressed similar ideas in a letter to his bishop and protector:

> Tell me, for the sake of your life [*per vostra vita*], do you not know that a great dispute is raging among the astronomers whether it might be a fixed star or comet? If it is a real star (something that will soon be ascertained by way of demonstrative observations), I do not understand why it should be called evil rather than good, since it has never been observed before, just because of what it shows through the beauty of the radiation of its bright rays, it being a star that assembles the nature of Jupiter and Venus together. But if it is a comet (something that undoubtedly conforms more to the doctrine of Aristotle, and perhaps even to the reality of truth), it need not be an indicator of evil, since comets have occurred at other times with a rather contrary signification.[101]

---

pdf; Cesare Catà, "*Lumen requirunt lumine*: Marsilio Ficino, Nicola Cusano e l'iconologia dei Magi nel Rinascimento," *Philosophical Readings*, 2015, 7: 110–131; Owen Gingerich, "Kepler's *De Vero Anno* (1614)," in *The Star of Bethlehem and the Magi: Interdisciplinary Perspectives from Experts on the Ancient Near East, the Greco-Roman World, and Modern Astronomy*, ed. Peter Barthel and Georg van Kooten (Leiden-Boston: Brill, 2015), pp. 3–16.

99 See Colin R. Nicholl, *The Great Christ Comet: Revealing the True Star of Bethlehem* (Wheaton: Crossway, 2015); Costantino Sigismondi, "Mira Ceti and the Star of Bethlehem," *Quodlibet*, 2004, 4.

100 Giostra (ref. 70, n. 2), pp. 326–335.

101 Pompeo Caimo, *Lettera* [*del*] *4 Dicembre 1604 che tratta ancora sopra la costellazione del cielo* [Archivio della Curia Arcivescovile di Udine, Fondo Caimo-Dragoni, envelope 115, manuscripts of Pompeo Caimo, f. n.n.], in Giovannini (ref. 47), Appendix 3: "Ditemi per vostra vita, non sapete voi, ch'è gran contesa fra gli Astronomi, s'ella sia stella fissa, ò Cometa? Quando sia stella reale (il che tosto si conoscerà per dimostrative osservanze)

Caimo added that the possible negative effect of the nova would be active only in those places where the astrological sign in which it appeared was dominant. For instance, he said, the city of Rome was considered under the sign of Libra for its old part and Leo for its new part, but the nova appeared between Ophiuchus and Sagittarius, so none of its negative effects would ever take place in Rome. As a result, Caimo concluded that he and his patron could consider themselves safe and secure.[102]

Expectations of dramatic change, and even the desire for a radical improvement of living conditions, ran deep not only at the popular level. Despite his scepticism towards various eschatological views, Dal Monte evinced in his text the pervasiveness of such popular beliefs, just as they were announced widely after the appearance of the nova.

9.8 Astrological Elements of the New Star According to the Testimony of Dal Monte

In his letter of January 1605, Dal Monte thanked Giordani for further reference material that he had enclosed with what Dal Monte had originally requested. One component of this material, vaguely mentioned by Giordani, was a Latin treatise of predictions written by an astrologer in skilful, contradictory terms. This promptly drew the ire of Dal Monte, who suggested sarcastically that such a wise prognosticator would have easily foreseen the future as he did, since he predicted periods of both war *and* peace, both epidemics *and* good health, both concord *and* discord, etc. We do not know the identity of the 'sagacious'

non sò io vedere perché rea piu tosto debba chiamarsi, che buona, non sendo mai più stata osservata, e per quello, che mostra col rotare i suoi raggi lucenti, e bella, assembrando stella di natura Gioviale, e Venerea. Ma s'ella è cometa (il che senza dubbio più conformemente si dice à la dottrina del Filosofo, e forse anco à la realtà del vero) non è neccessario, che sia indicatrice di male, che pura altre volte sono state comete di contrario significamento."

102 Giovannini (ref. 47), Appendix 3: "Il Caimo, riguardo la possibilità che la 'stella nova' potesse essere realmente una nuova stella nel cielo, preferì sospendere il giudizio nell'attesa di quelle più precise 'dimostrative osservanze' che si stavano concretizzando tra i contrasti e le collaborazioni degli astronomi sopra menzionati. Sbilanciarsi sarebbe stato avventato, la nova era apparsa da appena un mese; tuttavia, la visione d'insieme di Pompeo era evidentemente tendente a salvare la tradizione. Infatti, l'idea che la SN1604 potesse essere una cometa gli sembrava più conforme alla dottrina di Aristotele e, tra parentesi, 'forse anco à la realtà del vero.' [...] con l'ipotizzare <poi> che la 'stella nova' fosse una cometa, il Caimo cercò di dimostrare come questi mirabolanti eventi celesti non siano sempre stati nefasti."

soothsayer, whose credentials are not recorded and are probably destined to remain anonymous among the many venturesome fortunetellers of that time. However, it is reasonable to imagine that he was perhaps Alessandro Padovani from Forlì, who authored a short Latin treatise entitled *De novo sydere, causis, duratione, ac portentis discursus* (1604). While the text remained in manuscript, it was polished enough for informal distribution, perhaps for later publication.[103] This short treatise can be traced to an area that is chronologically and geographically compatible with the exchanges that took place between Dal Monte and Giordani. In it, we find the few elements described by Dal Monte in his letter, and there is no doubt that the author "practices the profession of the judicial [astrologer]."[104] Beyond the specific identification, however, such a pamphlet should be taken as further proof of the unscrupulous activity of so many astrologers who sought to prey on the general concerns and credulity of a society stunned by the new star.

Dal Monte briefly returned to astrology in a letter of 21 January 1605, the last of his known correspondence with Giordani. Here, Dal Monte apologized to Giordani for not considering more closely all of the notes that Giordani had collected and sent to him. In particular, Dal Monte had overlooked a text that, on second thought, he now considered "the most beautiful discourse,"[105] basically an astrological analysis of the phenomenon. It is not clear whether the text in question belonged to Giordani or another author, but it displayed astrological conjectures by way of "figures" as well.[106] Works of judicial astrology, in fact, often made use of charts and diagrams to visualize what the sky was like at a certain moment in time, and subsequently presented a horoscope.[107] If Dal Monte did not refer to a lost work by Giordani, such an act of interpretation backed by a series of illustrations may have been the *De stella nova anni 1604* by Francesco Ingoli (1578–1649). An unpublished manuscript that might have come from their Roman channels of intellectual commerce, the *De stella nova anni 1604* contained four or five astrological diagrams to complement the written text.

103 Alessandro Padovani, *De novo sydere, causis, duratione, ac portentis discursus Anni MDCIIII*, Biblioteca Malatestiana, Ms. 166.17, 1r–21r.

104 Guidobaldo Dal Monte to Pier Matteo Giordani, 20 January 1604, in Arrighi (ref. 69), no. 7, p. 195.

105 Guidobaldo Dal Monte to Pier Matteo Giordani, 21 January 1604, in Arrighi (ref. 69), no. 8, p. 196.

106 *Ibid.* (See the following image.)

107 On the use of diagrams in Renaissance judicial astrology, see Giuseppe Bezza, "Le tecniche astrologiche," in Ernst and Giglioni (ref. 50), pp. 53–70.

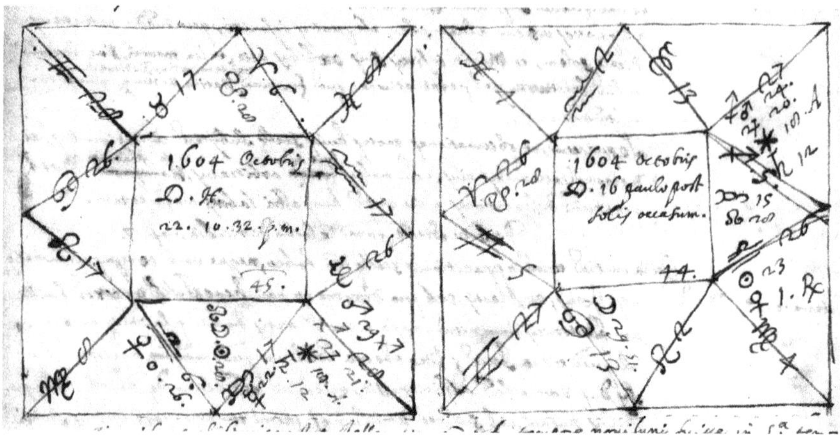

FIGURE 9.3 Franciscus Ingolus, *De stella anni 1604 tractatus* (1604–1605), Biblioteca Apostolica Vaticana, MS Ott.lat.2484 pt. 1, c. 143v (selection): astrological diagrams.

Despite his high regard for the "beautiful discourse," not even in this case was Dal Monte moved by the absence of parallax measured by the author to change his mind concerning the cometary nature of the great outburst.

At the beginning of their correspondence, Dal Monte mentioned to Giordani that astronomical controversies regarding the overall interpretation of the nova had soon arisen together with astrological disputes that were no less passionate, or no less contradictory, among authors as closely located as Bologna and Rimini:

> ... while in Bologna it is held that the nova signifies an increase of religion, in Rimini there is a friar astrologer who maintains the contrary, namely that it signifies a detriment to religion.[108]

It is interesting to note that an astronomer with an academic background and a noble upbringing such as Dal Monte paid some attention to apparently marginal astrologers and their ideas. But perhaps, as we often find, those almanac compilers and other authors were *not* so marginal at the time.[109] For the

108 Dal Monte (ref. 72), p. 193: "[...] se à Bologna tengano che significhi augmento di religione, in Rimini ci è un Frate Astrologo che tiene il contrario, cioè che significhi detrimento di Religione."

109 See Elide Casali, *Le spie del cielo: Oroscopi, lunari e almanacchi nell'Italia moderna* (Torino: Einaudi, 2003).

fact that neither Dal Monte nor the 'official' history of science mentions their names does not mean they were of negligible importance or their predictions remained unheard. Tracing these sources will perhaps prove worth the effort, for if they are identified, they can provide a more accurate outline of the various currents of early modern culture and the community of astronomically minded intellectuals operating in Italy at the beginning of the seventeenth century.

The aforementioned "friar astrologer" from Rimini has been identified as Casamatta Faentino,[110] whom Elide Casali has recently recognized as a pseudonym for Girolamo Casale.[111] This bizarre character was the author of a *Discorso astrologico*, a short Latin-vernacular bilingual essay that contained a section with "*a judgement on the new star that appeared in 1604*."[112] In this work, we find only a reference to "abscissioni," or 'schisms,'[113] which may be an indicator of what Dal Monte read as a "detriment to religion." In general, however, Casamatta pronounced a negative prediction of impending events announced by the star, which would have inaugurated an inauspicious period for all of Western civilization. In particular, the planet Mercury would apparently exercise an aggravating influence, and only the closeness of Jupiter in that area of the conjunction might mitigate the negative effects. The new star would indicate many disasters in the following order: strong winds, storms, droughts, discord and death of noblemen, conflicts and wars of succession, political divisions, schisms ("*abscissioni*") and segregation of populations.[114] All of these ill-fated consequences would be accompanied by a period of dryness in the soil immediately before and after the appearance of the novelty. According to a tradition that dated back (at least) to the *Meteorologica*, these signs heralded the passage of comets or shooting stars,[115] exactly as Dal Monte

110 Giostra (ref. 70, n. 2), pp. 326–335.

111 See Elide Casali, "Il numero fisico matematico e gli almanacchi di Carlo Cesare Scaletta da Faenza (1666–1748) tra astrologia, enciclopedismo e nuova scienza," in *Belle le contrade della memoria: Studi e libri in onore di M. G. Tavoni*, ed. Federica Rossi and Paolo Tinti (Granarolo dell'Emilia: Pàtron, 2009), pp. 81–98, on p. 84; Casali (ref. 109), p. 54.

112 Casamatta Faentino [*alias* Girolamo Casale], *Discorso astrologico, nel quale oltre le mutazioni de' Tempi, si scoprono varij naturali avvenimenti, non solo dell'Anno presente MDCV ma anco per tutto l'Anno MDCXI. Con un giudicio sopra la nova stella apparsa l'Anno 1604 del Mese d'Ottobre. Del famosissimo Sig. Casamatta Faentino Professor d'Astrologia* (Rimini: S. Simbeni, 1605).

113 *Ibid.*, c. 6v. This term, however, does not occur in the Latin version of the text, but only in the vernacular part.

114 *Ibid.*, esp. cc. 5r–6v; 9r–10r.

115 I have considered such a recurrent pattern of correlation and/or causation in Matteo Cosci, "Aristotelian Cometary Theory in Italian: Effects of Comets from Mid-Sixteenth-Century

expected. More precisely, Casamatta did not define the phenomenon as a new star as such, but as an "*assub*,"[116] a term transliterated from the Arabic astronomical language basically meaning any temporary nocturnal light that could appear in the sky.[117] Casamatta claimed that he had found such a notion in the *Summa* of "the *Perscrutator*," namely Robert of York, as well as in Clavius' commentary on the *Sphere*.[118] If he had actually consulted Clavius' commentary, Casamatta could have arrived at the idea that it might have been a new (fixed) star, as Clavius concluded, so it may be the case that he consciously rejected this option from the range of hypotheses. Again, this was exactly as Dal Monte (who was reading both authors) expected: since the nova was classified as a temporary event, it could not be a fixed eternal star.

At the same time, there were other self-styled experts who interpreted the sign in different ways. Following the same testimony from Dal Monte, we find that in Bologna, not far from Rimini, there was another individual who understood the new star as an omen with an opposite meaning, namely heralding an "increase of religion" (obviously the *Roman Catholic* religion, probably *to the detriment of others*). However, the identity of the astrologer is no easier to uncover than the former one, and no attempt has been made yet. In fact, it is difficult to find anything that was printed in Bologna on this topic before November 1604, precisely when the Marquis made his reference. In any case, it is still not clear whether the two works were published,[119] and it is improbable that such a prognostication was conceived, written, printed and circulated beyond its area of origin just a month and a few days after the nova first appeared. Therefore, these ideas were probably circulated either orally or by letters, perhaps by way of private consultations, ephemerides, or astrological 'instant essays' numbering a few sheets, which usually enjoyed a limited and local distribution. (Similar conjectures could be printed later, when more careful observations were made, as was often the case.) Once updated, these preliminary texts were frequently presented under fictitious identities in order to protect the reputation of the astrologer, in case his predictions turned out to be plainly wrong.[120] This happens to hold true for Casamatta/Casale, who

Vernacular Prose to the Reception of Galileo Galilei," *Rivista di Storia della Filosofia* (special issue), ed. David Lines and Anna Laura Puliafito, 2019, 74: 343–360.

116 Casamatta (ref. 112), c. 5r.

117 See Carlo Alfonso Nallino, "Etimologia araba e significato di '*asub*' e di '*azimut*,'" *Rivista degli studi orientali*, 1919, 8: 369–388.

118 Casamatta (ref. 112), c. 5r.

119 Dal Monte simply reported that Bologna was the place of origin of the astrological rumor ("[...] *à Bologna tengono che significhi* [...]," Dal Monte (ref. 72), p. 193.)

120 See Casali (ref. 109).

perfectly fits the picture, since Dal Monte made reference to his work while it still remained unpublished.

If we broaden our analysis to the works published in Bologna in the following years, we find, among others, the four-page *Pronostico o discorso* (1606) under the name of Laudadio Diodato.[121] Perhaps it was this work to which Dal Monte referred? Unfortunately, we cannot discern any clear reference to the impending boost in religiosity that Dal Monte recounted in his report. It is more probable to think, then, that Dal Monte was referring to the ideas that were going to be published in another text.[122] Incidentally, we can trace to the same publisher in Bologna a *Discorso astrologico delle mutationi de' tempi, e de i più notabili accidenti sopra l'anno 1607*. This was a short press copy with astrological announcements for the year 1607 in the light of earlier events, including the nova of Ophiuchus. Authored by an enigmatic "Lodovico Bonhombra," the work enjoyed great interest from the public and was reprinted in a slighty extended form a few months later. The identity of the author has been determined from a manuscript reference found in a copy of the text held at the Biblioteca Marciana in Venice: Giovanni Antonio Magini (1555–1617), professor of mathematics and astronomy at the University of Bologna.[123] The

121 Laudadio Diodato, *Pronostico o discorso* (Bologna: Giovanni Batista Bellagamba, 1606), Biblioteca Universitaria Bologna, no. invent. A 2764.

122 Another candidate for the identification of the source mentioned by Guidobaldo might be the *Discorso sopra la nuova stella apparsa l'anno 1604 alli 8 d'ottobre* written by the astrologer Giuseppe Rosaccio, who was active in Bologna in those years as well. That text is apparently no longer extant, but one can find a reference to the *Discorso* in one of Rosaccio's later works, where the prognostication (in terms of religious "detriment") is compatible with the one referred to by the Marquis' letter: " … as I have said many times in regard to that condensed exhalation, namely the light that was not a star that appeared under the sign of Sagittarius (the sign and house of Jupiter), that it did not mean anything other than the exaltation of the Holy Church, and detriment and damage to its enemies" ("*come molte volte ho detto sopra di quella esalatione unita, cioè lampada, e non stella, apparsa sotto il segno di Sagittario, segno e casa di Giove, ch'altro dinotar non voleva, se non esaltatione di santa Chiesa, e detrimento e danno a gl'inimici suoi*"). Giuseppe Rosaccio Cosmografo, *Discorso astronomico e cristiano* (Florence and Bologna: Girolamo Discepolo, 1614), c. 5. See Elide Casali, *Astrologia cristiana e nuova scienza: pronostici astrologici sulle comete (1577–1618)*, in *Celestial Novelties on the Eve of the Scientific Revolution 1540–1630*, ed. Dario Tessicini and Patrick J. Boner (Florence: Olschki, 2013), pp. 105–131, on pp. 117 and 122.

123 Lodovico Bonhombra [alias G. A. Magini], *Discorso astrologico delle mutationi de' tempi, e de i più notabili accidenti sopra l'anno 1607* (Bologna: Giovanni Batista Bellagamba, 1607). Biblioteca Nazionale Marciana, no. invent. OLD 6367. For the identification of Bonhombra as Magini, see, for instance: Emmanuele A. Cicogna, *Delle iscrizioni veneziane* (Venice: G. Piccotti, 1834), vol. 4, where he says on p. 436: "Bonhombra, Lodovico, *Discorso astrologico*. Bologna 1607, in 4°. L'autore è Giannantonio Magino, padovano dottore di filosofia

identification is made more plausible by the fact that Dal Monte knew Magini and was closely familiar with his scientific work, although he did not always hold it in high esteem.[124] After all, Magini often published astrological speculations that may have displeased the Marquis, including his series of *taccuini* that Magini published annually under his own name.[125] Thus, Dal Monte probably did not endorse the content of the *Discorso*, not only for the mathematical matters it explored,[126] but also on account of the passages involving the great planetary conjunction that preceded the appearance of the luminary. In his account, Magini wrote:

> On this [conjunction], it is believed that a great alteration of the world could occur, with the effect of good and optimal consequences in general, namely peace, tranquility, and the exaltation of our [Roman] Catholic religion, with the repression of infidels and heretics, and perhaps also some monarchy; despite the fact, however, that the above conditions will only come after great turmoil and disorder, including rebellions, insurrections, wars, and other negative effects, as it seems to show from a certain inclination of the stars assembled together at the same time as the said great conjunction of 1603 …[127]

In fact, Bonhombra/Magini specified that, although the planetary conjunction was benefical and favourable, other astronomical variables entered the account that might have interfered with or compromised the generally positive outcome of his prognostication. Those factors were: first, the approach of the planet Mercury, 'enemy' of Sagittarius, where the conjunction took place;

e pubblico primario matematico dello studio di Bologna, come da una nota mss. in un esemplare dell'operetta nella Biblioteca Marciana."

124 See Favaro (ref. 26), pp. 11, 13, 22.

125 See Ugo Baldini, *s.v.* "Magini, Giovanni Antonio," in *Dizionario Biografico degli Italiani*, vol. 67, 2006, online edition.

126 *Ibid.*: "[…] in questo testo il Magini superò i canoni usuali, *affermando la necessità di calcoli esatti* e del confronto tra figure astrali ed eventi climatici per serie prolungate di anni"; my italics.

127 Bonhombra, *Discorso astrologico* (ref. 123), c. 8: "[…] in questa [concomitanza] credono che potrebbe accadere grand'alteratione nel Mondo, con produttione però di buoni e ottimi effetti in universale, cioè di pace, tranquillità, e essaltatione della nostra Catholica religione, con depressione de gl'infedeli, e heretici, e forse qualche monarchia; non ostante che prima habbino da seguire commotioni grandi e disordini, con ribellioni, sedittioni, guerre, e altri mali effetti, come pare che dimostri certa abitudine di stelle, fattasi all'istesso tempo della detta gran congiontione del 1603 […]"

second, the proximity of the 'martial star,' namely Antares or *Alpha Scorpionii* (also known as *Calbalacrab* or 'heart of the Scorpion' in Arabic), an evil star that was often considered to bring misfortune; and finally, the fact that in the year 1604 the planet Mars, broadly associated with bloody wars, burning riots and general belligerency, conjoined with Jupiter and Saturn within the same constellation of Sagittarius.

The astrological circumstances were thus highly complex even before the nova suddenly appeared in the sky. Its incredible closeness to the convergence of Mars was thought by Magini and many others to accentuate the effects that were 'inclined by' the red planet ("rebellions, insurrections, wars"). However, these disasters were expected to take place at the end of a negative period of time that, in turn, would foreshadow the subsequent and long-awaited great peace of the Western world and the Christian religion in its entirety (that is, the "increase of religion" to which Dal Monte was probably referring).

Three years after the appearance of the star and two years after it had faded from view, the author reported that no sign of improvement could be seen in Europe, still less epochal ones, but only recurrent waves of bloodshed that continued to spill across the continent. "Who knows," the author asked, "whether blessed God has not created the said star from anew as a premonition of the punishment that he wishes to give us?".[128]

Magini then turned to consider sources of support for his "highly probable astrological conjecture." In the end, he lamented that the only reasonable thing to hope for was that "the Divine Lord, by his mercy, shall not allow the ongoing conflicts to carry on longer."[129] Fortunately, a good sign in that direction was witnessed by the end of the Thirteen Years' War (1593–1606), considering in particular the final years of the campaign, which had been the most cruel and violent.[130] From an astrological point of view, this trend was (hopefully)

128 *Ibid.*, c. 9: "[…] nel grado 19. del medesimo segno di Sagittario <vi fu> l'apparitione d'una nuova Stella nell'istesso luogo, che durò molto risplendente per parecchi mesi, doppo la quale, per quanto s'è veduto, non sono seguiti troppo buoni effetti, ma sì bene ribellioni, commottioni e moti grandi, con molta effusione di sangue, così in Ongaria, come altrove, e chi sa, che Dio benedetto non habbia prodotto da nuovo la detta Stella per presagio del castigo che ci vuol dare, e che significhi ancora i presenti disturbi suscitati in Italia, acciò che ci rivediamo dalle nostre colpe, e che ricorriamo all'aiuto di sua divina maestà […]"

129 *Ibid.*

130 See Rhoads Murphey, *Ottoman Warfare: 1500–1700* (London: UCL Press, 1999), pp. 173–174. In the opening chapter of *De stella nova* (1606), Kepler referred to the relation between the growing number of political predictions and the Fiery Trigon shortly before the appearance of the new star; *De stella nova*, *JKGW*, 1, 157.12–15: "Cum extarent jam a multis annis variae Astrologorum praedictiones de revolvente trigono igneo, et ingentibus

confirmed by Bonhombra through the approach of the Moon, which normally exerted a protective influece, as well as the approach of the Sun and the "beneficial" presence of Venus.[131]

The reason why Magini sought to share the above account under a false name may be explained by his attempt to avoid risking his reputation as a practitioner and professor through the promulgation of uncertain predictions. In addition, Magini was practicing judicial astrology at a difficult time (as we discussed above), when it was increasingly condemned and demonized under the Index and prosecuted by the Inquisition.[132] This happened often in the local milieu of Magini, as a contemporary procedural protocol of censure and condemnation tells us today.[133] In the following section, we turn from his pseudonymous account to his epistolary correspondence, where he exchanged ideas and information under his real name in his various roles as astronomer, mathematician and professor at the University of Bologna.

9.9 An Anonymous Consultation with Magini on the New Star

The modern editor of Magini's correspondence located after a letter from Giovanni di Guevara a document that bears no trace of its author nor when it was written. The title of the text is editorial and simply reads: *Consultation on the new star* [*Consulto intorno alla stella nuova*].[134]

For a start, if the text was actually a consultation, it was an answer *to* Magini rather than a request *for* his consultation, as can be gathered from the assertive, not interrogative, tone of the document. Morover, it came already enclosed with the *imprimatur* or licence for printing. Therefore, Magini was probably asked to examine the text, provide his professional opinion and give his authorization for publication as well. The fact that it was never printed suggests that there may have been some kind of interruption in publication, a request for revision or some other intervention made by Magini, or perhaps a completely different circumstance. Judging by the content of the text, the

prodigiis, rerumque publicarum motibus ex hoc anno 1604 consecuturis; cumque ideo omnium animi in res novas essent erecti."

131 Bonhombra (ref. 123), c. 9.

132 See Baldini (ref. 49); Casali (ref. 106), pp. 61–80.

133 *Breve informazione del modo di trattare le cause del S. Officio perli molto reverendi Vicarij della Santa Inquisitione, Istituiti nelle Diocesi di Modona, di Carpi, di Nonantola e della Garfagnana* (Modena: G. Cassiani, 1608), cc. 12, 52.

134 Anonymous, *Consulto intorno alla stella nuova*, in Favaro (ref. 25), no. 34, pp. 291–294.

author was probably a clergyman who expressed his proposals conservatively in a style of exposition that was characteristically scholastic. Magini, for his own part, was increasingly disposed to breaking with the traditional interpretation of the phenomenon. At that time, it seems that Magini was collecting a wide variety of reference material to publish a possible study under his own name (rather than rely on Bonhombra).[135] The anonymous document may thus have landed on his desk as further source material for his work in progress.

The author articulated his reflections in seven points on the "marvellous splendour" that will be briefly recounted in the following summary.[136]

(1) First, the idea of a heavenly body transported perpetually by another orb was dismissed entirely, since it had never been seen before, either at past or future points, along any part of its hypothetical path. It had to be something that, as such, could not be pre-existent.

(2) It had to be located in the region of the fixed stars where it shared the motion of the final sphere (rather than in the elementary region, as Dal Monte wished), for the altitude or distance should have been the same as what Clavius had assigned to the great novelty of 1572.

(3) What was observed could not have been a real star, since stars are considered to be perpetual by their very nature. The bright light had begun to show at a certain moment in time, so it could not be eternal; therefore, the author concluded, it could not be a star. Although he conceded that it had shone at night in a way that strongly resembled a star (in terms of dimension, form, splendour, "and also for some other hidden features that have been acknowledged to belong to real stars"),[137] the author could not commit to accepting it as a fixed star because it could not be considered perennial.

(4) There was no need to attribute the origin of the novelty to an act of free and creative divine power, as previously proposed by Clavius; consequently, there was no need to abandon the principle of celestial incorruptibility. The change took place in a region of space—the ether—that, in terms of its own structure, could not allow for any modification, so it had to be a kind of change of its own. As a result, the phenomenon was studied from the aetiological point of view of the four Aristotelian causes

135 According to Favaro (ref. 26), p. 110.

136 *Ibid.*, p. 291: "mirabile fulgore."

137 *Ibid.*, p. 292: "[…] et forte etiam quad aliquas virtutes occultas quas in veris sideribus inesse expertum est."

(efficient, material, formal and final) to understand its peculiar typology of change.

At the level of material cause, the bright light could be neither a simple alteration (which was considered to be possible only in the sublunar region) nor generated from nothing (*ex nihilo*). Rather, the author argued for a *different* kind of change, which he called, employing scholastic terminology, *alteratio accidentalis perfectiva*, namely a change that never implies a deterioration or worsening of state, but what occurs only when change serves to preserve a fully acquired form; and since it does not require a passage between contrary qualities, such a change could also take place in the celestial region, where it was thought that tensions between opposite physical polarities were absent.[138]

As for the formal cause, the astronomical event was considered to be, on the one hand, a necessary optical consequence of the universe as it was, and, on the other, a sort of maintenance measure of its matter. With regard to the former, some authors such as Bartholomeus Amicus (1562–1649) explained that the overlaps between the celestial crystalline spheres, differing in vitreous density, could reflect the sidereal light in such a way that, from our point of view on earth, the nova would exhibit optical properties like those stemming from the fixed stars.[139] With regard to the 'maintenance of matter,' we may recall the famous position of Kepler, who claimed that the cosmos possessed a sort of soul and 'natural faculty' that was responsible for self-preservation, both by way of the creation of something new and by way of the destruction of excess matter. In

138 The remote origin of the doctrine of "alteratio accidentalis perfectiva," which became widespread by the late Renaissance, can be traced back to the work of Jean Buridan. In his commentary on *De caelo*, Buridan argued the following: "De alteratione dicendum est, quod caeloum est alterabile alteratione improprie dicta vel communiter dicta, quae est receptio alicuius qualitatis sine corruptione contrariae qualitatis et sine resistentia; quia est illuminabile, ut apparet de luna, et oportet quod spherae inferiores recipiant influentias solis et astrorum superiorum, aliter non possent illae influentiae multiplicari ad ista inferiora. Sed dicendum est quod caeloum non est alterabile alteratione proprie dicta, quae fit cum corruptione qualitatis contrariae et cum resistentia qualitatum praeexistentium; talis enim alteratio ordinatur in generationem et corruptionem, ideo non invenitur in caelo cum caelum non sit generabile." Jean Buridan, *Quaestiones super libris quattuor de caelo et mundo*, ed. Ernest Addison Moody (Cambridge, MA: Medieval Academy of America, 1942), Book 1, quaestio 10, 82.14–25. On this point, see Grant (ref. 29), esp. on pp. 103–104, 107–108.

139 Edward Grant, *Planets, Stars, and Orbs: The Medieval Cosmos, 1200–1687* (Cambridge: Cambridge University Press, 1994), pp. 208–209.

De stella nova (1606), Kepler supposed that the new star burned in such a way as to provide a form of physiological purification, which restored old matter to a state of purity that he equated with the transparency of the ether.[140]

At the level of efficient cause, the object was said to have some repercussions on the alignment and disposition of the other celestial bodies, from whose convergence or 'conspiration' it was thought to be made.

As for the final cause, the author assigned the appearance of the new luminary to the maintenance of balance between the composite influences that were emanating from the stars (according to their various dispositions) upon the inferior bodies. Hence, the new convergence of light that now included the stellar novelty would contribute to fostering and perpetuating events on earth.

(5) The author of the consultation did not venture to pronounce what precisely the nova announced or foretold. The sudden appearance had changed the entire configuration of the heavens, and so, he argued, had all of the physical characteristics and qualities that stemmed from them. "Those things that shall lead to future change [*mutatio*]," he lamented, "remain unknown to us."[141]

(6) The author then asked why the nova had diminished in brightness so considerably since it had first appeared. He reckoned in response that, since the light of the 'optical nova' ultimately relied on the irradiating planets and stars and the contingency of their convergence, the apparent brightness could not remain constant and uniform over time.

(7) Although it was believed that the light would not disappear at once and would last a while longer (just as the nova in Cassiopeia had remained visible for roughly sixteen months), the author claimed that it could not endure forever. The reason, he adduced, was that the relative periodicity of the planets and stars, which projected the image of the phenomenon on the celestial vault through their temporary alignment (like an interplay of mirrors and reflections) could not last forever. In a similar fashion, Francesco Maurolico and Cornelio Frangipane (1553–1643) had

140 Miguel Á. Granada, "Johannes Kepler and David Fabricius: Their Discussion on the Nova of 1604," in *Change and Continuity in Early Modern Cosmology*, ed. Patrick J. Boner (Dordrecht: Springer, 2011), pp. 67–92, on pp. 76–77: "The purpose, Kepler thought, was to preserve as clean and proper the transparent and fluid '*campos liquentes*' or ethereal matter, always encumbered by 'perspiration,' fatty excesses, and putridity exhaled by the stars."

141 Anonymous (ref. 134), p. 293.

> maintained that the nova of 1572 was actually a focalized concentration of light coming from the surrounding planets and stars, while Girolamo Cardano had suggested the same thing about the Milky Way, assigning the expanse to the convergence of other light sources in the sky.[142]

All of this, concluded the author, was argued "*rationibus physicis*" and submitted for the approval of professors of natural philosophy and other academic authorities. However, despite his best effort, the author may have encountered some resistance from Magini, who might have expressed reservations about finally approving the account for publication. The ideas were original in some ways, but on the whole they remained conservative and ran contrary to the research of Magini, who never hesitated to consider the nova as a new star from the start.

9.10 Magini's Correspondence on the New Star

The following four letters to Magini all deal, in part or entirely, with the nova of 1604. Bartolomeo Cristini, a scientific correspondent serving as ducal librarian for the Savoia family, wrote these letters during the first months of 1605. Prefect of the palace and grand master of ceremonies for the same Piedmontese house, Cristini expressed a keen interest in mathematics and astronomical calculations. Although we do not possess any replies from Magini, it is possible to reconstruct the essential topics and turns of their epistolary exchange. In addition, it is clear from a letter from Federico Cesi to Jan van Heeck in Prague that Magini first thought the nova originated as a light-reflecting cloud of vapours,[143] an opinion highly similar to the one held by Galileo in his lectures, which, apparently, he later abandoned.

Regarding the location of the star, Cristini reported that he had received observational coordinates from Tengnagel by way of Giovanni Stefano Ferrero, the Bishop of Vercelli (from 1599 to 1610), a personal friend and apostolic *nuncio* at the imperial court.[144] Moreover, Cristini noted that another nova, in addition to the one in Ophiuchus, had been recently observed from Rome. The

142 See Tessicini (ref. 58), p. 52.

143 Federico Cesi to Jan van Heeck, 19 March 1605, in Gabrieli (ref. 33), p. 56: "Maginus solus suam opinionem edidit, absurdissimam ex vaporum materia conflatam censes; reliqui et maxime Peripatetici magno pelago obruti videntur, et iam Aristotelis dogmata denegant celestia."

144 Bartolomeo Cristini to Giovanni Antonio Magini, various letters (see below), in Favaro (ref. 26), pp. 295, 302, 307. Cfr. Ricci (ref. 4, n. 2) p. 221.

more recent nova was apparently located in the constellation of Gemini, at the level of the Pleiades, or rather close to the eye of Taurus, according to sources. Cristini confessed, however, that in spite of careful examination he could not see anything,[145] and Magini never made reference to such a novelty in his extant correspondence. We may imagine that the discovery was a mistake, or that someone in Rome had been dazzled by visual conditioning, deceptively moved by excessive enthusiasm regarding the resounding appearance of the nova of 1604.

Turning to the bright light in Ophiuchus, Cristini and Magini agreed that it was actually an alteration "in the ether … contrary to Aristotle [*contra Aristotelem*]."[146] However, Cristini wrote that none of the scholars in his vicinity was willing to accept such evidence, since it directly opposed the principle of celestial immutability. The agreement between the two astronomers stopped here; for Cristini was not persuaded that it was actually a *new* star, but rather a star that had already been acknowledged before. He even proposed tracing the star back to the one found in the right shinbone of the Serpent Bearer ("*in tibia dextera Ophiuchi*," known today as *ϑ Ophiuci*, a star slightly greater than the third magnitude). Following one of Clavius' lesser endorsed hypotheses for the nova of 1572, Cristini inferred that the pre-existing dormant star "awoke" and suddenly began shining much more intensely ("*aucta lumine*," he wrote, "*quae in pristinum nunc reddita est*").[147] By contrast, Magini came to the conclusion that the nova was a completely new stellar formation, since there had been nothing previously at that point in the night sky, nor in *ϑ Ophiuci*. In affirming this, Magini benefited from consulting the stellar catalogue found in the first posthuomous version of Tycho's *Progymnasmata* (1602),[148] where in fact no star was listed at the actual position of the nova (see Fig. 9.4). Cristini was left dumbfounded by the divergence of opinion and their correspondence came to a natural end.

Among the last letters addressed to Magini concerning this controversy, we may recount a couple from Giovanni Strassoldo (1547–1610) in Udine at the end of 1605. Besides other minor matters, the occasion for their exchange

145 Bartolomeo Cristini to Giovanni Antonio Magini, 26 February 1605, in Favaro (ref. 26), no. 36, p. 303.

146 *Ibid.*, p. 298; see also p. 110.

147 Bartolomeo Cristini to Giovanni Antonio Magini, 8 April 1605, in Favaro (ref. 26), no. 37, p. 304. See also the following notice by Bartolomeo Cristini to Giovanni Antonio Magini, 8 May 1604, in *ibid.*, no. 38, p. 307.

148 See the first part of Frank Verbunt and Robert H. van Gent, "Three Editions of the Star Catalogue of Tycho Brahe," *Astronomy & Astrophysics*, 2010, 516: 1–25.

was the recent election of Camillo Borghese (1550–1621) as Pope Paul V; for Strassoldo asked Magini for some details in order to interpret the papal "genitura" based on the astrological circumstances of his birth.[149] Strassoldo then cast a horoscope for the neo-papal profile on the basis of the disposition of the heavenly bodies as he determined them for the start of the divine mandate (as per standard procedure). According to Strassoldo, the appearance of the new star, "so resplendent and admirable … provides enough for a powerful geniture and promises great honours … and for [the Pope] to make the greatest impact, which actually occurs."[150] (At the same time, we know from a letter written by Federico Cesi to Jan van Heeck that the same sign was considered at the court in Rome to be an omen of death for one of the previous popes, Clement VII, which occurred on March 1605 "*ex portentu syderis novi*").[151]

Despite the precarious art of prediction, it is interesting to note how long the new star loomed on the horizon of prognostication. Visible for nearly eighteen months after it first appeared, it remained an enduring anomaly to be accounted for, even when the meaning of the marvel escaped the imagination of astrologers. For his own part, Magini merged avant-garde astronomical insight with imaginative astrological speculation, which he advanced under the veil of Bonhombra. Dal Monte made reference to this pseudonymous account, but by way of negative mention. Clavius, by contrast, did not enter these aspects of the dispute, although he was certainly aware that astrological discussions—and disagreements—raged all around him.

9.11 Clavius' Correspondence on the New Star (2)

Meanwhile, Clavius continued to receive one letter after another, primarily from other prelates, requiring him to resolve more and more details about the nova. As a central link and source of authority in a sprawling correspondence

149 Giovanni Strassoldo to Giovanni Antonio Magini, 3 June 1605, in Favaro (ref. 26), no. 34, pp. 309–312. On the Renaissance practice of providing horoscopes, see H. Darrel Rutkin, "Various Uses of Horoscopes: Astrological Practices in Early Modern Europe," in *Horoscopes and Public Spheres: Essays on the History of Astrology*, ed. Guenther Oestmann, H. Darrel Rutkin and Kocku von Stuckrad (Berlin-New York: De Gruyter: 2005), pp. 167–182.

150 Giovanni Strassoldo to Giovanni Antonio Magini, 8 December 1605, in Favaro (ref. 26), p. 315: "[…] la nuova stella, così splendente e riguardevole … havesse bastato in una genitura potente et atta a prometter honori grandi … a far sortir l'effetto grandissimo, che succede."

151 Ricci (ref. 4, n. 3), p. 223.

network, Clavius strove to answer just as many inquiries as he could. Despite the fact that around Christmas the star descended in altitude closer to the horizon where it became increasingly more difficult to discern, interest in its nature and physical features did not wane among observers. Before concluding, let us touch briefly on these later forms of contact with Clavius.

The Benedictine monk Don Antonio da Mantova wrote to Clavius in December 1604 claiming that the star had already been observed from the monastery at Montecassino around 3 October, first as a light of the fourth magnitude, and then increasing to second magnitude before finally diminishing to the current dimension at which it could then be observed.[152]

Bernardino Salino, lecturer of mathematics at the University of Genova, wrote to Clavius on 14 January 1605 asking him for the coordinates of the star.[153]

Jean Chastellier, theologian and dean of the Collège de le Flèche, wrote from France to ask for his opinion on the new star or that held by his fellow Jesuit Jacques Sirmond.[154]

Giovanni Giacomo Staserio replied to Clavius from Naples, noting that the brightness of the nova was gradually declining. In his letter, Staserio included a comparative diagram explaining where the stellar magnitude had already dropped below that of Antares, though it was not yet inferior to one of the surrounding stars (probably *ϑ Ophiuchi*). Finally, Staserio referred to a separate consultation with Christoph Grienberger S.J. on the matter.[155]

Soon after hearing from Salino, Clavius was contacted by Giovanni Bartolomeo Biamino, teacher of mathematics at the college in Brera.[156] Biamino addressed Clavius as a guide and mentor, and informed him of the scrupulous and prolonged observations that he had made in Milan. Biamino mentioned the opinion of those who believed it was a comet, and their indecision as to whether it would have occasioned a clear sky or produced dryness in the soil. Although he could not calculate the parallax due to disturbed visibility, Biamino understood the relevance of this missing piece of evidence, and consequently asked Clavius to provide him with the fundamental measurement he lacked. Finally, Biamino informed his mentor that he intended to organize

152 Don Antonio da Mantova to Christoph Clavius, 28 December 1604, in Clavius (ref. 60), no. 242, p. 141.

153 Bernardino Salino to Christoph Clavius, 14 January 1605, in *ibid.*, no. 244, pp. 143–144.

154 Jean Chastellier to Christoph Clavius, 26 February 1605, in *ibid.*, no. 246, pp. 148–149.

155 Giovanni Giacomo Staserio to Christoph Clavius, 1 April 1605, in *ibid.*, no. 247, pp. 149–150.

156 Giovanni Bartolomeo Biamino to Christoph Clavius, 19 January 1605, in *ibid.*, no. 245, pp. 145–148.

a public lecture on the position of the new star at his college. Mellowing after a more formal opening, Biamino finally wrote:

> Now that I have spoken seriously about the star, let us have a bit of fun. For what figments of judgement, as is common in those cases, will they dream up without any judgement at all? Shall they foretell with a stern brow poisoning, pestilence, arrows, arms, wars, blood and destruction from an ill-fated position, just because they realize that it occupies the sign of Sagittarius, remaining in roughly the same place in the foot of Serpentarius? Whatever they may blather, the wise man shall always prefer to laugh at their judgements rather than be worried by them.[157]

Biamino ended by noting the inconsistency between such an unfortunate prediction and the original planetary conjunction that, with the exception of ominous Mars, was believed by the same astrologers to be benign and perhaps even favourable. Even though Biamino the mathematician poked fun at the astrologers, however, he proved in his letter to be perfectly well informed about their art: a shrewd display of rationalism, a manneristic expression of complicity or a critical (and common) response to other stargazers? In any case, his letter concluded with a thinly concealed *captatio benevolentiae*. Biamino bid farewell to Clavius by claiming that he preferred the favour of those, like Clavius, who were engaged in the study of the stars, rather than the favour of the stars themselves, whatever the magnificent nova might entail.

Bartolomeo Cristini, whose correspondence with Magini we briefly examined above, wrote to Clavius as well.[158] Cristini noted that the new heavenly body, measured in relation to several other stars chosen as points of reference, had never budged. Cristini then articulated the same opinion he had conveyed (above) to Magini, namely that the 'new' star had already existed and simply grown in radiance and splendour for a certain length of time before returning to obscurity. By holding this position, Cristini implied the dismissal of celestial immutability, but preserved the principle of ingenerability of celestial bodies. According to him, then, the new star was actually *old*, and precisely the same

157 *Ibid.*, p. 147: "Serio sum de stella locutus, iocari liceat. Quaenam, ut assolent, figmenta iudiciarii absque iudicio somniabunt? An quod in Serpentarii pede fixa medium fere Sagittarii signum occupare cernitur, venena, pestes, sagittas, arma, bella, caedes, exitia ex infausto situ severiore supercilio pronunciabunt? Quidquid garriant, iudicia eorum ridere mavult sapiens quam timere."

158 Bartolomeo Cristini to Christoph Clavius, 18 April 1605, in Clavius (ref. 60), no. 250, pp. 154–155.

OPHIVCHVS.								
DENOMINATIO STELLARVM.	LONGITVDO.				LATITVDO.			MAGN.
	S.	G.	M.		G.	M.		
In capite	♐	16	50		35	57	B	3
In dextro humero	♐	19	45		28	1	B	3
Inferior & ſequens in dextro humero	♐	21	5		26	11	B	3.
Præcedens in ſiniſtro humero	♐	4	59½		32	35½	B	4.
Sequens in eodem humero	♐	6	16		31	56	B	4
Quæ in ſiniſtro cubito	♐	0	3		23	39½	B	4
In ſiniſtra manu borealior	♏	26	44½		17	19	B	3
Sequens & auſtralior	♏	27	57		16	30½	B	3.
In dextro ancone	♐	19	3		15	19	B	4
Auſtralior & præcedens in dextra manu	♐	24	13½		13	47	B	4
Borealior & ſequens in eadem manu	♐	25	14½		15	20	B	5
In dextro genu	♐	12	24		7	18	B	3
Quæ in ſiniſtro genu	♐	3	39		11	30	B	3
In dextra tibia	♐	12	24		7	18	B	3
Quinta informium in via lactea	♐	26	31		33	2½	B	4.

FIGURE 9.4 Tycho's posthumous star chart for Ophiuchus. Tycho Brahe, *Astronomiae instauratae progymnasmata* (Prague, 1602), cc. 267v–268r (composite image).

star found near the shin of the Serpent Bearer, a known star that had been catalogued by Copernicus as third-magnitude in the same constellation of Ophiuchus.[159] (Again, this thesis was basically a reiteration of the first option concerning the nova of 1572 that Clavius dismissed in his commentary on the *Sphere*). Cristini was newly persuaded by the fact that, even though the star had once surpassed the first magnitude after appearing, it soon 'returned' to being no brighter than its 'original' third magnitude. Cristini accepted the decline in light and luster as compelling evidence that the star had returned to an earlier or original state, and that the object under observation must have always been the same star, namely *Tibia Ophiuchi*, the third-magnitude star that according

159 Nicolaus Copernicus, *De revolutionibus orbium celestium* (Nuremberg: Johannes Petreius, 1543), Book 2, "Signorum stellarumque descriptio canonica," f. 50r.

to Cristini stood in exactly the same position as where the nova had appeared (and remained). To strengthen his case, Cristini even suggested an imprecision in the stellar catalogue of Brahe. (Magini may have made him aware of this conceit or curious ommission in their separate correspondence.)[160] It is clear, however, that the imprecision was Cristini's *own* interpretation, since Magini had referred to the same catalogue and come to the *opposite* conclusion, namely that it was a new star that *did not* correspond to the star in *Tibia Ophiuchi*, and thus they did not share the same coordinates.

Cristini told Clavius that he had consulted with Magini and Tengnagel (evidently without finding support from either author), and he was now hoping to hear from Clavius, in case he could provide Cristini with a source of support for his more traditional interpretation. It is quite possible that Clavius never replied to this letter, however, because he had already adopted a completely different premise, having long abandoned the idea of saving the 'old' system of stellar astronomy at any cost. Contrary to Cristini, Clavius had turned to another set of assumptions to ascertain the celestial origins of the star. At the conference on the new star at the Collegio Romano (mentioned above), where Clavius probably oversaw the proceedings and may have even established the official position proposed by his colleagues, it had been formally concluded that the nova could *not* have been a pre-existing star, and that this had been confirmed by the careful comparison of countless observations.[161]

The judgement of the Jesuits did not enjoy a broad reception beyond Rome and apparently remained relatively obscure. It certainly did not stem the tide of debate sweeping across the continent. On 17 July 1605, the Jesuit professor of ethics and mathematician in Mainz, Johann Reinhard Ziegler (1569–1636), wrote to Clavius to notify him that debates over the novelty were as lively as ever in Germany. Many essays had been published on it, Ziegler observed, although from his personal point of view there were few details from the argument that actually deserved any deeper attention.[162]

160 It should be recalled that the positions of the fixed stars provided by Tycho in his catalogue were originally calculated for the purpose of determining precisely the position of the nova of 1572. See C. Doris Hellman, "The Rôle of Measurement in the Downfall of a System: Some Examples from 16th Century Comet and Nova Observations," in *New Aspects in the History and Philosophy of Astronomy*, ed. Arthur Beer (Oxford: Pergamon Press, 1967), pp. 43–52, on p. 50.

161 Baldini (ref. 91), p. 80.

162 Johann Reinhard Ziegler to Christoph Clavius, 17 July 1605, in Clavius (ref. 60), no. 254, pp. 160–161.

Two years later, when the star had vanished from sight some months before, another Jesuit, Johann Lantz, professor of mathematics at Ingolstadt,[163] wrote to Clavius condemning his account as "too easy."[164] Why had Clavius accepted the famous luminary as a fixed star, Lantz asked, which had never appeared before? Why could it not have been something else, such as an exhalation that had risen from the earth and set on fire in the upper reaches of the air from friction caused by the crystalline sphere of the Moon? For that was, in fact, the traditional view commonly held to explain the fiery origin of comets. To fuel his doubt, Lantz claimed that, once the 'star' had been extinguished, a sort of slight shift in its remnants had been observed. If a proper shift had truly taken place, Lantz asked rhetorically, how could it ever be the product of a *fixed* star? At this point, Lantz believed that it would have been more reasonable for Clavius to consider the novelty a comet. It thus appears that Lantz shared some of the same concerns as Dal Monte long after the latter first formulated his opinion across the Alps. Moreover, Lantz did not accept the divine nature of the novelty and complained that Clavius was wrong to leave open the hypothesis of a free act of spontaneous creation when the phenomenon could be explained entirely in physical terms. Echoing the opinion of others, Lantz argued that the bright light had originated from the great conjunction of the superior planets that had converged in the same region of the sky shortly before the 'star' appeared. In the course of conveying his thesis, Lantz expressed close familiarity with Kepler's *De stella nova*, recently published and widely received, and where his theory, shared with other astrologers, was discussed at length.[165] Clavius, for his own part, seems to have paid scant attention to his opinion.[166]

163 Lantz is better known for having taught Christoph Scheiner, the author of *Rosa ursina* (1626) and the bitter rival of Galileo. On the career of Scheiner, see Luigi Ingaliso, *Filosofia e cosmologia in Christoph Scheiner* (Soveria Mannelli: Rubettino, 2005).

164 Johann Lantz to Christoph Clavius, 10 July 1607, in Clavius (ref. 42), no. 271, pp. 59–60.

165 *De stella nova*, *JKGW*, 1, Chapters 2–12, pp. 165–208.

166 Johann Lantz to Christoph Clavius, 10 July 1607, in Clavius (ref. 42), no. 271, p. 60: "Iam quod ad novas stella dicemus. Certe stellam Sagittarij naturalem, non miraculosam fuisse, argumento est magna coniunctio, ad quam exorta fuit. Sed R.a V.a. facile stellas novas in caelo esse concedit. Quod si ita est, cur stella Sagittarij 30 diebus promota est 36'? Cur non potius stella fixa, aut aliquem ex planetis secuta?". See Galileo Galilei & Christoph Scheiner, *On Sunspots*, ed. Eileen Reeves and Albert Van Helden (Chicago: University of Chicago Press, 2010), p. 41: "If Clavius answered Lantz's barrage of questions and speculations, that letter has not survived."

9.12 Conclusion: A World (of Letters) Mirrored in the New Star

The demand for data on the nova of 1604 ignited an international exchange that extended far beyond the boundaries of modern Italy to every corner of the continent. The push for parallactic measurements in particular was aimed at establishing whether the novelty could be observed more precisely from different points on earth, allowing observers to conclude once and for all whether it was a comet or a fixed star. The answer, in turn, would suggest an ordinary event or the occurrence of an exceptional anomaly in the world system. Although no resounding agreement was reached on this point in the short or middle term, the written letter served a wide range of scholars as the medium of choice for discussing their many concerns and queries.

The account of Clavius on the nova of 1572, widely read and warmly received, remained a popular reference point for many scholars long after the novelty had disappeared. On the one hand, Clavius dared to define a new star in the sky, while on the other he cautioned that it had been "a physically unpredictable divine creation." Blending innovation and tradition, Clavius' account informed the work of many who observed the new star of 1604.

In the present study, I have identified Andrea Argoli as one of the first known Italian observers of the new star, along with the Franciscan Ilario Altobelli. Their opening roles in the debate can now be restored as a result. We have also recovered the intermediary roles of individuals such as Giovanni Mellei and Homero Tortora, who promoted the exchange of information between their patrons and other scholars, often by serving as 'on site' middle men who served as sources of mediation for various motives.

The Dal Monte-Giordani correspondence reveals the many complex forms of cognitive and cultural bias that accompanied the interpretation of the nova, while at the same time offering a taste of the many astrological accounts that arose in response to the new luminary, including the prognostications of Casamatta (alias Casale) and Bonhombra (alias Magini).

At the same time, astronomers such as Magini asserted other, more mathematically oriented accounts on the basis of Brahe's accurate observations. As we have seen, Magini concluded that the nova could not be one of the 'old' stars observed already, even though he was well aware of the existence of other, more conservative opinions, as extant documents in his possession confirm today regarding various aspects of content and formulation.

Despite attempts to identify the nova with a pre-existing comet or star, a growing number of scholars gradually accepted that they were witnessing a new star never before seen in the superlunary realm. This was evident from the many parallactic measurements that converged toward a "*diversitas aspectus*" close to zero, though there was no lack of hesitation regarding how to receive and interpret the data, especially since it ran contrary to convention and the standard foundations of astronomy. It all depended on whose eyes were watching: the new luminary could serve as the source of a startling horoscope, a pretext for erudite dispute or a persuasive piece of evidence for confirming a new cosmological system, as Galileo had endeavoured in the working notes for his public lectures.[167]

The names of prominent astronomers that appear in the present study—Clavius, Altobelli and Magini—were highly regarded sources of reference in the epistolary exchanges of Galileo and Kepler during these years. Through the exercise of their authority and the application of their skill, they played a key role in fostering the circulation of ideas and information. We would probably find the name of the Marquis Dal Monte more frequently, were it not for his aristocratic inclination to underexpose himself in the first person.

The above letters do not simply articulate a scientific network that linked a broad spectrum of scholars and their separate ideas across long distances in a relatively short amount of time; they also speak to a massive intellectual mobilization that, when the new star suddenly appeared, far surpassed the scope of the written letter, extending to public squares, princely courts, private residences, university lecture halls, pubs, taverns, and printing presses, among other places. Concern over the conspicuous astrological aspects, in particular, encouraged debate over the great novelty in every sphere of society, from the common populace to high court councils.

Finally, the appearance of the nova served as a sort of measure of resilience (and resistance) of the cosmological theories that then held sway. While many of these theories would eventually falter in the long run, they proved highly robust in the short term. As conceptual tools, they patched many openings in the old system caused by the succession of celestial novelties over the sixteenth and seventeenth centuries.[168]

---

167 See Massimo Bucciantini, *Galileo e Keplero: Filosofia, cosmologia e teologia nell'Età della Controriforma* (Torino: Einaudi, 2000), esp. Chapter 6, "*Supernova*: Galileo lettore di Tycho e di Keplero," pp. 117–143; William R. Shea, "Galileo and the Supernova of 1604," in *1604–2004: Supernovae as Cosmological Lighthouses*, ed. Massimo Turatto, Stefano Benetti, Luca Zampieri and William R. Shea (San Francisco: Astronomical Society of the Pacific, 2005), pp. 13–20; Michele Camerota, *Galileo Galilei e la cultura scientifica nell'età della Controriforma*, vol. 1 (Milan: il Giornale Biblioteca Storica, 2006 [Rome: Salerno, [1]2004]), esp. Ch. III.5, "*False imposture, fraudolenti inganni, temerari usurpamenti*," pp. 138–148; Cosci (ref. 10).

168 In parallel with the present chapter, see Matteo Cosci, "The Correspondence of Galileo Galilei and Other Italian Astronomers on the Nova of 1604," *forthcoming*.

Acknowledgements

The present contribution was made possible by the ERC Starting Grant 2013–335949: "Aristotle in the Italian Vernacular: Rethinking Renaissance and Early-Modern Intellectual History (c. 1400–c. 1650)." Special thanks to Patrick Boner for his careful editorial advice.

CHAPTER 10

The Scientific Legacy of Kepler's 'Stella Nova'

William P. Blair

10.1 Introduction

The "new star" observed by Kepler and others in 1604, and Kepler's analysis of its characteristics, had an obvious and enormous impact on the direction of modern science, even though none of those who observed it had an inkling about the real mechanism involved in creating it.[1] Those observers could be forgiven for thinking that once the object faded from view, it would be gone forever.

However, with the advent of large optical and radio telescopes in the mid-twentieth century and space-based telescopes in the late twentieth century, the expanding remains of that titanic stellar explosion have been rediscovered and studied in great detail. Despite this intense scrutiny with telescopes across the electromagnetic spectrum (Fig. 10.1), the object we now know as "Kepler's supernova remnant" has only recently begun to yield some of its closely held secrets. In doing so, it has only become more relevant to astronomers as we strive to put this object into its proper astrophysical context. To understand the scientific import of Kepler's supernova remnant to modern science, some background is needed.

10.2 Background and Context

Initially, supernovae were classified based on simple observable characteristics in their dispersed light spectra—either they showed evidence of hydrogen in their spectra (Type II) or they did not (Type I). Some systematic behavior in the light curves (or changes in intensity over time) were also observed as more observations became available. This was strictly an observational classification scheme, as no real understanding of the underlying physical mechanisms was in place yet.

But as more supernovae were observed, and as different techniques were applied to their analysis, it quickly became clear that more complexity was needed to classify the range of observed characteristics. Type I's came in several

1 Physics and Astronomy Department, Johns Hopkins University.

 | DOI:10.1163/9789004437272_012

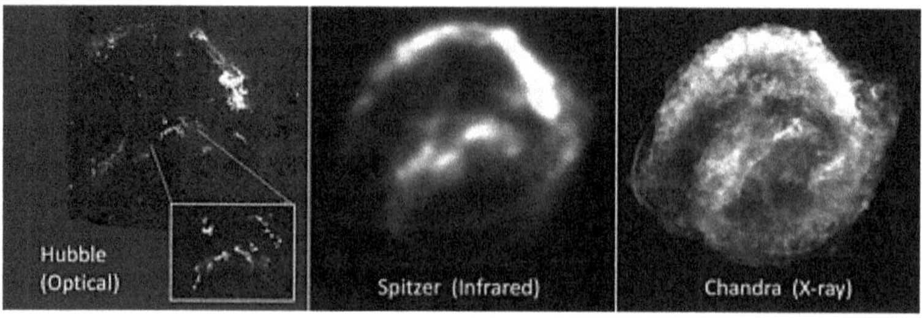

FIGURE 10.1 Three views of the remnant of Kepler's supernova, from NASA's Great Observatories: (left) an optical light view from Hubble Space Telescope; (middle) an infrared light view from the Spitzer Space Telescope; (right) the appearance in X-ray light from the Chandra X-ray Observatory. Hubble's camera shows details in the densest knots of circumstellar material, as shown in the enlargement. Spitzer shows hot dust that has been heated by the expanding supernova shock wave. The chemical composition of the X-ray emitting gas is what clinched the identification of Kepler's supernova as a Type Ia.
FIGURE COURTESY OF THE AUTHOR

varieties designated as Ia, Ib, and Ic, based on other observed characteristics in their spectra in addition to no evidence of hydrogen. Type II's took on additional sub-classes based on the shapes of their light curves, with Type II-L supernovae showing a linear decline after the peak, and Type II-P supernovae exhibiting an extended flat plateau prior to the linear decline. Other sub-types have continued to be added as more and more supernovae have been observed and additional sub-groupings of characteristics have been identified.[2]

As the observations began to yield to interpretation of the kinds of stars responsible for the explosions and their observed characteristics, aspects of the original observational classification system have not translated smoothly, leaving us with a rather awkward nomenclature that is still used in the scientific literature today. We now know that supernovae arise in two different genres of stars: low-mass exploding white dwarf stars with 1.4 times the mass of the sun, and relatively massive stars (approximately nine or more times the mass of the sun, and perhaps as much as 100 solar masses or more in some extreme cases). Unfortunately, these physical properties do not translate into "Type I" and "Type II." The exploding white dwarf variety corresponds to the Type Ia classification, while Type Ib, Ic, and all of the Type II varieties arise from massive stars of various kinds and evolutionary state.

2 Some original early classification schemes included SN Types III, IV, and V; see Fritz Zwicky, "Supernovae," *Handbuch der Physik*, 1958, 51: 766. Over time, these classes were subsumed into various sub-types of the Type II classification scheme.

To understand a bit more about what is going on physically, we need to go slightly deeper than the types of stars, into the explosion mechanisms involved and the evolutionary state of the stars when the explosion is triggered. The low-mass white dwarf supernovae do not show hydrogen in their spectra because the chemical composition of the white dwarf stars that spawn them are dominated by carbon and oxygen (and sometimes other heavy chemical elements), with no surface hydrogen present. In contrast, massive stars lose mass as they evolve, stripping their outer hydrogen layers over time and exposing their interiors, where the heavier elements lie. Hence, a massive star that has lost its hydrogen atmosphere and subsequently explodes also does not show hydrogen in its spectrum, and is classified as Type Ib. The Type Ic class goes even further, not showing evidence of hydrogen or helium (the next heaviest element) in its spectrum, meaning of course that even more material has been lost by the star prior to its explosion. The Type II classes that do show hydrogen in their spectra are massive stars that have exploded earlier in their evolution, before all of their surface hydrogen has been stripped away. Collectively, these massive star supernovae are termed "core-collapse supernovae" because the explosion mechanism is related to the loss of interior pressure support as their nuclear fuel runs out, causing the stellar core to collapse and then rebound outward. If this rebound has sufficient energy, it causes the star to be ripped apart, thus creating the supernova explosion.

Not surprisingly, the supernovae arising from the large range of massive star progenitors show a broad range of observable characteristics both in their spectra and in their photometric light curves. However, the Type Ia supernovae arise from a very uniform set of progenitors, namely 1.4 solar mass carbon-oxygen white dwarf stars, and as such, the observed supernova characteristics for this class are quite similar. For example, observations of relatively nearby Type Ia supernovae (where distances can be determined from other methods) have established that the absolute peak brightness of Type Ia supernovae is very nearly constant,[3] establishing them as "standard candles." It is therefore possible to calculate the distance to other Type Ia supernovae if their peak *apparent* brightness can be measured for comparison, by simple application of the inverse square law. The huge value of this peak luminosity, coupled with the very large telescopes available to today's astronomers, makes it possible to detect and measure Type Ia supernovae at very great distances indeed.[4]

3 Bradley E. Schaefer, "Peak Brightnesses of Historical Supernovae and the Hubble Constant," *Astrophysical Journal*, 1996, 459: 438–454.

4 In reality, with the large number of carefully observed Type Ia supernovae available today, we now know that they are not strictly speaking standard candles. Some are slightly brighter or

It is thus the Type Ia supernovae that have become so useful for judging distances in astronomy, and hence, they have become a crucial element of establishing the cosmic distance ladder that is used in cosmology. It is the accuracy of this very distance ladder that has permitted one of the most astounding conclusions about our universe to be established: that not only is the universe expanding, but that this expansion is actually *accelerating* over time,[5] a finding that resulted in a Nobel prize in 2011 for its discoverers.

10.3 The Type Ia Progenitor Conundrum

With this background, it is somewhat embarrassing to admit that modern astronomers do not understand the progenitor systems or exact explosion mechanisms that produce the Type Ia supernovae. White dwarf stars are held up against gravity by electron degenerate pressure (rather than thermal pressure as in normal stars). Subramanyan Chandrasekhar was the first to calculate that a white dwarf star with more than a certain amount of mass would not be able to hold itself up against the inward pull of gravity, and hence, the star would collapse in on itself.[6] Others refined the argument and concluded that 1.4 solar masses was the magic number. Under the assumption that this is an unstable process, there are any number of possible triggering mechanisms that would cause such a star to release a tremendous burst of energy that would tear the star apart, thus creating a Type Ia supernova. Since the trigger happens when the progenitor white dwarf star is always 1.4 solar masses, the physical conditions at the time of the explosion are so similar that the energy released (and hence the peak brightness of the supernova) is nearly uniform. The problem is, how does Nature make 1.4 solar mass white dwarf stars? The vast majority of isolated white dwarf stars with mass estimates populate a fairly narrow range from about 0.5–0.7 solar masses.

---

fainter than the mean. However, these brightness variations vary systematically with other characteristics that can be observed, such as the width of the brightness peak in the light curve, allowing the variations to be calibrated and improving the use of these objects for calculating distances. Hence, it would be more appropriate to call these objects "standardizable candles."

5 Adam G. Riess, Alexei V. Filippenko, Peter Challis, et al., "Observational Evidence from Supernovae for an Accelerating Universe and a Cosmological Constant," *Astronomical Journal*, 1998, 116: 1,009–1,038.

6 Subrahmanyan Chandrasekhar, "The Maximum Mass of Ideal White Dwarfs," *Astrophysical Journal*, 1931, 74: 81.

There are two popular and competing pathways that in principle could produce a white dwarf star with the requisite 1.4 solar masses. The first pathway involves having two white dwarf stars in a tight binary orbit that lose energy and merge into a single object with enough mass to surpass the threshold. This is called the double degenerate model. The second pathway would be to have a white dwarf in a binary system with a normal star; as the system evolves, the normal star expands and transfers sufficient material onto the white dwarf star to push it over the Chandrasekhar limit, thus causing it to explode. Each of these models has its pros and cons, and it has not been clear whether one dominates or whether both models are operative in the real universe, and if so for what fractions of the observed supernovae. Many recent studies in the nearby universe, where the remnants of suspected Type Ia supernovae can be observed, seem to favor the double degenerate pathway.

10.4 The Classification of Kepler's Supernova

The optical remnant of SN 1604 was first discovered by Baade as a handful of faint emission line filaments in a very crowded star field toward the center of our Galaxy.[7] Over the ensuing decades, as better and better observational data have been obtained,[8] a debate has raged about the type of the 1604 supernova. The reconstructed light curve has been used as evidence for both Type Ia and Type II designations.[9] The position of the object well off the galactic plane seemed to point to a possible massive runaway star from an earlier binary system.[10] Deeper optical observations and spectroscopy revealed a partial shell of emitting material with strong nitrogen lines, consistent with the idea of mass loss from a massive progenitor star, tipping the scale toward core-collapse. Initial X-ray observations by satellite telescopes were alternatively modeled with both Type Ia and core-collapse models.[11]

7 Walter Baade, "Nova Ophiuchi of 1604 as a Supernova," *Astrophysical Journal*, 1943, 97: 119.

8 Rino Bandiera and Sidney van den Bergh, "Changes in the Optical Remnant of Kepler's Supernova during the Period 1942–1989," *Astrophysical Journal*, 1991, 374: 186–201; William P. Blair, Knox S. Long, and Olaf Vancura, "A Detailed Optical Study of Kepler's Supernova Remnant," *Astrophysical Journal*, 1991, 366: 484–494; Ravi Sankrit, William P. Blair, Tracey Delaney, et al., "HST/ACS Imaging of a Balmer-Dominated Shock in Kepler's Supernova Remnant," *Advances in Space Research*, 2005, 35: 1,027–1,030.

9 Baade (ref. 6); Jesse B. Doggett and David Branch, "A Comparative Study of Supernova Light Curves," *Astrophysical Journal*, 1985, 90: 2,303–2,311.

10 Bandiera and Van den Bergh (ref. 7).

11 A. Decourchelle and J. Ballet, "Modeling the X-Ray Spectrum of Kepler's Supernova Remnant," *Astronomy and Astrophysics*, 1994, 287: 206–218; R. Rothenflug, B. Magne,

It was not until 2007 that two papers combined to definitively nail down Kepler's supernova as a Type Ia. The first insight came when Blair et al. pointed out that the apparent nitrogen enhancement in optical spectra was consistent with *normal* interstellar abundances at the progenitor's position more than halfway toward the center of our Galaxy.[12] This removed the need to attribute the enhanced abundance to material from a massive progenitor star, which would be expected to be enriched in nitrogen. The most solid determination, however, came from interpretation of a deep X-ray observation with the *Chandra* satellite,[13] which was able to determine more accurate chemical abundances of the hot ejecta material from the exploded star. The highly enhanced abundances of iron and silicon were directly aligned with expectations for an exploded carbon-oxygen white dwarf, and inconsistent with enriched abundances of oxygen, sulfur, and other elements seen from young core-collapse remnants. Kepler's supernova must have been a Type Ia.

10.5 Ramifications

Returning to the uncertain progenitor systems of Type Ia's, the importance of knowing the type of Kepler's supernova becomes clear. In the double degenerate model, the two purported white dwarf stars are not expected to be surrounded by circumstellar material because the time it takes for them to spiral in and merge means any such material would have long since dissipated. Indeed, there are at least two other historical supernova remnants in our Galaxy that seem to be consistent with this picture: SN 1572 (Tycho's) and SN 1006. In the single degenerate model, the normal star in the binary system transfers mass to the white dwarf companion, but it is a 'sloppy' process. Some material spirals in onto the white dwarf, but some material also has to be expelled in this process in order to balance the angular momentum of the system. As such, one might expect substantial material to surround the site of single degenerate Type Ia

J. P. Chieze, and J. Ballet, "Hydrodynamic Model of Kepler's Supernova Remnant Constrained by EINSTEIN and EXOSAT X-Ray Spectra," *Astronomy and Astrophysics*, 1994, 291: 271–282; Kenzo Kinugasa and Hiroshi Tsunemi, "ASCA Observation of Kepler's Supernova Remnant," *Publications of the Astronomical Society of Japan*, 1999, 51: 239–252; G. Cassam-Chenai, A. Decourchelle, J. Ballet, et al., "XMM-Newton Observation of Kepler's Supernova Remnant," *Astronomy and Astrophysics*, 2004, 414: 545–558.

12 William P. Blair, Parviz Ghavamian, Knox S. Long, et al., "Spitzer Space Telescope Observations of Kepler's Supernova Remnant," *Astrophysical Journal*, 2007, 662: 998–1,013.

13 Stephen P. Reynolds, Kazimierz J. Borkowski, Una Hwang, et al., "A Deep Chandra Observation of Kepler's Supernova Remnant: A Type Ia Event with Circumstellar Interaction," *Astrophysical Journal*, 2007, 668: L135–L138.

supernovae. This is the exact situation we seem to have confirmed for Kepler's supernova, and hence, it becomes the best example we know of that points to a single degenerate progenitor for a Type Ia supernova. Taken in context with the others, this seems to confirm that at least some Type Ia supernovae occur via both of the progenitor paths that have been identified.

With at least one confirmed example of the single degenerate progenitor, one can look for other examples that might also be consistent with this pathway. There is one young supernova remnant in our neighboring galaxy, the Large Magellanic Cloud, that is so similar to the remnant of Kepler's supernova that is has been called Kepler's "cousin."[14] There is also a very small sub-category of supernovae that are initially classified as Type Ia from their early spectra and light curve development, but which subsequently develop narrow hydrogen lines and are classified with a new moniker: Type IIn. (Recall that Type II's show hydrogen lines, and "n" in this case stands for narrow.) It is not too hard to imagine that these objects are cases where the blast wave from a single degenerate supernova has swept out into surrounding circumstellar material lost from the binary prior to explosion, thus exciting the hydrogen emission in this material at late times. The identification of Kepler's supernova remnant as a bona fide Type Ia is largely what places this sort of reasoning on a firm footing.

While little uncertainty remains in determining the type of Kepler's supernova, it still maintains some of its secrets and refuses to give them up without more work. One of the theoretical expectations of the single degenerate model is that the companion star to the supernova should remain intact, although possibly affected by the explosion of its close companion in one of several ways. One observable of the erstwhile companion might be a peculiar velocity (or depending on the direction of motion, a proper motion) relative to other nearby stars. Another possibility might be a peculiar atmospheric chemical composition under the assumption that the companion star intercepts some of the supernova debris during the explosion, and that this would be manifest in the spectrum of the star. Both of these effects have been searched for in the stars that are seen in projection near the location of the expected supernova position in Kepler's supernova remnant, and no smoking guns have been identified.[15] However, since no surviving companion to a Type Ia supernova has

---

14 Brian J. Williams, Kazimierz J. Borkowski, Stephen P. Reynolds, et al., "*Spitzer* Observations of the Type Ia Supernova Remnant N103B: Kepler's Older Cousin?," *Astrophysical Journal*, 2014, 790: 139–149.

15 Pilar Ruiz-Lapuente, Francesco Damiani, Luigi Bedin, et al., "No Surviving Companion in Kepler's Supernova," *Astrophysical Journal*, 2018, 862: 124.

ever been clearly identified, it is difficult to know how much credence to give to these theoretical ideas.

10.6 Epilogue

Johannes Kepler was an intriguing personality. He came from humble beginnings, possessed a fierce and staunch religious faith that was entwined with his world view but at odds with contemporary culture, and was a scientific visionary. While understanding the new star of 1604 was only one of his many interests, I think he would be secretly pleased to know that his name had been attached to this equally intriguing supernova remnant, and that it continues to have a prominent and relevant place in modern astrophysics.

Bibliography

Primary Literature

Original Editions

Andreae, Johann Valentin, *FAMA FRATERNITATIS R. C. Das ist, Gerucht der Bruderschafft des Hochlöblichen Ordens R. C. An alle Gelehrte und Heupter Europas.* Kassel, 1615.

Apian, Peter, *Ein kurtzer Bericht & Observation unnd Urtels des jüngst erschinnen Cometen.* Ingolstadt, 1532.

Argoli, Andrea, *Problemata astronomica triangulorum ope demonstrata per sinus, tangentes et secantes ex sola multiplicatione absque divisione.* Rome, 1604.

Bartsch, Jacob, *Planisphaerium stellatum sive vice-globus coelestis in plano delineatus.* Strasbourg, 1624.

Bayer, Johann, *Uranometria: Omnium asterismorum continens schemata, nova methodo delineata, aereis laminis expressa.* Augsburg, 1603.

Bonhombra, Lodovico [alias Giovanni A. Magini], *Discorso astrologico delle mutationi de' tempi, e de i più notabili accidenti sopra l'anno 1607.* Bologna, 1607.

Brahe, Tycho, *Astronomiae instauratae progymnasmata.* Prague, 1602.

Büttner, Balthasar, *Cometen Stundbüchlein.* Eisleben, 1605.

Cellarius, Andreas, *Harmonia macrocosmica seu atlas universalis et novus totius universi creati cosmographiam generalem, et novam exhibens.* Amsterdam, 1661.

Cesalpino, Andrea, *De metallicis libri tres.* Rome, 1596.

Clavius, Christoph, *Geometria practica.* Rome, 1604.

Clavius, Christoph, *Compendium brevissimum describendorum horologiorum horizontalium ac declinantium.* Rome, 1603.

Clavius, Christoph, *Horologiorum nova description.* Rome, 1599.

Clavius, Christoph, *Astrolabium.* Rome, 1593.

Copernicus, Nicolaus, *De revolutionibus orbium coelestium.* Nuremberg, 1543.

Dal Monte, Guidobaldo, *De stella magorum*, B. U. V. cod. Urb. lat. 1743, cc. 65r–69r.

De Boodt, Anselmus Boëtius, *Gemmarum et lapidum historia.* Hanover, 1609.

Diodato, Laudadio, *Pronostico o discorso delle mutationi de' tempi, e de i più notabili accidenti sopra l'anno 1607.* Bologna, 1606.

Du Bartas, Guillaume, *La seconde Sepmaine de Guillaume de Saluste Seigneur du Bartas /Die andere Woche*, trans. Tobias Hübner. Köthen, 1622.

Fabricius, David, *Kurtzer und Gründtlicher Bericht von Erscheinung und Bedeutung deß grossen Newen WunderSterns.* Hamburg, 1605.

Faentino, Casamatta, *Discorso astrologico, nel quale oltre le mutazioni de' Tempi, si scoprono varij naturali avvenimenti, non solo dell'Anno presente MDCV ma anco per tutto l'Anno MDCXI*. Rimini, 1605.

Galilei, Galileo, *Le operazioni del compasso geometrico et militare*. Padua, 1606.

Gemma, Cornelius, *De prodigiosa specie naturaque cometae*. Antwerp, 1578.

Gemma, Cornelius, *De naturae divinis characterismis seu raris et admirandis spectaculis, indiciis, proprietatibus rerum in partibus singulis universi*. Antwerp, 1575.

Hagecius, Thaddaeus, *Dialexis de novae et prius incognitae stellae inusitatae magnitudinis et splendidissimi luminis apparitione*. Frankfurt, 1574.

Herlicius, David, *Astronomische und historische Erklerung des Newen Sterns oder ungeschwänzten Cometen, so Anno 1604 im ende des Septembris, und folgendem Octobri, auch Novembri erschienen*. Stettin, 1605.

Ingoli, Francesco, *De novo sydere, causis, duratione, ac portentis discursus Anni MDCIIII*, Biblioteca Malatestiana, Ms. 166.17, 1r–21r.

Kircher, Athanasius, *Iter exstaticum coeleste*. Würzburg, 1661.

Krabbe, Johannes, *Alter und Newer Schreibkalendar*. Erfurt, 1613.

Krabbe, Johannes, *Newes Astrolabium, Sampt dessen Nutz und Gebrauch*. Frankfurt, 1609.

Krabbe, Johannes, *Cometa eo anno 1604. den 3. Tag Octobris / am Himmel eschienen*. Magdeburg, 1605.

Krabbe, Johannes, *Gründliche observation des newen Cometen lauff / höhe / und grösse, welcher in diesem 1596. Jahre / im Monat Julio am Himmel erschienen*. Magdeburg, 1596.

Krabbe, Johannes, *Practica astrologica*. Berlin, 1591.

Mazzoni, Jacopo, *De triplici hominum vita, activa nempè, contemplativa, et religiosa methodi tres, quaestionibus quinque millibus, centum et nonagintaseptem distinctae. Cesena*, 1576.

Melanchthon, Philipp, *Initia doctrinae physicae*. Wittenberg, 1549.

Moller, Albin, *Gründlicher und warer Bericht / Von den newen Cometstern / so in der Lufft unter dem Himmel gesehen / nach dem Niedergang der Sonnen / in October und Novemb. des 1604. Jahrs*. Eisleben, 1605.

Neuhauser, Wilhelm Eo, *Tractatus de nova stella: oder von dem newen Abent Stern Scheinende*. Strasbourg, 1604.

Palau, Marco Antonio, *Discurso astronomico sobre el cometa: Estrella que de nuevo ha aparecido en el Zodiaco, en 18. grados del signo de Sagitario, cerca de la Ecliptica, con poca latitud septentrional, a los 8. de Octubre 1604*. Valencia, 1604.

Placentinus, Johannes, *Physicalischer und Astrologischer Bericht Von Denen erschrecklichen, ungewöhnlichen … und schädlichen Winden*. Frankfurt, 1661.

Riccioli, Giovanni Battista, *Almagestum novum*. Bologna, 1651.

Roeslin, Helisaeus, *Mitternächtige Schiffarth Von den Herrn Staden inn Niderlanden vor XV. Jaren vergebenlich fürgenommen, wie dieselbige anzustellen, daß man daselbst herumb in Orient und Chinam kommen möge, zu sonderen der Christenheit, sonderlich Teutschlands Nutzen und Wolfart*. Oppenheim, 1611.

Roeslin, Helisaeus, *Historischer / Politischer und Astronomischer naturlicher Discurs Von heutiger zeit Beschaffenheit / Wesen und Standt der Christenheit und wie es ins künfftig in derselben ergehn werde / Aus anleitung dero von Anno 1600. her am hohen Himmel erschienenen grossen Wunderzeichen / sonderlichen dess Cometens anno 1607. genommen*. Strasbourg, 1609.

Roeslin, Helisaeus, *Tractatus meteorastrologiphysicus*. Strasbourg, 1597.

Scaliger, Julius Caesar, *Exotericarum exercitationum liber XV*. Frankfurt, 1582.

Schickard, Wilhelm, *Astroscopium: Pro facillima stellarum cognitione excogitatum, & commentariolo illustratum*. Stuttgart and Leipzig, 1698.

Schickard, Wilhelm, *Cometen Beschreibung In zwen underschidliche Partes abgetheilt, deren Erster Von denselbigen ins gemein: der Ander Von allen Insonderheit, sonderlich aber denen drey Jüngsten, In abgeloffenen 1618 Jahr erschienen, aussführlich handelt*. Tübingen, 1619.

Schiller, Julius, *Coelum stellatum christianum, ad maiorem Dei Omnipotentis, Sanctaeque Eius tam Triumphantis, quam Militantis Ecclesiae gloriam, Obductis Gentilium Simulachris, Eidem Domino et Creatori Suo, postliminio quasi restitutum*. Augsburg, 1627.

Sifard, Johann, *Cometen Spiegel*. Leipzig, 1605.

Tortora, Homero, and Pier Matteo Giordani, *Lettere di Homero Tortora a P.M. Giordani*, Biblioteca Oliveriana di Pesaro, Ms. 415, cc. 35–41.

Van Heeck, Jan, *De nova stella disputatio*. Rome, 1605.

Weidner, Johannes, *Acclamatio ad stellam novam prodigiosam, anno labente 1604, mensib. Octobri & Novembri conspectam, post Solis occasum, in dodecatemorio Sagittarii*. Göltiz: Johannes Rhamba, 1605.

Modern Editions and Translations

Agrippa, Heinrich Cornelius, *Three Books of Occult Philosophy*. Ed. Donald Tyson, trans. James Freake. St. Paul, MN: Llewellyn Worldwide, 1993.

Andreae, Johann Valentin, *The Fame and Confession of the Fraternity of R: C: Commonly, of the Rosie Cross*. Trans. Thomas Vaughan. London: Giles Calvert, 1652.

Arago, François, *Oeuvres complètes*, vol. 4. Ed. J. A. Barral. Paris: Gide, 1858.

Aristotle, *The Complete Works of Aristotle: The Revised Oxford Translation*. Ed. Jonathan Barnes. Princeton: Princeton University Press, 1984.

Aristotle, *On the Heavens*. Trans. W. K. C. Guthrie. Cambridge, MA: Harvard University Press, 1939.

Brahe, Tycho, *Opera omnia*. Ed. J. L. E. Dreyer. 15 vols. Copenhagen: Swets & Zeitlinger, 1972.

Bruno, Giordano, *Dialoghi filosofici italiani*. Ed. Michele Ciliberto. Milan: Mondadori, 2000.

Bruno, Giordano, *De immenso, et innumerabilibus*. In: *Jordani Bruni Nolani Opera Latine conscripta*, ed. Francesco Fiorentino (Naples: Morano, 1884; facsimile ed., Stuttgart: Frommann, 1962), I, 2.

Buridan, Jean, *Quaestiones super libris quattuor de caelo et mundo*. Ed. Ernest Addison Moody. Cambridge, MA: Medieval Academy of America, 1942.

Cardano, Girolamo, *De Subtilitate*. Trans. John M. Forrester. Tempe: Arizona Center for Medieval and Renaissance Studies, 2013.

Cicero, *De natura deorum*. Trans. H. Rackham. Cambridge, MA: Harvard University Press, 1933.

Clavius, Christoph, *Corrispondenza*, 6 vols. Ed. Ugo Baldini and Pier Daniele Napolitani (Pisa: University of Pisa: Department of Mathematics, 1992).

Copernicus, Nicolaus, *On the Revolutions*. Trans. and commentary by Edward Rosen. Baltimore: Johns Hopkins University Press, 1992.

Copernicus, Nicolaus, *Erster Entwurf seines Weltsystems sowie eine Auseinandersetzung Johannes Keplers mit Aristoteles über die Bewegung der Erde nach den Handschriften*. Ed. Fritz Rossmann. Darmstadt: Wissenschaftliche Buchgesellschaft, 1974.

Crasso, Lorenzo, *Elogii d'huomini letterati*, vol. 2. Venice, 1666.

Descartes, René, *The World and Other Writings*. Ed. Stephen Gaukroger. Cambridge: Cambridge University Press, 2004.

Descartes, René, *Principles of Philosophy*. Trans. Valentine Rodger Miller and Reese P. Miller (Dordrecht-Boston-London: Kluwer, 1982).

Descartes, René, *Oeuvres de Descartes*. Ed. Charles Adam and Paul Tannery (Paris: Cerf, 1905), vol. 7.

Ficino, Marsilio, *Three Books on Life*. Ed. Carol V. Kaske and John R. Clark. Binghamton: Renaissance Society of America, 1989.

Galilei, Galileo, *Le Opere di Galileo Galilei*, 20 vols. Ed. Antonio Favaro. Florence: Barbèra, 1936.

Gassendi, Pierre, *Opera omnia* (Lyon, 1658; facsimile ed., Stuttgart: Frommann, 1964), vol. 1.

Gilbert, William, *On the Magnet*. Trans. Silvanus Phillips Thompson. New York: Basic Books, 1958.

Hill, Nicholas, *Philosophia epicuraea democritiana theophrastica proposta simpliciter non edocta*. Ed. Sandra Plastina. Pisa-Rome: Fabrizio Serra, 2007.

Kepler, Johannes, *The Six-Cornered Snowflake: A New Year's Gift*. Trans. Jacques Bromberg. Philadelphia: Paul Dry Books, 2010.

Kepler, Johannes, *Über den Neuen Stern im Fuß des Schlangenträgers*. Trans. Otto and Eva Schönberger and Eberhard Knobloch. Würzburg: Königshausen & Neumann, 2006.

Kepler, Johannes, *Optics: Paralipomena to Witelo and Optical Part of Astronomy*. Trans. William H. Donahue. Santa Fe: Green Lion Press, 2000.

Kepler, Johannes, *The Harmony of the World*. Trans. E. J. Aiton, A. M. Duncan, and J. V. Field. Philadelphia: American Philosophical Society, 1997.

Kepler, Johannes, *New Astronomy*. Trans. William H. Donahue. Cambridge: Cambridge University Press, 1993.

Kepler, Johannes, *L'étoile nouvelle dans le pied du Serpentaire suivi de L'étoile inconnue du Cygne*. Trans. Jean Peyroux. Paris: Blanchard, 1988.

Kepler, Johannes, "A Thorough Description of an Extraordinary New Star Which First Appeared in October of This Year, 1604." Trans. J. V. Field and Anton Postl. *Vistas in Astronomy* 20 (1977): 333–339.

Kepler, Johannes, *L'étrenne ou la neige sexangulaire*. Trans. Robert Halleux. Paris: Vrin, 1975.

Kepler, Johannes, *Nova Kepleriana: Wiederaufgefundene Drucke und Handschriften von Johannes Kepler*. Ed. Walther von Dyck. Munich: Bayerische Akademie der Wissenschaften, 1934.

Melanchthon, Philipp, *Melanchthons Briefwechsel: Kritische und kommentierte Gesamtausgabe*. Ed. Christine Mundhenk. Stuttgart-Bad Cannstatt: Frommann-Holzboog, 2010.

Ptolemy, *Almagest*. Trans. G. J. Toomer. Princeton: Princeton University Press, 1998.

Rothmann, Christoph, *Christoph Rothmann's Discourse on the Comet of 1585: An Edition and Translation with Accompanying Essays*. Ed. Miguel A. Granada, Adam Mosley and Nicholas Jardine. Leiden: Brill, 2014.

Schickard, Wilhelm, *Briefwechsel*. Ed. Friedrich Seck. Stuttgart-Bad Cannstatt: Frommann-Holzboog, 2002.

Seneca, *Natural Questions*. Trans. Harry M. Hine. Chicago: University of Chicago Press, 2010.

Simplicius, *On Aristotle On the Heavens 1.2–3*. Trans. Ian Mueller. London-New York: Bloomsbury, 2011.

Secondary Literature

Almási, Gábor, "Rethinking Sixteenth-Century 'Lutheran Astronomy.'" *Intellectual History Review* 24 (2013): 1–16.

Almási, Gábor, "Tycho Brahe and the Separation of Astronomy from Astrology: The Making of a New Scientific Discourse." *Science in Context* 26 (2013): 3–30.

Almási, Gábor, "Humanistic Letter-Writing." *European History Online* (2010): http://www.ieg-ego.eu/almasig-2010-en.

Ariew, Roger, "Censorship, Condemnations, and the Spread of Cartesianism." In: *Cartesian Empiricism*, ed. Mihnea Dobre and Tammy Nyden (Dordrecht: Springer, 2013), 25–46.

Arrighi, Gino, "Un grande scienziato italiano: Guidobaldo Dal Monte in alcune carte inedite della Biblioteca Oliveriana di Pesaro." *Atti dell'Accademia Lucchese di scienze, lettere ed arte* 12 (1965): 183–199.

Aterini, Barbara, "Guidobaldo Dal Monte (1545–1607)." In: *Distinguished Figures in Descriptive Geometry and Its Applications for Mechanism Science: From the Middle Ages to the 17th Century*, ed. Michela Cigola (Cham-Heidelberg: Springer, 2016), 153–180.

Aurini, Raffaele, "Argoli, Andrea." In: *Dizionario bibliografico della gente d'Abruzzo*, vol. 1 (Teramo: Ars et Labora, 1952), 422–429.

Azzolini, Monica, *The Duke and the Stars: Astrology and Politics in Renaissance Milan*. Cambridge, MA: Harvard University Press, 2013.

Baade, Wilhelm H. Walter, "Nova Ophiuchi of 1604 as a Supernova." *The Astrophysical Journal* 97 (1943): 119–127.

Baldini, Ugo, "The Academy of Mathematics of the Collegio Romano from 1553 to 1612." In: *Jesuit Science and the Republic of Letters*, ed. Mordechai Feingold (Cambridge, MA: MIT Press, 2003), 47–98.

Baldini, Ugo, "The Roman Inquisition's Condemnation of Astrology: Antecedents, Reasons and Consequences." In: *Church, Censorship and Culture in Early Modern Italy*, ed. Gigliola Fragnito (Cambridge: Cambridge University Press, 2001), 79–110.

Baldini, Ugo, and Juan Casanovas, "La sfera celeste di Cristoforo Clavio." In: *Osservatorio astronomico di Capodimonte* (Naples: Almanacco, 1996), 175–182.

Baldini, Ugo (ed.), *Christoph Clavius e l'attività scientifica dei Gesuiti nell'età di Galileo*. Rome: Bulzoni, 1995.

Baldini, Ugo, *Legem impone subactis: Studi su filosofia e scienza dei Gesuiti in Italia, 1540–1632*. Rome: Bulzoni, 1992.

Baldini, Ugo, and George V. Coyne, S.J. (eds.), *The Louvain Lectures (Lectiones Lovanienses) of Bellarmine and the Autograph Copy of his 1616 Declaration to Galileo*. Vatican City: Vatican Observatory, 1984.

Baldini, Ugo, "La nova del 1604 e i matematici e filosofi del Collegio Romano." *Annali dell'Istituto e Museo di Storia della Scienza di Firenze* 6 (1981): 63–98.

Bandiera, Rino, and Sidney van den Bergh, "Changes in the Optical Remnant of Kepler's Supernova during the Period 1942–1989." *Astrophysical Journal* 374 (1991): 186–201.

Barker, Peter, Hanne Andersen, and Xiang Chen, *The Cognitive Structure of Scientific Revolutions*. Cambridge: Cambridge University Press, 2006.

Barker, Peter, "How Rothmann Changed His Mind." *Centaurus* 46 (2004): 41–57.

Barker, Peter, “Brahe and Rothmann on Atmospheric Refraction.” In: *Optics and Astronomy: Proceedings of the XXth International Congress of History of Science*, ed. Gérard Simon and Simone Débarbat (Turnhout: Brepols, 2001), 43–50.

Barker, Peter, and Bernard R. Goldstein, “Theological Foundations of Kepler’s Astronomy.” *Osiris* 16 (2001): 88–113.

Barker, Peter, “The Role of Religion in the Lutheran Response to Copernicus.” In: *Rethinking the Scientific Revolution*, ed. Margaret J. Osler (Cambridge: Cambridge University Press, 2000), 59–88.

Barker, Peter, “Kepler’s Epistemology.” In: *Method and Order in Renaissance Natural Philosophy*, ed. Daniel A. Di Liscia, Eckhard Kessler, and Charlotte Methuen (Aldershot: Ashgate, 1997), 355–368.

Barker, Peter, “The Optical Theory of Comets from Apian to Kepler.” *Physis* 30 (1993): 1–25.

Barnes, Robin B., *Astrology and Reformation*. Oxford: Oxford University Press, 2016.

Barreca, Francesco, and Patrick J. Boner, “A Perfect Similitude: Science and Politics in Kepler’s Dedicatory Letters to *De stella nova* and the *Astronomia nova*.” In: *Celestial Novelties on the Eve of the Scientific Revolution, 1540–1630*, ed. Dario Tessicini and Patrick J. Boner (Florence: Olschki, 2013), 209–229.

Bellucci, Dino, *Science de la Nature et Réformation: La physique au service de la Réforme dans l’enseignement de Philippe Mélanchthon*. Rome: Edizioni Vivere, 1998.

Beltrán Marí, Antonio, *Cometas, ciencia y religión: La polémica Galileo-Grassi*. Madrid: Tecnos, 2016.

Bezza, Giuseppe, “Le tecniche astrologiche.” In: *Il linguaggio dei cieli: Astri e simboli nel Rinascimento*, ed. Germana Ernst and Guido Giglioni (Rome: Carocci, 2012), 53–70.

Biagioli, Mario, *Galileo, Courtier: The Practice of Science in the Culture of Absolutism*. Chicago: University of Chicago Press, 1993.

Biagioli, Mario, *Galileo’s Instruments of Credit: Telescopes, Images, Secrecy*. Chicago: University of Chicago Press, 2006.

Bialas, Volker, *Johannes Kepler: Astronom und Naturphilosoph* (Linz: Trauner, 2013).

Blair, William P., Parviz Ghavamian, Knox S. Long, et al., “Spitzer Space Telescope Observations of Kepler’s Supernova Remnant.” *Astrophysical Journal* 662 (2007): 998–1,013.

Blair, William P., Knox S. Long, and Olaf Vancura, “A Detailed Optical Study of Kepler’s Supernova Remnant.” *Astrophysical Journal* 366 (1991): 484–494.

Blumenberg, Hans, *The Genesis of the Copernican World*. Cambridge, MA: MIT Press, 1987.

Boner, Patrick J., *Kepler’s Cosmological Synthesis: Astrology, Mechanism and the Soul*. Leiden: Brill, 2013.

Boner, Patrick J., “Beached Whales and Priests of God: Kepler and the Cometary Spirit of 1607.” *Early Science and Medicine* 17 (2012): 589–603.

Boner, Patrick J. (ed.), *Change and Continuity in Early Modern Cosmology*. Dordrecht: Springer, 2011.

Boner, Patrick J., "Kepler's Copernican Campaign and the New Star of 1604." In: *Change and Continuity in Early Modern Cosmology*, ed. Patrick J. Boner (Dordrecht: Springer, 2011), 93–114.

Boner, Patrick J., "Kepler's Imprecise Astrology." In: *Kepler: La physique céleste*, ed. Édouard Mehl (Paris: Les Belles Lettres, 2011), 115–132.

Boner, Patrick J., "Kepler on the New Star: *De stella nova*, Chapters 7–9." *Culture and Cosmos* 14 (2010): 209–234.

Boner, Patrick J., "Finding Favour in the Heavens and Earth: Stadius, Kepler and Astrological Calendars in Early Modern Graz." In: *Studia Copernicana*, vol. 42, ed. Richard L. Kremer and Jaroslaw Włodarczyk (Warsaw: Instytut Historii Nauki, 2009), 159–177.

Boner, Patrick J., "A Tenuous Tandem: Patrizi and Kepler on the Origins of Stars." *Journal for the History of Astronomy* 40 (2009): 381–391.

Boner, Patrick J., "Life in the Liquid Fields: Kepler, Tycho and Gilbert on the Nature of the Heavens and Earth." *History of Science* 46 (2008): 275–297.

Boner, Patrick J., "Kepler v. the Epicureans: Causality, Coincidence and the Origins of the New Star of 1604." *Journal for the History of Astronomy* 38 (2007): 207–221.

Boner, Patrick J., "Kepler's Living Cosmology: Bridging the Celestial and Terrestrial Realms." *Centaurus* 48 (2006): 32–39.

Boner, Patrick J., "Kepler on the Origins of Comets: Applying Earthly Knowledge to Celestial Events." *Nuncius* 21 (2006): 31–47.

Boner, Patrick J., "Soul-Searching with Kepler: An Analysis of Anima in His Astrology." *Journal for the History of Astronomy* 36 (2005): 7–20.

Boockmann, Friederike, Daniel A. Di Liscia, and Hella Kothmann (eds.), *Miscellanea Kepleriana: Festschrift für Volker Bialas zum 65. Geburtstag*. Augsburg: Rauner, 2005.

Bouman, José, and Cis van Heerten (eds.), *Divine Wisdom, Divine Nature: The Message of the Rosicrucian Manifestoes in the Visual Language of the Seventeenth Century* (Amsterdam: In de Pelikaan, 2014).

Brancaccio, Giovanni, "Tagliacozzo e il confine tra il Regno di Napoli e lo Stato della Chiesa." In: *Tagliacozzo e la Marsica tra Antico regime e Risorgimento*, ed. Franco Salvatori (Roma: Abilgraf, 2005), 101–111.

Brethauer, Karl, "Johannes Krabbe Mundensis." In: *Braunschweigisches Jahrbuch*, vol. 55, ed. J. König (Braunschweig: Braunschweigischer Geschichtsverein, 1974), 72–89.

Brosseder, Claudia, "The Writing in the Wittenberg Sky: Astrology in Sixteenth-Century Germany." *Journal of the History of Ideas* 66 (2005): 557–576.

Brosseder, Claudia, *Im Bann der Sterne: Caspar Peucer, Philipp Melanchthon und andere Wittenberger Astrologen*. Berlin: Akademie Verlag, 2004.

Brucker, Jakob, *Historia critica philosophiae*. Leipzig: Christoph Breitkopf, 1766.

Bucciantini, Massimo, *Galileo e Keplero: Filosofia, cosmologia e teologia nell'Età della Controriforma*. Turin: Einaudi, 2007.

Burke-Gaffney, Michael W., *Kepler and the Jesuits*. Milwaukee: Bruce Publishing Company, 1944.

Burke-Gaffney, Michael W., "Kepler and the Star of Bethlehem." *Journal of the Royal Astronomical Society of Canada* 31 (1937): 417–425.

Camerota, Michele, *Galileo Galilei e la cultura scientifica nell'età della Controriforma*. Rome: Salerno, 2004.

Camerota, Michele, and Patrizia Ruffo, "Le lezioni di Galileo sulla nuova stella del 1604 nel resoconto di Antonio Alberti (17 dicembre 1604)." *Galilaeana* 12 (2015): 193–201.

Canone, Eugenio, and Germana Ernst (eds.), *Enciclopedia Bruniana e Campanelliana*. Pisa-Rome: Fabrizio Serra, 2010.

Caredda, Barbara, *Aspetti e momenti del dibattito astronomico nella prima Accademia dei Lincei (1603–1616)*. Ph.D. Dissertation, University of Cagliari, 2008.

Casali, Elide, "Il diavolo dal mantello stellato e la condanna dell'astrologia." In: *Il linguaggio dei cieli: Astri e simboli nel Rinascimento*, ed. Germana Ernst and Guido Giglioni (Rome: Carocci, 2012), 153–167.

Casali, Elide, "Il numero fisico matematico e gli almanacchi di Carlo Cesare Scaletta da Faenza (1666–1748) tra astrologia, enciclopedismo e nuova scienza." In: *Belle le contrade della memoria: Studi e libri in onore di M. G. Tavoni*, ed. Federica Rossi and Paolo Tinti (Granarolo dell'Emilia: Pàtron, 2009), 81–98.

Cassam-Chenai, Gamil, A. Decourchelle, J. Ballet, et al., "XMM-Newton Observation of Kepler's Supernova Remnant." *Astronomy and Astrophysics* 414 (2004): 545–558.

Catà, Cesare, "*Lumen requirunt lumine*: Marsilio Ficino, Nicola Cusano e l'iconologia dei Magi nel Rinascimento." *Philosophical Readings* 7 (2015): 110–131.

Chandrasekhar, Subrahmanyan, "The Maximum Mass of Ideal White Dwarfs." *Astrophysical Journal* 74 (1931): 81.

Chapman, Allan, "Tycho Brahe—instrument designer, observer and mechanician." *Journal of the British Astronomical Association* 99 (1989): 70–77.

Chen-Morris, Raz, *Measuring Shadows: Kepler's Optics of Invisibility*. University Park: Penn State University Press, 2016.

Christianson, John Robert, *Tycho Brahe and the Measure of the Heavens*. London: Reaktion, 2020.

Christianson, John Robert, *On Tycho's Island: Tycho Brahe, Science, and Culture in the Sixteenth Century*. Cambridge: Cambridge University Press, 2000.

Cicogna, Emmanuele A., *Delle iscrizioni veneziane*, vol. 4. Venice: Giuseppe Piccotti, 1834.

Ciliberto, Michele, *La ruota del* tempo. Rome: Editori Riuniti, 1986.

Coope, Ursula, "Aristotle on the Infinite." In: *Oxford Handbook of Aristotle*, ed. Christopher Shields (Oxford: Oxford University Press, 2012), 267–286.

Cosci, Matteo, "Aristotelian Cometary Theory in Translation: The Effects of Comets from Mid-Sixteenth-Century Vernacular Prose to the Reception of Galileo Galilei." *Giornale di Storia della Filosofia Italiana* (special issue), ed. David Lines and Anna Laura Puliafito, forthcoming.

Cosci, Matteo, "Le fonti di Galileo Galilei per le *Lezioni e studi sulla stella* nuova del 1604." *Archives Internationales d'Histoire des Sciences* 68 (2018): 6–70.

Cosci, Matteo, "Galileo *alias* Astolfo Arnerio Marchiano e la disputa padovana sulla Stella Nuova." *Atti e memorie dell'Accademia Galileiana di Scienze, Lettere ed Arti in Padova* 131 (2019): 35–84.

Crowther, Kathleen, and Peter Barker, "Training the Intelligent Eye: Understanding Illustrations in Early Modern Astronomy Texts." *Isis* 104 (2013): 429–470.

Daston, Lorraine, "The Empire of Observation, 1600–1800." In: *Histories of Scientific Observation*, ed. Lorraine Daston and Elizabeth Lunbeck (Chicago: University of Chicago Press, 2011), 81–113.

Daston, Lorraine, and Katherine Park, *Wonders and the Order of Nature, 1150–1750*. New York: Zone Books, 1998.

Dear, Peter, *Discipline and Experience: The Mathematical Way in the Scientific Revolution*. Chicago: University of Chicago Press, 1995.

Debus, Allen G., *The Chemical Philosophy: Paracelsian Science and Medicine in the Sixteenth and Seventeenth Centuries*. New York: Science History Publications, 1977.

Decourchelle, Anne, and J. Ballet, "Modeling the X-Ray Spectrum of Kepler's Supernova Remnant." *Astronomy and Astrophysics* 287 (1994): 206–218.

Del Prete, Antonella, *Universo infinito e pluralità dei mondi: Teorie cosmologiche in età moderna*. Naples: Città del Sole, 1998.

Doggett, Jesse B., and David Branch, "A Comparative Study of Supernova Light Curves." *Astrophysical Journal* 90 (1985): 2,303–2,311.

Donahue, William H., "Astronomy in the Service of Cosmology? Kepler's Arguments for a Finite Universe." In: *Astronomy as a Model for the Sciences in Early Modern Times*, ed. Menso Folkerts and Andreas Kühne (Augsburg: Rauner, 2006), 297–306.

Donahue, William H., *The Dissolution of the Celestial Spheres, 1595–1650*. New York: Arno Press, 1981.

Dossena, Marina, and Gabriella Del Lungo Camiciotti (eds.), *Letter Writing in Late Modern Europe*. Amsterdam-Philadelphia: John Benjamins, 2012.

Drake, Stillman, and Charles D. O'Malley (trans.), *The Controversy on the Comets of 1618*. Philadelphia: University of Pennsylvania Press, 1960.

Dreyer, J. L. E., *Tycho Brahe: A Picture of Scientific Life and Work in the Sixteenth Century*. Edinburgh: A. & C. Black, 1890.

Dupré, Louis, *Passage to Modernity: An Essay in the Hermeneutics of Nature and Culture*. New Haven: Yale University Press, 1995.

Dupré, Sven, "Wonder and Experiment in Kepler's *Optics* and *Dioptrics*." In: *Studia Copernicana*, vol. 42, ed. Richard L. Kremer and Jaroslaw Włodarczyk (Warsaw: Instytut Historii Nauki, 2009), 223–237.

Dygebauer, Peter, "Clavius und die Nova von 1572." *Sterne und Weltraum* 1968: 7–22.

Evans, Joan, *Magical Jewels of the Middle Ages and the Renaissance, Particularly in England*. Oxford: Clarendon Press, 1922.

Evans, R. J. W., *Rudolf II and His World: A Study in Intellectual History, 1576–1612*. Oxford: Oxford University Press, 1984.

Fabbri, Natacha, *Cosmologia e armonia in Kepler e Mersenne: Contrappunto a due voci sul tema dell'Harmonice mundi*. Florence: Leo S. Olschki, 2003.

Favaro, Antonio, "I successori di Galileo nello studio di Padova fino alla caduta della Repubblica." *Nuovo Archivio Veneto* 33 (1917): 96–182.

Favaro, Antonio, *Carteggio inedito di Ticone Brahe, Giovanni Keplero e di altri celebri astronomi e matematici dei secoli XVI e XVII con Giovanni Antonio Magini tratto dall'Archivio Malzezzi de' Medici in Bologna*. Bologna: Zanichelli, 1886.

Fernie, J. D., "The Historical Search for Stellar Parallax." *Journal of the Royal Astronomical Society of Canada* 69 (1975): 153–161.

Field, Judith V., *Kepler's Geometrical Cosmology*. London: Athlone, 1988.

Field, Judith V., "A Lutheran Astrologer: Johannes Kepler," *Archive for History of Exact Sciences* 31 (1984): 189–272.

Findlen, Paula (ed.), *Empires of Knowledge: Scientific Networks in the Early Modern World*. London: Routledge, 2018.

Findlen, Paula, *Possessing Nature: Museums, Collecting, and Scientific Culture in Early Modern Italy*. Berkeley: University of California Press, 1994.

Findlen, Paula, "Jokes of Nature and Jokes of Knowledge: The Playfulness of Scientific Discourse in Early Modern Europe." *Renaissance Quarterly* 43 (1990): 292–331.

Folkerts, Menso, "Der Astronom David Fabricius (1564–1617): Leben und Wirken." *Berichte zur Wissenschaftsgechichte* 23 (2000): 127–142.

Fox, Robert (ed.), *Thomas Harriot: An Elizabethan Man of Science*. Aldershot: Ashgate, 2000.

Freudenthal, Gad, "Theory of Matter and Cosmology in William Gilbert's *De Magnete*." *Isis* 74 (1983): 22–37.

Gabrieli, Giuseppe (ed.), *Il Carteggio Linceo della vecchia accademia di Federico Cesi (1603–1630)*. Rome: Memorie dell'Accademia Nazionale dei Lincei, 1938.

Gal, Ofer, and Raz Chen-Morris, *Baroque Science*. Chicago: University of Chicago Press, 2013.

Galluzzi, Paolo, *Tra atomi e indivisibili: La materia ambigua di Galileo*. Florence: Olschki, 2011.

Gamba, Antonio, "Andrea Argoli, il terzo successore di Galileo Galilei." *Galileo* 36 (1996): 4–7.

Gamba, Enrico, and Vico Montebelli, *Le scienze a Urbino nel tardo Rinascimento*. Urbino: Quattroventi, 1988.

Gilly, Carlos, "Las novas de 1572 y 1604 en los manifiestos rosacruces y en la literatura teosófica y escatológica alemana anterior a la Guerra de los Treinta Años." In: *Novas y Cometas entre 1572 y 1618: Revolución cosmológica y renovación política y religiosa*, ed. Miguel A. Granada (Barcelona: University of Barcelona, 2012), 275–332.

Gilly, Carlos, *Cimelia Rhodostaurotica: Die Rosenkreuzer im Spiegel der zwischen 1610 und 1660 entstandenen Handschriften und Drucke*. Amsterdam: In de Pelikaan, 1995.

Gingerich, Owen, "Kepler's *De Vero Anno* (1614)." In: *The Star of Bethlehem and the Magi: Interdisciplinary Perspectives from Experts on the Ancient Near East, the Greco-Roman World, and Modern Astronomy*, ed. Peter Barthel and Georg van Kooten (Leiden-Boston: Brill, 2015), 3–16.

Gingerich, Owen, "Kepler Then and Now." *Perspectives on Science* 10 (2002): 228–240.

Giostra, Alessandro, "Guidobaldo Dal Monte e i nuovi corpi celesti." In: *Guidobaldo Dal Monte (1545–1607): Theory and Practice of the Mathematical Disciplines from Urbino to Europe*, ed. Antonio Becchi, Domenico Bertoloni Meli, and Enrico Gamba (Berlin: Edition Open Access, 2013), 317–338.

Giostra, Alessandro, "Ilario Altobelli e il cosmo medieval." In: *Atti del V Convegno Internazionale su Francesco d'Appignano*, ed. Domenico Priori (Jesi: Terra dei Fioretti, 2011), 263–277.

Giostra, Alessandro, "La *'stella o cometa'* nelle lettere di Guidobaldo Dal Monte a Pier Matteo Giordani." *Giornale di astronomia* 3 (2003): 46–53.

Giostra, Alessandro, Francesco Merletti, and William R. Shea (eds.), *Ilario Altobelli: Scienziato, teologo, corrispondente di Galileo Galilei*. Camerano: Empatiabooks, 2011.

Giovannini, Michel, *Accademie e astrologia: Ambiente culturale e relazioni erudite attorno a Pompeo Caimo (1568–1631), tra Udine e Roma*. Michel Giovannini, 2014.

Giusti, Anna Maria, *Pietre Dure: The Art of Semiprecious Stonework*. Trans. Fabio Barry. Los Angeles: Getty Publications, 2006.

Gliozzi, Mario, "Argoli, Andrea." In: *Dizionario biografico degli italiani*, vol. 4 (Roma: Enciclopedia Treccani, 1962), 132–134.

Grafton, Anthony, *Worlds Made by Words: Scholarship and Community in the Modern West*. Cambridge, MA: Harvard University Press, 2009.

Grafton, Anthony, *Bring Out Your Dead: The Past as Revelation*. Cambridge, MA: Harvard University Press, 2001.

Grafton, Anthony, *Cardano's Cosmos: The Worlds and Works of a Renaissance Astrologer*. Cambridge, MA: Harvard University Press, 1999.

Grafton, Anthony, *Defenders of the Text: The Traditions of Scholarship in an Age of Science, 1450–1800*. Cambridge, MA: Harvard University Press, 1991.

Granada, Miguel A., "Conflagración en el Renacimiento." Bruniana & Campanelliana 26 (2020): 125–146.

Granada, Miguel A., "Bartholomaeus Keckermann and Christoph Hunichius on Novas and Comets at the Beginning of the 17th Century: Two Opposing Views on the Relation Between Natural Philosophy and Mathematics." In: *Natural Knowledge and Aristotelianism at Early Modern Protestant Universities*, ed. Pietro D. Omodeo and Volkhard Wels (Wiesbaden: Otto Harrassowitz, 2019), 209–234.

Granada, Miguel A., Patrick J. Boner, and Dario Tessicini (eds.), *Unifying Heaven and Earth: Essays in the History of Early Modern Cosmology*. Barcelona: Universitat de Barcelona, 2016.

Granada, Miguel A., "Helisaeus Röslin y la libertad de religión." *Anales del Seminario de Historia de la Filosofía* 31 (2014): 69–88.

Granada, Miguel A., "Michael Maestlin and His Unpublished Treatise on the Nova of 1604." *Journal for the History of Astronomy* 45 (2014): 91–122.

Granada, Miguel A. (ed.), *Novas y cometas entre 1572 y 1618: Revolución cosmológica y renovación política y religiosa* (Barcelona: Universitat de Barcelona, 2012).

Granada, Miguel A., "After the Nova of 1604: Roeslin and Kepler's Discussion on the Significance of the Celestial Novelties (1607–1613)." *Journal for the History of Astronomy* 43 (2011): 353–390.

Granada, Miguel A., "Johannes Kepler and David Fabricius: Their Discussion on the Nova of 1604." In: *Change and Continuity in Early Modern Cosmology*, ed. Patrick J. Boner (Dordrecht: Springer, 2011), 66–92.

Granada, Miguel A., "'A quo moventur planetae?': Kepler et la question de l'agent du mouvement planétaire après la disparition des orbes solides." *Galilaeana* 7 (2010): 111–141.

Granada, Miguel A., "Novelties in the Heavens between 1572 and 1604 and Kepler's Unified View of Nature." *Journal for the History of Astronomy* 40 (2009): 393–402.

Granada, Miguel A., "Kepler and Bruno on the Infinity of the Universe and of Solar Systems." *Journal for the History of Astronomy* 39 (2008): 469–493.

Granada, Miguel A., "The Defence of the Movement of the Earth in Rothmann, Maestlin and Kepler: From Heavenly Geometry to Celestial Physics." In: *Mechanics and Cosmology in the Medieval and Early Modern Period*, ed. Massimo Bucciantini, Michele Camerota, and Sophie Roux (Florence: Olschki, 2007), 95–119.

Granada, Miguel A., "Did Tycho Eliminate the Celestial Spheres Before 1586?" *Journal for the History of Astronomy* 37 (2006): 125–145.

Granada, Miguel A., "Kepler v. Roeslin on the Interpretation of Kepler's Nova: (1) 1604–1606." *Journal for the History of Astronomy* 36 (2005): 299–319.

Granada, Miguel A., "Aristotle, Copernicus, Bruno: Centrality, the Principle of Movement and the Extension of the Universe." *Studies in History and Philosophy of Science* 35 (2004): 91–114.

Granada, Miguel A., Jürgen Hamel, and Ludolf von Mackensen (eds.), *Christoph Rothmanns Handbuch der Astronomie von 1589: Kommentierte Edition der Handschrift Christoph Rothmanns "Observationum stellarum fixarum liber primus," Kassel 1589*. Frankfurt: Harri Deutsch, 2003.

Granada, Miguel A., *Sfere solide e cielo fluido: Momenti del dibattito cosmologico nella seconda metà del Cinquecento*. Milan: Guerini, 2002.

Granada, Miguel A., "Cálculos cronológicos, novedades cosmológicas y expectativas escatológicas en la Europa del siglo XVI." *Rinascimento* 37 (1998): 357–435.

Granada, Miguel A., "Il rifiuto della distinzione tra *potentia absoluta* e *potentia ordinata* di Dio e l'affermazione dell'universo infinito in Giordano Bruno." *Rivista di Storia della Filosofia* 49 (1994): 495–532.

Graney, Christopher M., Mathematical Disquisitions*: The Booklet of Theses Immortalized by Galileo*. Notre Dame: University of Notre Dame Press, 2017.

Graney, Christopher M., *Setting Aside All Authority: Giovanni Battista Riccioli and the Science Against Copernicus in the Age of Galileo*. Notre Dame: University of Notre Dame Press, 2015.

Graney, Christopher M., and Timothy P. Grayson, "On the Telescopic Disks of Stars: A Review and Analysis of Stellar Observations from the Early 17th through the Middle 19th Centuries." *Annals of Science* 68 (2011): 351–373.

Grant, Edward, *A History of Natural Philosophy: From the Ancient World to the Nineteenth Century*. Cambridge: Cambridge University Press, 2007.

Grant, Edward, *The Foundations of Modern Science in the Middle Ages: Their Religious, Institutional and Intellectual Contexts*. Cambridge: Cambridge University Press, 1996.

Grant, Edward, *Planets, Stars, and Orbs: The Medieval Cosmos, 1200–1687*. Cambridge: Cambridge University Press, 1994.

Grant, Edward, "Celestial Incorruptibility in Medieval Cosmology 1200–1687." In: *Physics, Cosmology and Astronomy, 1300–1700: Tension and Accommodation*, ed. Sabetai Unguru (Dordrecht: Springer, 1991), 101–127.

Günther, Ludwig, *Kepler und die Theologie: Ein Stück Religions- und Sittengeschichte aus dem XVI. und XVII. Jahrhundert*. Giessen: Alfred Töpelmann, 1905.

Håkansson, Håkan, "Tycho the Prophet: History, Astrology, and the Apocalypse in Early Modern Science." In: *The Word and the World: Biblical Exegesis and Early Modern Science*, ed. Peter J. Forshaw and Kevin Killeen (Dordrecht: Springer, 2007), 137–156.

Hall, A. Rupert, *The Scientific Revolution 1500–1800*. Boston: Beacon Press, 1962.

Halleux, Robert, "L'œuvre minérologique d'Anselme Boèce de Boodt (1550–1632)." *Histoire et Nature* 14 (1979): 63–78.

Hallyn, Fernand, *The Poetic Structure of the World: Copernicus and Kepler*. Trans. Donald M. Leslie. New York: Zone Books, 1990.

Hamel, Jürgen, "Die Begründung der modernen astronomischen Beobachtungstechnik an der Kasseler Sternwarte im 16. Jahrhundert: Der Kasseler Sternkatalog 1586—eine Fehleranalyse der ihm zugrunde liegenden Positionsmessungen." In: *Nouveau ciel, nouvelle terre: La révolution copernicienne dans l'Allemagne de la Réforme, 1530–1630*, ed. Miguel Á. Granada and Édouard Mehl (Paris: Les Belles Lettres, 2009), 67–84.

Hamel, Jürgen, "Die Vorstellung von den Kometen seit der Antike bis ins 17. Jahrhundert: Tradition und Innovation." In: *Georg Samuel Dörffel (1643–1688): Theologe und Astronom*, ed. Johannes Richter (Plauen: Vogtland, 1994), 97–122.

Hamel, Jürgen, *Die astronomischen Forschungen in Kassel unter Wilhelm IV.* Frankfurt: Harri Deutsch, 1989.

Han, Giora, and Bernard R. Goldstein, "Symmetry in Copernicus and Galileo." *Journal for the History of Astronomy* 25 (2004): 273–292.

Harrison, Peter, *The Bible, Protestantism, and the Rise of Natural Science*. Cambridge: Cambridge University Press, 2001.

Hatch, Robert A., "Discovering Mira Ceti: Celestial Change and Cosmic Continuity." In: *Change and Continuity in Early Modern Cosmology*, ed. Patrick J. Boner (Dordrecht: Springer, 2011), 153–176.

Hatch, Robert A., "Between Friends: Huygens & Boulliau." *De Zeventiende Eeuw: Cultuur in de Nederlanden in interdisciplinair perspectief* 12 (1996): 106–113.

Healy, Margaret, "A Most Troublesome and Dangerous Ailment: Encounters with the Stone in Early Modern Europe." *Journal de la Renaissance* 3 (2005): 207–216.

Heidarzadeh, Tofigh, *A History of Physical Theories of Comets, from Aristotle to Whipple*. New York: Springer, 2009.

Hellman, C. Doris, "The Rôle of Measurement in the Downfall of a System: Some Examples from Sixteenth Century Comet and Nova Observations." *Vistas in Astronomy* 9 (1967): 43–52.

Hellman, C. Doris, "Maurolyco's 'Lost' Essay on the New Star of 1572." *Isis* 51 (1960): 322–336.

Henderson, Judith R., "Humanist Letter Writing: Private Conversation or Public Forum?" In: *Self-Presentation and Social Identification: The Rhetoric and Pragmatics of Letter Writing in Early Modern Times*, ed. Toon Van Houdt, Jan Papy, Gilbert Tournoy, and Constant Matheeussen (Leuven: Leuven University Press, 2002), 17–38.

Henry, John, "Thomas Harriot and Atomism: A Reappraisal." *History of Science* 20 (1982): 267–303.

Hirai, Hiro, *Medical Humanism and Natural Philosophy: Renaissance Debates on Matter, Life and the Soul*. Leiden: Brill, 2011.

Hirai, Hiro, "*Logoi Spermatikoi* and the Concept of Seeds in the Mineralogy and Cosmogony of Paracelsus." *Revue d'histoire des sciences* 61 (2008): 245–264.

Hirai, Hiro, *Le concept de semence dans les théories de la matière à la Renaissance: De Marsile Ficin à Pierre Gassendi*. Turnhout: Brepols, 2005.

Hoskin, Michael, *Stellar Astronomy: Historical Studies*. Chalfont St Giles: Science History Publications, 1982.

Howell, Kenneth, *God's Two Books: Copernican Cosmology and Biblical Interpretation in Early Modern Science*. Notre Dame: University of Notre Dame Press, 2002.

Hübner, Jürgen, *Die Theologie Johannes Keplers zwischen Orthodoxie und Naturwissenschaft*. Tübingen: Mohr, 1975.

Hübner, Wolfgang, "Die Christianisierung der Sternbilder in Schickards *Astroscopium*." In: *Zum 400. Geburtstag von Wilhelm Schickard: Zweites Tübinger Schickard-Symposion 25. bis 27. Juni 1992*, ed. Friedrich Seck (Sigmaringen: Thorbecke, 1995), 131–150.

Hübner, Wolfgang, *Zodiacus Christianus: Jüdisch-christliche Adaptationen des Tierkreises von der Antike bis zur Gegenwart*. Königstein: Hain, 1983.

Ingaliso, Luigi, *Filosofia e cosmologia in Christoph Scheiner*. Soveria Mannelli: Rubettino, 2005.

Jardine, Nicholas, "Kepler=Koestler: On Empathy and Genre in the History of the Sciences." *Journal for the History of Astronomy* 45 (2014): 271–286.

Jardine, Nicholas, "God's 'Ideal Reader': Kepler and His Serious Jokes." In: *Johannes Kepler: From Tübingen to Żagan*, ed. R. L. Kremer and J. Wlodarczyk, *Studia Copernicana* 42 (2009): 41–51.

Jardine, Nicholas, "Kepler, God, and the Virtues of Copernican Hypotheses." In: *Nouveau ciel, nouvelle terre: La révolution copernicienne dans l'Allemagne de la Réforme*, ed. Miguel A. Granada and Édouard Mehl (Paris: Belles Lettres, 2009), 269–281.

Jardine, Nicholas, and Alain-Philippe Segonds, *La guerre des astronomes: La querelle au sujet de l'origine du système géo-héliocentrique à la fin du XVIe siècle*. 2 vols. Paris: Les Belles Lettres, 2008.

Jardine, Nicholas, "Koyré's Kepler/Kepler's Koyré." *History of Science* 38 (2000): 363–376.

Jardine, Nicholas, "The Places of Astronomy in Early Modern Culture." *Journal for the History of Astronomy* 29 (1998): 49–62.

Jardine, Nicholas, *The Scenes of Inquiry: On the Reality of Questions in the Sciences*. Oxford: Clarendon Press, 1991.

Jardine, Nicholas, *The Birth of History and Philosophy of Science: Kepler's* A Defence of Tycho against Ursus *with Essays on Its Provenance and Significance*. Cambridge: Cambridge University Press, 1984.

Jerratsch, Anna, "Celestial Phenomena in Early Modernity: The Integrated Image of Comets." In: *Natural Knowledge and Aristotelianism at Early Modern Protestant Universities*, ed. Pietro D. Omodeo and Volkhard Wels (Wiesbaden: Otto Harrassowitz, 2019), 187–208.

Kargon, Robert H., *Atomism in England from Harriot to Newton*. Oxford: Clarendon, 1966.

Karpenko, Vladimír, and Ivo Purš, "Tycho Brahe: Between Astronomy and Alchemy." In: *Rudolf II and Alchemy: Exploring the Secrets of Nature in Central Europe in the 16th and 17th Centuries*, ed. Ivo Purš and Vladimír Karpenko (Prague: Artefactum, 2016), 459–88.

Kelly, Mary S., *Celestial Motors: 1543–1632*. Ph.D. Dissertation, University of Oklahoma, 1964.

Kinugasa, Kenzo, and Hiroshi Tsunemi, "ASCA Observation of Kepler's Supernova Remnant." *Publications of the Astronomical Society of Japan* 51 (1999): 239–252.

Kjellgren, Martin, *Taming the Prophets: Astrology, Orthodoxy, and the Word of God in Early Modern Sweden*. Lund: Sekel, 2011.

Kodera, Sergius, "The Art of the Distillation of 'Spirits' as a Technological Model for Human Physiology: The Cases of Marsilio Ficino, Joseph Duchesne and Francis Bacon." In: *Blood, Sweat and Tears: The Changing Concepts of Physiology from Antiquity into Early Modern Europe*, ed. Manfred Horstmanshoff, Helen King, and Claus Zittel (Leiden: Brill, 2012), 139–170.

Koestler, Arthur, *The Sleepwalkers*. New York: Macmillan, 1959.

Koyré, Alexandre, *From the Closed World to the Infinite Universe*. Baltimore: Johns Hopkins University Press, 1957.

Kruger, Justin, and David Dunning, "Unskilled and Unaware of It: How Difficulties in Recognizing One's Own Incompetence Lead to Inflated Self-Assessments." *Journal of Personality and Social Psychology* 77 (1999): 1121–1134.

Kurihara, Ken, *Celestial Wonders in Reformation* Germany. Abingdon: Routledge, 2016.

Kusukawa, Sachiko, *The Transformation of Natural Philosophy: The Case of Philip Melanchthon*. Cambridge: Cambridge University Press, 1995.

Lach, Donald F., *Asia in the Making of Europe*, vol. 2. Chicago: University of Chicago Press, 1970.

Lanuza Navarro, Tayra M. C., "From Intense Teaching to Neglect: The Decline of Astrology at the University of Valencia and the Role of the Spanish *Novatores*." *Early Science and Medicine* 22 (2017): 410–437.

Lanuza Navarro, Tayra M. C., "La astrología como explicación científica de la historia: Los pronósticos españoles del siglo XVII." In: *Synergia: Primer Encuentro de Jóvenes Investigadores en Historia de la Ciencia*, ed. Néstor Herrán, Josep Simon, Ximo Guillem-Llobat, Tayra Lanuza Navarro, Pedro Ruiz Castel, and Jaume Navarro (Madrid: CSIC, 2007), 303–323.

Lattis, James M., *Between Copernicus and Galileo: Christoph Clavius and the Collapse of Ptolemaic Cosmology*. Chicago: University of Chicago Press, 1994.

Leopold, John H., *Astronomen, Sterne, Geräte: Landgraf Wilhelm IV. und seine sich selbst bewegenden Globen*. Lucerne: Fremersdorf, 1986.

Lerner, Michel-Pierre, "L'entrée de Tycho Brahe chez les jésuites ou le channt du cygne de Clavius." In: *Les jésuites à la Renaissance: Système éducatif et production du savoir*, ed. Luce Giard (Paris: Presses Universitarires de France, 1995), 145–185.

Lovejoy, Arthur O., *The Great Chain of Being: A Study of the History of an Idea*. Cambridge, MA: Harvard University Press, 1936.

Machamer, Peter K., "Aristotle on Natural Place and Natural Motion." *Isis* 69 (1978): 377–387.

Mandrou, Robert, *From Humanism to Science 1480–1700*, trans. Brian Pearce. Bungay: Penguin, 1978.

Martens, Rhonda, *Kepler's Philosophy and the New Astronomy*. Princeton: Princeton University Press, 2000.

Martin, Craig, *Renaissance Meteorology: Pomponazzi to Descartes*. Baltimore: Johns Hopkins University Press, 2011.

Mascella, Raffaele, and Danilo Pelusi, "Uno scienziato abruzzese: Andrea Argoli." *Ratio Mathematica* 17 (2006): 49–52.

Mauder, Horst, "Die Kometenschriften von Schickard und Kepler." In: *Zum 400. Geburtstag von Wilhelm Schickard: Zweites Tübinger Schickard-Symposion 25. bis 27. Juni 1992*, ed. Friedrich Seck (Sigmaringen: Thorbecke, 1995), 151–166.

Mazzuchelli, Giammaria, *Gli Scrittori d'Italia, cioè notizie storiche e critiche intorno alle vite e agli scritti dei letterati italiani*, vol. 1. Brescia: Bossini, 1753.

Mehl, Édouard, *Descartes et la fabrique du monde: Le problème cosmologique de Copernic à Descartes*. Paris: Presses Universitaires de France, 2019.

Mehl, Édouard, "Kepler's Second Copernican Campaign: The Search for an Annual Stellar Parallax after the Roman Decree (1616)." In: *Galileo Banned: The Entangled Matter of the Anti-Copernican Decree of 1616*, ed. Natacha Fabbri and Federica Favino (Florence: Olschki, 2018), 191–209.

Mehl, Édouard, "'Novum struam mundum': Kepler's Rebuilding of the Copernican 'symmetria mundi.'" In: *Unifying Heaven and Earth: Essays in the History of Early Modern Cosmology.*, ed. Miguel A. Granada, Patrick J. Boner, and Dario Tessicini (Barcelona: Universitat de Barcelona, 2016), 197–216.

Mehl, Édouard, "Théorie physique et optique des comètes de Kepler à Descartes." In: *Novas y cometas entre 1572 y 1618: Revolución cosmológica y renovación política y religiosa*, ed. Miguel A. Granada (Barcelona: Universitat de Barcelona, 2012), 255–274.

Methuen, Charlotte, *Kepler's Tübingen: Stimulus to a Theological Mathematics*. Brookfield: Aldershot, 1998.

Methuen, Charlotte, "'This Comet or New Star': Theology and the Interpretation of the New Star of 1572." *Perspectives on Science* 5 (1997): 499–515.

Miller, David Marshall, *Representing Space in the Scientific Revolution*. Cambridge: Cambridge University Press, 2014.

Monaco, Giuseppe, *Andrea Argoli: Astronomo e medico del Seicento*. Tagliacozzo: Scacco Matto, 2007.

Moran, Bruce T., *The Alchemical World of the German Court: Occult Philosophy and Chemical Medicine in the Circle of Morris of Hessen (1572–1632)*. Stuttgart: Franz Steiner Verlag, 1991.

Moran, Bruce T., "Patronage and Institutions: Courts, Universities, and Academies in Germany: An Overview, 1550–1750." In: *Patronage and Institutions: Science, Technology, and Medicine at the European Court, 1500–1750*, ed. Bruce T. Moran (Bury St Edmunds: Boydell Press, 1991), 167–183.

Moran, Bruce T., "Christoph Rothmann, the Copernican Theory, and Institutional and Technical Influences on the Criticism of Aristotelian Cosmology." *Sixteenth Century Journal* 13 (1982): 85–108.

Mosley, Adam, "The Cosmographer's Role in the Sixteenth Century: A Preliminary Study." *Archives internationes d'histoire des sciences* 59 (2009): 423–439.

Mosley, Adam, *Bearing the Heavens: Tycho Brahe and the Astronomical Community of the Late Sixteenth Century*. Cambridge: Cambridge University Press, 2007.

Mosley, Adam, "Objects of Knowledge: Mathematics and Models in Sixteenth-Century Cosmology and Astronomy." In: *Transmitting Knowledge: Words, Images, and Instruments in Early Modern Europe*, ed. Sachiko Kusukawa and Ian Maclean (Oxford: Oxford University Press, 2006), 193–216.

Mosley, Adam, Nicholas Jardine, and Karin Tybjerg, "Epistolary Culture, Editorial Practices, and the Propriety of Tycho's Astronomical Letters." *Journal for the History of Astronomy* 34 (2003): 421–451.

Murphey, Rhoads, *Ottoman Warfare: 1500–1700*. London: UCL Press, 1999.

Nallino, Carlo Alfonso, "Etimologia araba e significato di '*sub*' e di '*azimut*.'" *Rivista degli studi orientali* 8 (1919): 369–388.

Navarro Brotons, Víctor, *Jerónimo Muñoz: Matemáticas, cosmología y humanismo en la época del Renacimiento*. Valencia: Universitat de Valencia, 2019.

Navarro Brotons, Víctor, *Disciplinas, saberes y prácticas: Filosofía natural, matemáticas y astronomía en la sociedad española de la época moderna*. Valencia: Universitat de Valencia, 2014.

Navarro Brotons, Víctor, "Las novedades celestes en España entre 1572 y 1618." In: *Novas y cometas entre 1572 y 1618: Revolución cosmológica y renovación política y religiosa*, ed. Miguel A. Granada (Barcelona: Universitat de Barcelona, 2012), 15–41.

Nicholl, Colin R., *The Great Christ Comet: Revealing the True Star of Bethlehem*. Wheaton: Crossway, 2015.

North, John, *Cosmos: An Illustrated History of Astronomy and Cosmology*. Chicago: University of Chicago Press, 2008.

Nothaft, C. Philipp E., *Dating the Passion of Christ: The Life of Jesus and the Emergence of Scientific Chronology (200–1600)*. Leiden: Brill, 2012.

Omodeo, Pietro D., and Volkhard Wels (eds.), *Natural Knowledge and Aristotelianism at Early Modern Protestant Universities*. Wiesbaden: Otto Harrassowitz, 2019.

Omodeo, Pietro D., and Jürgen Renn (eds.), *Science in Court Society: Giovanni Battista Benedetti's* Diversarum speculationum mathematicarum et physicarum liber *(Turin, 1585)* (Berlin: Edition Open Access, 2019).

Omodeo, Pietro D., "Central European Polemics over Descartes: Johannes Placentinus and His Academic Opponents at Frankfurt on Oder (1653–1656)." *History of Universities* 29 (2016): 29–64.

Omodeo, Pietro D., "Institutionalized Metaphysics of Astronomy at Early-Modern Melanchthonian Universities." In: *Wissen in Bewegung: Institution—Iteration—Transfer*, ed. Eva Cancik-Kirschbaum and Anita Traninger (Wiesbaden: Harrassowitz, 2016), 51–78.

Omodeo, Pietro D., "La contingente geometria del cosmo nella Dotta ignoranza cusaniana: Cusano e Keplero a confronto." In: *Filosofia Arte Scienza in Cusano e Leibniz*, ed. Antonio Dall'Igna and Damiano Roberi (Milan: Mimesis, 2014), 215–226.

Omodeo, Pietro D., *Copernicus in the Cultural Debates of the Renaissance: Reception, Legacy, Transformation*. Leiden: Brill, 2014.

Omodeo, Pietro D., "Minimum und Atom: Eine Begriffserweiterung in Brunos Rezeption des Cusanus." In: *Die "Modernitäten" des Nicolaus Cusanus: Debatten und Rezeptionen*, ed. Tom Müller and Matthias Vollet (Bielefeld: Transcript, 2013), 285–304.

Omodeo, Pietro D., "Astronomia, filosofia e teologia nel tardo Rinascimento tedesco: Heinrich Julius di Braunschweig e il soggiorno di Giordano Bruno in Germania." *Giornale critico della filosofia italiana* 92 (2011): 307–326.

Omodeo, Pietro D., "David Origanus's Planetary System (1599 to 1609)." *Journal for the History of Astronomy* 42 (2011): 439–454.

Omodeo, Pietro D., "Perfection of the World and Mathematics in Late Sixteenth-Century Copernican Cosmologies." In: *The Invention of Discovery, 1500–1700*, ed. James Douglas Fleming (Farnham-Burlington: Ashgate, 2011), 93–108.

Osler, Margaret J., *Reconfiguring the World: Nature, God, and Human Understanding from the Middle Ages to Early Modern Europe*. Baltimore: Johns Hopkins University Press, 2010.

Pantin, Isabelle, "Les fables des astres: Continuité et mutations de 'l'affichage céleste' à la Renaissance." In: *Savoirs et fiction au Moyen Âge et à la Renaissance*, ed. Dominique Boutet and Joëlle Ducos (Paris: Presses de l'Université Paris-Sorbonne, 2015), 363–378.

Pizzamiglio, Pierluigi, *L'astrologia in Italia all'epoca di Galileo Galilei (1550–1650): Rassegna storico-critica dei documenti librari custoditi nella Biblioteca 'Carlo Viganò.'* Milan: Vita e Pensiero, 2004.

Plastina, Sandra, "'Philosophia lucis proles verissima est.' Nicholas Hill lettore di Francesco Patrizi." *Bruniana & Campanelliana* 10 (2004): 175–182.

Plastina, Sandra, "Nicholas Hill: 'The English Campanella?'" *Bruniana & Campanelliana* 4 (1998): 207–221.

Premuda, Loris, "Medici Abruzzesi nello Studio di Padova." *Abruzzo: Rivista dell'Istituto di Studi Abruzzesi* 6 (1968): 505–512.

Principe, Lawrence M., *The Secrets of Alchemy*. Chicago: University of Chicago Press, 2013.

Purš, Ivo, "Rudolf II's Patronage of Alchemy and the Natural Sciences." In: *Rudolf II and Alchemy: Exploring the Secrets of Nature in Central Europe in the 16th and 17th Centuries*, ed. Ivo Purš and Vladimír Karpenko (Prague: Artefactum, 2016), 139–204.

Purš, Ivo, "Anselmus Boëtius de Boodt: Physician, Mineralogist and Alchemist." In: *Rudolf II and Alchemy: Exploring the Secrets of Nature in Central Europe in the 16th and 17th Centuries*, ed. Ivo Purš and Vladimír Karpenko (Prague: Artefactum, 2016), 535–537.

Rabin, Sheila J., "The Astrological Cosmos of Johannes Kepler." In: *World-Building and the Early Modern Imagination*, ed. Allison B. Kavey (New York: Palgrave Macmillan, 2010), 59–66.

Rabin, Sheila J., "Was Kepler's *Species Immateriata* Substantial?" *Journal for the History of Astronomy* 36 (2005): 49–56.

Rabin, Sheila J., "Kepler's Attitude Toward Pico and the Anti-Astrology Polemic." *Renaissance Quarterly* 50 (1997): 750–770.

Ragep, F. Jamil, "Tusi and Copernicus: The Earth's Motion in Context." *Science in Context* 14 (2001): 145–163.

Randles, William G. L., *The Unmaking of the Medieval Christian Cosmos, 1500–1760: From Solid Heavens to Boundless Aether*. Aldershot: Ashgate, 1999.

Redondi, Pietro, *Galileo Heretic*. Princeton: Princeton University Press, 1989.

Regier, Jonathan, "Reading Cardano with the Roman Inquisition: Astrology, Celestial Physics, and the Force of Heresy." *Isis* 110 (2019): 661–679.

Regier, Jonathan, "An Unfolding Geometry: Appropriating Proclus in the *Harmonice Mundi* (1619)." In: *Unifying Heaven and Earth: Essays in the History of Early Modern Cosmology*, ed. Miguel A. Granada, Patrick J. Boner, and Dario Tessicini (Barcelona: Universitat de Barcelona, 2016), 217–237.

Regier, Jonathan, "Kepler's Theory of Force and His Medical Sources." *Early Science and Medicine* 19 (2014): 1–27.

Reynolds, Stephen P., Kazimierz J. Borkowski, Una Hwang, et al., "A Deep Chandra Observation of Kepler's Supernova Remnant: A Type Ia Event with Circumstellar Interaction." *Astrophysical Journal* 668 (2007): L135–L138.

Riccardi, Pietro, *Biblioteca matematica italiana dalla origine della stampa ai primi anni del secolo XIX*, vol. 1. Modena: Società tipografica modenese, 1870.

Ricci, Saverio (ed.), *I primi lincei e il Sant'Uffizio: Questioni di scienza e di fede*. Rome: Bardi, 2005.

Ricci, Saverio (ed.), *"Una filosofica milizia": Tre studî sull'Accademia dei Lincei*. Udine, Campanotto: 1994.

Ricci, Saverio, *La fortuna del pensiero di Giordano Bruno*. Florence: Le Lettere, 1990.

Ricci, Saverio, "Federico Cesi e la *nova* del 1604: La teoria della fluidità del cielo e un opuscolo dimenticato di Joannes van Heeck." *Atti dell'Accademia Nazionale dei Lincei* 43 (1988): 111–133.

Riess, Adam G., Alexei V. Filippenko, Peter Challis, et al., "Observational Evidence from Supernovae for an Accelerating Universe and a Cosmological Constant." *Astronomical Journal* 116 (1998): 1,009–1,038.

Roelants, Nienke, *Lutheran Astronomers After the Fall (1540–1590): A Reappraisal of the Renaissance Dynamic of Religion and Astronomy*. Ph.D. Dissertation, Ghent University, 2013.

Rothenflug, Robert, B. Magne, J. P. Chieze, and J. Ballet, "Hydrodynamic Model of Kepler's Supernova Remnant Constrained by EINSTEIN and EXOSAT X-Ray Spectra." *Astronomy and Astrophysics* 291 (1994): 271–282.

Rothman, Aviva, *The Pursuit of Harmony: Kepler on Cosmos, Confession, and Community*. Chicago: University of Chicago Press, 2017.

Rothman, Aviva, "Forms of Persuasion: Kepler, Galileo, and the Dissemination of Copernicanism." *Journal for the History of Astronomy* 40 (2009): 403–419.

Rublack, Ulinka, *The Astronomer & the Witch: Johannes Kepler's Fight for His Mother*. Oxford: Oxford University Press, 2015.

Ruffner, James A., "The Curved and the Straight: Cometary Theory from Kepler to Hevelius." *Journal for the History of Astronomy* 2 (1971): 178–194.

Ruiz-Lapuente, Pilar, Francesco Damiani, Luigi Bedin, et al., "No Surviving Companion in Kepler's Supernova." *Astrophysical Journal* 862 (2018): 124.

Russo, Elena, *Styles of Enlightenment: Taste, Politics, and Authorship in Eighteenth-Century France*. Baltimore: Johns Hopkins University Press, 2007.

Rutkin, H. Darrel, "Astrology." In: *The Cambridge History of Science*, vol. 3, ed. Katharine Park and Lorraine Daston (Cambridge: Cambridge University Press, 2006), 541–561.

Rutkin, H. Darrel, "Various Uses of Horoscopes: Astrological Practices in Early Modern Europe." In: *Horoscopes and Public Spheres: Essays on the History of Astrology*, ed. Guenther Oestmann, H. Darrel Rutkin, and Kocku von Stuckrad (Berlin-New York: De Gruyter, 2005), 167–182.

Sachs, Abram J., and Christopher B. F. Walker, "Kepler's View of the Star of Bethlehem and the Babylonian Almanac for 7/6 BC." *Iraq* 46 (1984): 43–51.

Sakamoto, Kuni, "The German Hercules's Heir: Pierre Gassendi's Reception of Keplerian Ideas." *Journal of the History of Ideas* 70 (2009): 69–91.

Sankrit, Ravi, William P. Blair, Tracey Delaney, et al., "HST/ACS Imaging of a Balmer-Dominated Shock in Kepler's Supernova Remnant." *Advances in Space Research* 35 (2005): 1,027–1,030.

Schaefer, Bradley E., "Peak Brightnesses of Historical Supernovae and the Hubble Constant." *Astrophysical Journal* 459 (1996): 438–454.

Schechner, Sara J., *Comets, Popular Culture, and the Birth of Modern Cosmology*. Princeton: Princeton University Press, 1997.

Shea, William R., "Guidobaldo Dal Monte: Galileo's Patron, Mentor and Friend." In: *Guidobaldo Dal Monte (1545–1607): Theory and Practice of the Mathematical Disciplines from Urbino to Europe*, ed. Antonio Becchi, Domenico Bertoloni Meli, and Enrico Gamba (Berlin: Edition Open Access, 2013), 95–104.

Shea, William R., "Galileo and the Supernova of 1604." In: *1604–2004: Supernovae as Cosmological Lighthouses*, ed. Massimo Turatto, Stefano Benetti, Luca Zampieri, and William R. Shea (San Francisco: Astronomical Society of the Pacific, 2005), 13–20.

Seck, Friedrich (ed.), *Wilhelm Schickard 1592–1635: Astronom, Geograph, Orientalist, Erfinder der Rechenmaschine* (Tübingen: Mohr, 1978).

Segonds, Alain-Philippe, "Astronomie terrestre/Astronomie celeste chez Tycho Brahe." In: *Nouveau ciel, nouvelle terre: La révolution copernicienne dans l'Allemagne de la Réforme*, ed. Miguel A. Granada and Édouard Mehl (Paris: Belles Lettres, 2009), 109–142.

Segonds, Alain-Philippe, "Kepler as Reader and Translator of Aristotle." In: *Philosophy in the Sixteenth and Seventeenth Centuries: Conversations with Aristotle*, ed. Sachiko Kusukawa and Constance Blackwell (Aldershot: Ashgate, 1999), 205–32.

Seidengart, Jean, *Dieu, l'univers et la sphère infinie: Penser l'infinité cosmique à l'aube de la science classique*. Paris: Albin Michel, 2006.

Shackelford, Jole, "Providence, Power, and Cosmic Causality in Early Modern Astronomy: The Case of Tycho Brahe and Petrus Severinus." In: *Tycho Brahe and Prague: Crossroads of European Science*, ed. John R. Christianson, Alena Hadravová, Petr Hadrava, and Šolc Martin (Frankfurt: Harri Deutsch, 2002), 46–69.

Shackelford, Jole, "Tycho Brahe, Laboratory Design, and the Aim of Science: Reading Plans in Context," *Isis* 84 (1993): 211–30.

Siebert, Harald, "The Early Search for Stellar Parallax: Galileo, Castelli, and Ramponi." *Journal for the History of Astronomy* 36 (2005): 251–271.

Simon, Gérard, *Kepler astronome astrologue*. Paris: Gallimard, 1979.

Simon, Gérard, "Kepler's Astrology: The Direction of a Reform." *Vistas in Astronomy* 18 (1975): 439–448.

Smith, A. Mark, *From Sight to Light: The Passage from Ancient to Modern Optics*. Chicago: University of Chicago Press, 2014.

Smith, A. Mark, "Ptolemy's Theory of Visual Perception: An English Translation of the *Optics* with Introduction and Commentary." *Transactions of the American Philosophical Society* 86 (1996): iii–300.

Taub, Liba C., *Ptolemy's Universe: The Natural Philosophical and Ethical Foundations of Ptolemy's Astronomy*. Chicago: Open Court, 1993.

Tessicini, Dario, and Patrick J. Boner (eds.), *Celestial Novelties on the Eve of the Scientific Revolution, 1540–1630* (Florence: Olschki, 2013).

Tessicini, Dario, "Il dibattito italiano sulla nuova stella del 1572 (con il testo del *Discorso intorno a la stella* di Giuseppe Valdagno, ms Ambrosiana R 95 sup.)." In: *Novas y cometas entre 1572 y 1618: Revolución cosmológica y renovación política y religiosa*, ed. Miguel A. Granada (Barcelona: Universitat de Barcelona, 2012), 43–93.

Tessicini, Dario, "Cornelius Gemma and the New Star of 1572." In: *Change and Continuity in Early Modern Cosmology*, ed. Patrick J. Boner (Dordrecht: Springer, 2011), 51–65.

Tessicini, Dario, "'Vere gemmeum est?': Cornelio Gemma e la stella nuova del 1572." In *Cornelius Gemma: Cosmology, Medicine and Natural Philosophy in Renaissance Louvain*, ed. Hiro Hirai (Pisa-Rome: Fabrizio Serra, 2008), 51–64.

Tessicini, Dario, *I dintorni dell'infinito: Giordano Bruno e l'astronomia del Cinquecento*. Pisa-Rome: Fabrizio Serra, 2007.

Tessicini, Dario, "Bruno e Roeslin: Sulla presenza della *Theoria nova coelestium meteoron* nel *De immenso*." *Bruniana & Campanelliana* 2 (1998): 475–487.

Thoren, Victor E., *The Lord of Uraniborg: A Biography of Tycho Brahe*. Cambridge: Cambridge University Press, 1990.

Thorndike, Lynn, *A History of Magic and Experimental Science*, vol. 7. New York: Columbia University Press, 1958.

Thorndike, Lynn, *A History of Magic and Experimental Science*, vol. 6. New York: Columbia University Press, 1941.

Tredwell, Katherine A., "Michael Maestlin and the Fate of the *Narratio prima*." *Journal for the History of Astronomy* 35 (2004): 305–325.

Vanden Broecke, Steven, *The Limits of Influence: Pico, Louvain, and the Crisis of Renaissance Astrology*. Leiden: Brill, 2003.

Van der Krogt, Peter, *Globi Neerlandici: The Production of Globes in the Low Countries*. Utrecht: HES, 1993.

Van der Schoot, Albert, "Kepler's Search for Form and Proportion." *Renaissance Studies* 15 (2001): 59–78.

Van Gent, Rob, "De nieuwe sternen van 1572, 1600 en 1604 op de hemelglobes van Willem Jansz. Blaeu." *Caert-Thresoor: Tijdschrift voor de Geschiedenis van de Kartografie in Nederland* 12 (1993): 40–46.

Van Helden, Albert, *Measuring the Universe: Cosmic Dimensions from Aristarchus to Halley*. Chicago: University of Chicago Press, 1985.

Van Nouhuys, Tabitta, *The Age of Two-Faced Janus: The Comets of 1577 and 1618 and the Decline of the Aristotelian View in the Netherlands*. Leiden: Brill, 1998.

Vasconi, Paola (ed.), *Cristoforo Clavio e la cultura scientifica del suo tempo*. Rome: Gangemi, 2015.

Vasoli, Cesare, *I miti e gli astri*. Naples: Guida, 1977.

Verbunt, Frank, and Robert H. van Gent, "Three Editions of the Star Catalogue of Tycho Brahe." *Astronomy & Astrophysics* 516 (2010): 1–25.

Vermij, Rienk, "The Marginalization of Astrology among Dutch Astronomers in the First Half of the 17th Century." *History of Science* 52 (2014): 153–177.

Vermij, Rienk, *The Calvinist Copernicans: The Reception of the New Astronomy in the Dutch Republic, 1575–1750*. Amsterdam: Koninklijke Nederlandse Akadamie van Wetenschappen, 2002.

Vink, Jacco, "Supernova 1604, Kepler's Supernova, and Its Remnant." In: *Handbook of Supernovae*, ed. Athem W. Alsabti and Paul Murdin (Cham: Springer, 2017), 1–22.

Voelkel, James R., *The Composition of Kepler's* Astronomia nova. Princeton: Princeton University Press, 2001.

Von Mackensen, Ludolf, *Die erste Sternwarte Europas mit ihren Instrumenten und Uhren: 400 Jahre Jost Bürgi in Kassel*. Munich: Callwey, 1979.

Weichenhan, Michael, "Neugier und Furcht: Blicke auf Kometen in der frühen Neuzeit." In: Himmelsspektakel: Astronomie im Protestantismus der Frühen Neuzeit, ed. Sascha Salatowsky and Karl-Heinz Lotze (Gotha: Forschungsbibliothek Gotha, 2015), 59–71.

Weichenhan, Michael, *"Ergo perit coelom ...": Die Supernova des Jahres 1572 und die Überwindung des aristotelischen Kosmologie*. Stuttgart: Steiner, 2004.

Weinberg, Steven, *To Explain the World: The Discovery of Modern Science*. New York: Harper Collins, 2015.

Westman, Robert S., *The Copernican Question: Prognostication, Skepticism, and Celestial Order*. Berkeley: University of California Press, 2011.

Westman, Robert S., "Kepler's Early Physical-Astrological Problematic." *Journal for the History of Astronomy* 32 (2001): 227–236.

Westman, Robert S., "Nature, Art, and Psyche: Jung, Pauli, and the Kepler-Fludd Polemic. In: *Occult & Scientific Mentalities in the Renaissance*, ed. Brian Vickers (Cambridge: Cambridge University Press, 1984), 177–229.

Westman, Robert S., "The Astronomer's Role in the Sixteenth Century: A Preliminary Study." *History of Science* 18 (1980): 105–147.

Westman, Robert S., "The Comet and the Cosmos: Kepler, Maestlin and the Copernican Hypothesis." In: *The Reception of Copernicus' Heliocentric Theory*, ed. Jerzy Dobrzycki (Dordrecht: D. Reidel, 1973), 7–30.

Westman, Robert S., *Johannes Kepler's Adoption of the Copernican Hypothesis*. Ph.D. Dissertation, University of Michigan, 1971.

Williams, Brian J., Kazimierz J. Borkowski, Stephen P. Reynolds, et al., "*Spitzer* Observations of the Type Ia Supernova Remnant N103B: Kepler's Older Cousin?" *Astrophysical Journal* 790 (2014): 139–149.

Williams, Raymond, *Keywords: A Vocabulary of Culture and Society*. Oxford: Oxford University Press, 1985.

Woodey, Kennerly M., "Dante and the Doctrine of Great Conjunctions." *Dante Studies* 95 (1997): 119–134.

Zwicky, Fritz, "Supernovae." *Handbuch der Physik* 51 (1958): 766.

Index of Persons

Agrippa von Nettesheim, Heinrich Cornelius 112, 125
Altobelli (Altobelius), Ilario 210, 213, 215, 228, 256–257
Al-Tusi, Nasir al-Din 35
Ambrose, Aurelius 222
Amicus, Bartholomeus 246
Andreae, Johann Valentin 149n9
Apian, Peter 94, 143
Arago, François 107
Aratus 161
Argoli, Andrea 15, 214–218, 231, 235, 256
Aristarchus 52, 64, 66
Aristotle 13, 18, 23, 25, 29–30, 33, 35–37, 76, 89, 92, 186–187, 192, 208, 221, 226–227, 235, 249, 258
Arrighi, Gino 225, 230n85
Augustine 12
Averroes 36

Baade, Wilhelm H. Walter 215n34, 263
Baldini, Ugo 232
Barbaro, Ermolao 137
Barnaud, Nicolas 119
Barocci, Federico 228
Bartsch, Jacob 147, 173–174, 176, 179
Basil of Caesarea 222
Bayer, Johann 153–155, 163, 171, 174, 176
Bellarmine, Robert 222
Benedetti, Giovanni Battista 184
Biamino, Giovanni Bartolomeo 251–252
Blair, William P. 15, 264
Blaeu, Willem Janszoon 147, 150, 157n29, 163
Boethius, Anicius Manlius Severinus 130
Boner, Patrick J. 14, 20, 74, 131, 183, 209n7
Bonhombra, Lodovico (see also Magini, Giovanni Antonio) 241–250, 256
Borghese, Camillo (Pope Paul v) 250
Brahe, Tycho 6, 9–11, 13, 24n23, 26, 28, 37–38, 41, 54, 65n4, 68, 72, 81, 85, 87, 88n45, 98, 99n121, 101, 102n144, 105, 112n22, 113n26, 117, 132, 147–148, 152, 161, 181, 213, 220, 223, 229, 254n160
Brengger, Johann Georg 154n22, 213n32
Brotons, Víctor Navarro 2n9
Bruce, Edmund 21–22
Brucker, Jakob 196
Bruno, Giordano 13, 18, 31–34, 63, 73–79, 80, 184, 185–190, 192–194, 196, 202–203
Brunowsky, Johannes 231
Bürgi, Jost (Justus) 10, 93, 150–152, 163–164
Buridan, Jean 246n138
Büttner, Balthasar 4, 92

Caimo, Pompeo 235–236
Camerarius, Elias 23, 28
Capra, Baldassarre 218n51
Cardano, Girolamo 109n9, 112, 191, 243, 248
Casale, Girolamo (see also Faentino, Casamatta) 239–240, 256
Casali, Elide 239
Caspar, Max 154n22, 160n37
Cellarius, Andreas 179–180
Cesi, Federico 206–207, 214, 248, 250
Cesalpino, Andrea 122–123
Chandrasekhar, Subramanyan 262, 263
Charles v, Holy Roman Emperor 95
Chastellier, Jean 251
Christ, Jesus 6–7, 11, 15, 134–135, 145, 147–149, 156, 165–168, 172–178, 234
Cicero, Marcus Tullius 111
Clavius (Clavio), Christoph 15, 204, 209–233, 240, 245, 249, 250–255, 256–257
Copernicus, Nicolaus 13, 22, 25, 33–36, 42, 44n5–6, 45, 48, 49, 50, 51, 52, 54, 55, 56, 57, 58, 60, 65n4 and 5, 68, 70, 74, 76, 80, 132, 143, 253
Cosci, Matteo 15
Crasso, Lorenzo 217n45
Cristini, Bartolomeo 248–250, 252–254
Crüger (Krüger), Peter 44–45, 199
Cusanus (of Cusa), Nicholas 131, 234
Cysat, John Baptist 177n103

Da Mantova, Antonio 251
Dal Monte, Guidobaldo 204, 208, 224–244, 245, 250, 255–257
Daniel 135
De Boodt, Anselmus Boëtius 14, 107–108, 117–128

De' Medici, Ferdinando I (Grand Duke of Tuscany) 118
De' Medici, Giulio (Pope Clement VII) 250
Dee, John 28, 178
Descartes, René 15, 182, 184, 193–196, 197, 199, 201, 203
Digges, Thomas 28, 31, 47
Di Guevara, Giovanni 244
Diodato, Laudadio 241
Donahue, William H. 229
Du Bartas, Guillaume 173n90

Elizabeth, Queen of England and Ireland 115n34
Erastus, Thomas 118–119
Eriksen, Johan 152
Evans, Robert J. W. 125

Fabricius, David 1, 10, 11, 24, 36–38, 67, 84–85, 88n47, 90, 98–99, 102, 105–106, 146, 152, 157–161, 165
Faentino, Casamatta (see also Casale, Girolamo) 239–240
Ferrero, Giovanni Stefano 248
Ficino, Marsilio 14, 112, 125, 234
Field, Judith V. 32n51, 64, 77–78, 141
Findlen, Paula 136–137, 224
Fludd, Robert 143
Foucault, Michel 19
Frangipane, Cornelio 247
Franz Otto, Duke of Braunschweig-Lüneburg 95
Friedrich I, Duke of Württemberg 114, 115
Friedrich Wilhelm, Elector of Brandenburg 197
Frisch, Christian 152n17, 154n22, 160n37

Galilei, Galileo 2n8, 44, 50, 74, 79, 96, 104, 106, 113n26, 166, 168, 184, 193, 204, 209–213, 218, 224, 228, 248, 255n163, 257
Gassendi, Pierre 15, 184, 190–196
Gemma, Cornelius 9, 28, 147–148, 178n106
Gilbert, William 13, 31, 33, 34n55, 77, 112, 127
Giordani, Pier Matteo 225–226, 229–234, 236–238, 256
Giostra, Alessandro 225
Goldstein, Bernard 44n6, 74
Granada, Miguel Á. 15, 31
Graney, Christopher M. 13
Grassi, Orazio 193
Gregory of Nyssa 222
Grienberger, Christoph 211, 232, 251

Habrecht, Isaac 39
Hagecius, Thaddaeus 9–10
Halley, Edmond 182
Hamel, Jürgen 164n57
Harriot, Thomas 115, 188–189
Hatch, Robert A. 81, 197n56
Heinrich Julius, Duke of Braunschweig-Lüneburg 84, 93, 100
Henisch, Georg 154n22
Herlicius, David 1–2, 83, 90–92, 105
Herod, King of Judea 7
Herwart von Hohenburg, Johann Georg 22, 38, 70, 139, 153–154
Hevelius, Johannes 200
Hill, Nicholas 15, 184, 185–190, 192, 193, 202
Hipparchus 64, 66, 161
Hirai, Hiro 126
Hoffman, Johann Friedrich 158
Hon, Giora 74
Horrocks, Jeremiah 60

Ingoli, Francesco 44, 237
Isaiah 11, 170

Jardine, Nicholas 86n33, 138
John 170
Josephus, Titus Flavius 7
Jöstel (Jostelius), Melchior 15, 152–153

Keckermann, Bartholomaeus 99
Kircher, Athanasius 180
Koyré, Alexandre 17–19, 78
Krabbe, Johannes 14, 81–84, 88–106

Lansberg (Lansbergen), Philip 48, 199
Lantz, Johann 255
Liceti, Fortunio 130
Locher, Johann Georg 42, 44
Lorenzini, Ambrosio 67n11
Luke 166
Luna, Javier 13, 29n37, 104n157
Luther, Martin 129

Maestlin, Michael 13, 46, 55, 74, 87n39, 95, 97, 115–117, 131, 153n20, 158–159, 223
Magini (Maginus), Giovanni Antonio (see also Bonhombra, Lodovico) 8, 15, 204, 211, 212–215, 217, 228, 231, 241–250, 252, 254, 256–257
Martens, Rhonda 40
Matthew 166, 178n105, 234
Matthias, Holy Roman Emperor 122, 132
Maurice, Landgrave of Hesse-Kassel 152, 164
Maurice, Prince of Orange 152n12
Maurolico, Francesco 222, 247
Mayr (Marius), Simon 44, 60, 218n51
Mazzoni (Mazzone), Jacopo 227–228
Mehl, Édouard 74
Melanchthon, Philipp 23, 87, 94, 183
Mellei, Giovanni 219, 223–224, 230, 256
Micanzio, Fulgenzio 218n47
Mögling, Daniel (see also Schweighardt, Theophilus) 150
Moller, Albin 1, 89
Mosley, Adam 114 (nos. 27 and 28)
Müller, Philipp 39
Muñoz, Jerónimo 2–3, 9

Newton, Isaac 182
North, John 7n39

Omodeo, Pietro Daniel 15, 63n1

Padovani, Alessandro 237
Palau, Marco Antonio 2–3
Pantin, Isabelle 173n90
Peucer, Caspar 87n41
Pico della Mirandola, Giovanni 7, 19, 89, 126, 132–133, 137, 193
Placentinus, Johannes 15, 197–202, 203
Plato 227
Pliny the Elder 57, 197
Prester John 167
Ptolemy, Claudius 41, 49, 51–52, 54, 65–66, 68, 70–72, 113, 143, 159, 161–162

Raimondo, Annibale 9–10
Regier, Jonathan 14
Rheticus, Georg Joachim 23, 76, 82n6
Rhodius, Ambrosius 152
Riccioli, Giovanni Battista 45, 60, 161n44
Ripollés, Antonio Juan 2
Robert of York 240
Roeslin, Helisaeus 3, 88n4, 147, 165–169, 171, 172n85, 178n105
Rosaccio, Giuseppe 241n122
Rosen, Edward 36n64
Rosenkreutz, Christian 149
Rothman, Aviva 14–15, 126n72
Rothmann, Christoph 28n36, 47–48, 81–82, 95, 102, 117, 164, 181n2
Rubino, Francesco Antonio 232n90
Rudolf (Rudolph) II, Holy Roman Emperor 6n31, 8, 85, 107, 117–118, 122, 125n68, 127–128, 132, 163

Salino, Bernardino 251
Sasso, Camillo 218n51
Scaliger, Julius Caesar 107–109
Scheiner, Christoph 42, 44, 255n163
Schickard, Wilhelm 147, 154, 156–157, 169–174, 179
Schiller, Julius 147, 157n27, 174–180
Schweighardt, Theophilus (see also Mögling, Daniel) 150
Seneca, Lucius Annaeus 111
Severinus, Petrus 117
Shakespeare, William 115n34
Sifard, Johann 4
Simon, Gérard 19, 130
Sirmond, Jaques 251
Staserio, Giovanni Giacomo 212–213, 251
Strabo 197
Strassoldo, Giovanni 249–250
Suslyga, Lawrence (Laurentius) 6, 145

Tanckius, Joachim 83
Tengnagel (Tengnaglius), Franz Gansneb 10, 152, 213–214, 230–231, 234, 248, 254
Tessicini, Dario 12–13, 63n1, 178n106, 203n63, 220
Thorndike, Lynn 120n51, 126n70
Tortora, Homero 229–233, 256

Ursus, Nicolaus Reimers 113n26

Van Heeck (Heckius), Jan 206, 248, 250
Van Helden, Albert 42n1, 44, 46, 64, 99n122
Van Maelocote, Odo 211
Von Dyck, Walther 117
Vicke, Nikolaus 83–85
Virgil 24

Wacker von Wackenfels, Johannes Matthaeus 21, 122, 203n63
Weidner, Johannes 2n7
Welser, Markus (Marcus) 153–154
Westman, Robert S. 125
Wilhelm IV, Landgrave of Hesse-Kassel 164
Wolfgang Wilhelm, Count Palatine of Neuburg 148n6

Ziegler, Johann Reinhard 254

Index of Places

Abruzzo 215–216
Alatri 216
Amsterdam 180
Augsburg 97, 153–154, 174

Barcelona 12
Bavaria 153
Benátky nad Jizerou 118
Bethlehem 6, 234
Bologna 8, 212, 228, 238, 240–241, 244
Bohemia 10
Brandenburg 89, 185, 200
Braunschweig 93, 95, 98n116
Bruges 118

Calabria 211, 215, 218n51
Cossenza 215

Dresden 105

East Frisia 10, 90
Erfurt 97

Forlì 237
Frankfurt (Main) 122, 187
Frankfurt (Oder) 28, 185, 197
Frosinone 216

Gdańsk 200
Geneva 184, 190
Genova 251
Goa 232n90
Graz 93n78, 116–117, 141
Greifswald 91n65

Halberstadt 84
Hanau 107
Heidelberg 119
Helmstedt 93

Ingolstadt 255

Jerusalem 166, 173

Kassel 81, 150, 152, 164

L'Aquila 215
Leipzig 83, 196
Linz 92, 122
London 112
Louvain 28, 222

Magdeburg 97
Mainz 254
Messina 222
Milan 118, 251
Mombaroccio 224–225, 228
Montecassino 251
Münden 93

Naples 212, 216–217, 251

Padua 2n8, 14n78, 119, 204, 209–210, 217–218
Paterno di Avezzano 211, 215–216, 218
Paterno Calabro 215–216
Paris 108, 184, 189–190
Pesaro 225, 230
Prague 10, 12, 14, 67, 83, 98n116, 100, 105–106, 108, 112, 117–119, 122, 152, 164, 184, 202–203, 213, 233, 248

Rimini 238–240
Rome 44, 57, 122, 210–214, 217, 221, 224, 228–231, 235–236, 248–250, 254

Savoy 8
Strasbourg 172
Stuttgart 116, 153n20

Tagliacozzo 215–216, 218
Tübingen 113, 115, 159, 169, 173

Udine 235, 249
Uraniborg 117
Urbino 225, 228, 230

Valencia 2
Venice 217n45, 241
Vercelli 248
Verona 210

Wittenberg 91n65, 152
Wolfenbüttel 82–84, 88, 93, 95, 98, 102–103, 106
Württemberg 113–114, 116, 171

www.ingramcontent.com/pod-product-compliance
Lightning Source LLC
LaVergne TN
LVHW010641300926
849161LV00005B/1005